Neurodegenerative Diseases

This book delves into the complexities of brain health, aging, and neurodegeneration, exploring the varied roles that genetic, environmental, and lifestyle factors play in shaping brain health. It examines their contributions to the development and progression of neurodegenerative diseases. The opening chapters introduce the concept of the blood–brain barrier and its crucial role in maintaining brain health. It also discusses how this barrier is compromised with aging and in various neurodegenerative diseases, underscoring its importance in understanding disease progression. The book reviews specific conditions, including Alzheimer's disease and Parkinson's disease, analyzes their underlying mechanisms, and presents a range of pharmacological and nonpharmacological treatments, along with emerging therapies that offer hope for the future. It also explores other neurodegenerative diseases, such as ALS, Huntington's disease, and multiple sclerosis, detailing the unique pathways that lead to neurodegeneration and the latest approaches to treatment. The chapters provide insights into brain biomarkers and diagnostic techniques that help with early detection and diagnosis of neurodegenerative conditions. Toward the end, the book discusses the challenges faced by patients and caregivers and offers strategies for support and coping. This book serves as an essential resource for neuroscientists and neurologists.

Neurodegenerative Diseases

Pathologies, Mechanisms, and Treatment Strategies

Edited by
Kakarla Ramakrishna, Sachchida Nand Rai,
Santosh Kumar Singh, and Veer Singh

CRC Press
Taylor & Francis Group
Boca Raton London New York

CRC Press is an imprint of the
Taylor & Francis Group, an **informa** business

Designed cover image: Shutterstock

First edition published 2026
by CRC Press
2385 NW Executive Center Drive, Suite 320, Boca Raton FL 33431

and by CRC Press
4 Park Square, Milton Park, Abingdon, Oxon, OX14 4RN

CRC Press is an imprint of Taylor & Francis Group, LLC

ISBN: 978-1-032-87024-3 (hbk)
ISBN: 978-1-032-87025-0 (pbk)
ISBN: 978-1-003-53051-0 (ebk)

DOI: 10.1201/9781003530510

Typeset in Times
by KnowledgeWorks Global Ltd.

Contents

Preface

Neurodegenerative diseases, encompassing disorders such as Alzheimer's disease, Parkinson's disease, Huntington's disease, and amyotrophic lateral sclerosis (ALS), represent a growing global health challenge with profound personal, societal, and economic consequences. With aging populations worldwide, the incidence of these debilitating conditions continues to rise, emphasizing the urgent need for a comprehensive understanding of brain health and the underlying mechanisms of neurodegeneration.

This book, *Neurodegenerative Diseases: Pathologies, Mechanisms, and Treatment Strategies*, is a collaborative effort by researchers, clinicians, and scholars dedicated to advancing knowledge and fostering interdisciplinary discourse in the field of neurodegeneration. It brings together current scientific insights into the intricate interplay between genetic, environmental, and lifestyle factors that influence brain health and delves into the pathological processes that drive neuronal loss and cognitive decline.

The book begins with a foundational overview of factors influencing brain health and disease onset. Chapters 1 and 2 explore the impact of genetics, environment, lifestyle, and the critical role of the blood–brain barrier in maintaining central nervous system integrity. These sections lay the groundwork for understanding how systemic and molecular changes contribute to the pathophysiology of neurodegenerative disorders.

Chapters 3–8 provide an in-depth examination of the mechanisms of neurodegeneration in Alzheimer's, Parkinson's, and other neurodegenerative diseases, followed by comprehensive discussions on their pharmacological and nonpharmacological management strategies. These chapters also address emerging therapeutic interventions that hold promise for modifying disease progression and improving quality of life.

Diagnostic advancements are covered in Chapter 9, with a focus on brain biomarkers and their potential to detect diseases in early stages. Chapter 10 highlights cutting-edge research and technological innovations, including neuroimaging, computational tools, and regenerative medicine, which are reshaping our approach to diagnosis and treatment.

Finally, Chapter 11 gives voice to the human aspect of neurodegeneration, addressing the psychological, emotional, and practical challenges faced by patients and their caregivers. This chapter emphasizes the need for holistic care models that encompass not only medical treatment but also social support and mental health resources.

This book is intended for a diverse audience, including neuroscientists, medical professionals, students, and caregivers, as well as policy-makers engaged in healthcare planning. It aims to provide a balanced and integrative perspective, combining rigorous scientific content with real-world applications.

We hope this compilation serves as a valuable resource for understanding the complexities of brain health and neurodegeneration and inspires future innovations that improve outcomes for those affected by these formidable diseases.

List of Editors

Kakarla Ramakrishna is currently working as Assistant Professor at the KL College of Pharmacy, part of the Koneru Lakshmaiah Education Foundation (KL University), Andhra Pradesh, India. He completed his Ph.D. from the Department of Pharmaceutical Engineering and Technology at IIT (BHU), Varanasi, UP, in 2022. Dr. Kakarla has published research and review articles in international journals and authored four chapters on neurological diseases and their management. Currently, he is working on platelets and thrombosis, and neuronal damage to combat the ischemic stroke and its associated neuronal complications. Dr. Kakarla is also working on identifying the natural bioactive compounds to combat Alzheimer's disease and diabetes-associated brain damage through targeting oxidative stress, inflammation, and mitochondrial dysfunction.

Sachchida Nand Rai is currently an Assistant Professor at the University Centre for Research & Development (UCRD), Chandigarh University, Mohali, Punjab, India. He completed his Ph.D. in 2018 from the Department of Biochemistry, Institute of Science, Banaras Hindu University (BHU), Varanasi. Following his doctorate, he served as a Post-Doctoral Fellow at the Centre of Biotechnology, University of Allahabad, Prayagraj, India, from 2020 to 2023, and later as a Research Associate (ICMR-RA) at the Centre of Experimental Medicine and Surgery, Institute of Medical Sciences, BHU, from 2023 to 2025. Dr. Rai has published numerous research articles in high-impact journals and has authored three books. He serves as an editorial board member for several reputed journals and is a member of the Indian Academy of Neurosciences. His research focuses on investigating the neuroprotective activity of common Indian medicinal plants in toxin-induced Parkinsonian mouse models.

Santosh Kumar Singh is Senior Scientist at the Centre of Experimental Medicine and Surgery (CEMS), Faculty of Medicine, Institute of Medical Sciences, Banaras Hindu University, Varanasi-221005, India. He completed his Ph.D. from the Botany Department (Institute of Science), BHU in 1998. He was Visiting Scientist at the Department of Photosynthesis, Institute of Physical Biology, University of South Bohemia, Czech Republic, in 2003. He has published more than 75 research articles in high impact factor journals and has published several books. Drug Discovery and Development, Microbial Biotechnology, and Nanobiotechnology are some of his major areas of research. He is also a Lifetime Member of the Mycological Society of India, a Lifetime Member of the Society of Biological Chemists, India, and a Member of the Society for Ethnopharmacology, Kolkata, India (SFE-INDIA). He also participated as a resource person in the Karyashala Training Course under the Accelerate Vigyan Scheme, SERB, DST. He was awarded the "National Award for Outstanding Efforts in Science and Technology Communication through Innovative and Traditional Methods" (2021) conferred by the Department of Science and Technology, Ministry of Science and Technology, Government of India. He has received the Order of Academic Excellence in Publication (2020–2021), Institute of Medical Sciences, Banaras Hindu University, Varanasi.

Veer Singh is currently working as Assistant Professor (Research) in the Department of Biosciences at Manipal University Jaipur, Jaipur, India. Prior to joining MUJ, he served as a Young Scientist at the ICMR–Rajendra Memorial Research Institute of Medical Sciences, Patna, India. He completed his Ph.D. in Biochemical Engineering from IIT (BHU), Varanasi. He also received funding/grant from the Department of Health Research (DHR), Government of India of ₹48.65 lakhs. His research is focused on wastewater treatment, development of eco-friendly adsorbents, material characterization, bioremediation, adsorption, environmental toxicity and human health, and nanomaterial for wastewater treatment. He has published 36 research/review articles in SCI/Scopus-indexed journals, five patents, and contributed 17 book chapters in books. He serves as an editor for several reputed journals and has edited multiple books published by reputed international publishers.

1 Genetic, Environmental, and Lifestyle Factors in Brain Health and Neurodegeneration

*Akanksha Mishra**
Department of Pharmacology, Institute of Pharmaceutical Sciences, University of Lucknow, Lucknow, U.P., India

Arun Pratap Mishra
Office of the Registrar General of India, Ministry of Home Affairs, Government of India, New Delhi, India

1.1 INTRODUCTION

The term "brain health" indicates overall health and wellness along with intellectual fitness in the environment which consists of different positive and negative elements. Any irretrievable and progressive structural or functional deprivation of neuronal cells is defined as neurodegeneration which leads to various neurodegenerative diseases based on the location of affected neuronal cell groups in certain brain regions. Taken together, brain health and neurodegeneration indicate the consequences of living in a specific environment and following particular health practices on deprivation of neuronal cells. Its understanding helps to enhance cognition abilities throughout the lifespan. Despite having a large number of research studies, neurodegenerative diseases, including but not limited to Alzheimer's disease (AD), Huntington's disease (HD), motor neuron disease, Parkinson's disease (PD), amyotrophic lateral sclerosis (ALS), Lewy body disease (LBD), progressive supranuclear palsy (PSP), are still incurable (Ayeni et al., 2022; Petersen et al., 2015).

Disturbances in the excitatory and inhibitory effects of neurotransmitters lead to neurodegeneration. Chemical messengers present inside the human body take part in the transmission of neuron-secreted signal molecules across synapses and are known as neurotransmitters. AD involves the accumulation of β amyloid on the neuronal surface and the occurrence of neurofibrillary tangles due to the deposition of hyperphosphorylated Tau protein. The main pathophysiological feature of AD is low levels of neurotransmitter acetylcholine concentration in the brain. Additionally, amyloid precursor proteins (APP) and presenilin are also involved in AD onset (Ayeni et al., 2022).

PD, mainly a motor behavioral disorder, creates an impact on both motor and non-motor neurons. Progressive deficiency of neurotransmitter dopamine (DA) in the striatum and substantia nigra region of the brain is primarily involved in the symptoms which include bradykinesia, tremors, gait, and muscle stiffness. The presence of Lewy bodies, which are the aggregates of α-synuclein (SNCA) protein in the cytoplasmic areas of interneurons and genetic mutations in PTEN-induced putative kinase (PINK1), leucine-rich repeat kinase 2 (LRRK2), and SNCA, are some of the pathophysiological features of PD (Ayeni et al., 2022). Pluripotent stem cells-dopamine neurons-derived 3D organoids and various genetic models also played an important role in extracting the information about etiology and therapeutics of PD (Ramakrishna, Asokwah, et al., 2025; Ramakrishna,

*Corresponding Author. E-mail: mishra_akanksha@lkouniv.ac.in

DOI: 10.1201/9781003530510-1

Mohammed, et al., 2025). HD, a neurodegenerative inherited disorder occurs due to the specific degeneration of neurons in the striatum and cortex region of the brain. It further causes cognitive impairment, dementia, and dystonia. Genetic abnormalities include unusual extensions of cytosine, adenine, and guanine (repeat code) in the huntingtin gene. The decline in psychological abilities is also associated with neurotransmitters gamma-aminobutyric acid (GABA), DA, and serotonin (Ayeni et al., 2022). Table 1.1 provides an outline of neurotransmitters related to the pathophysiology of neurodegeneration.

Neurodegenerative diseases have been affecting people worldwide with a two-fold increase since 1990. Available drug treatments only provide short-term relief by reducing the symptoms of the disease. However, low bioavailability, toxicity, and neuroinflammation-like adverse effects worsen the brain efficiency. Physical activity and other body exercises are highly effective against neurodegenerative diseases by improving the quality of life of affected individuals. A permanent cure for neurodegenerative diseases is still a dream to be achieved (Ayeni et al., 2022). There is a need to understand brain health in a new paradigm to find more targeted therapy for the disease-modifying treatment of neurodegeneration. Evolutionary biology and epigenomics showed that neurodegeneration may be one of the consequences of altered genes in the body, but they don't give the whole picture of all the responsible pathways leading to neurodegenerative diseases. Social, environmental, economic, and lifestyle-related factors play an important role in the occurrence of neurodegeneration which somewhere is the reason why there is a requirement to shift the scientific way of thinking of researchers toward cultural and public health prevention way of thinking. It has been found that public awareness, good care, increased education, and inculcation of human values into the lives of patients suffering from neurodegenerative diseases are proving beneficial (Petersen et al., 2015). Environmental factors including pesticides, pollution, detergents, chemicals, and industry waste materials are greatly linked to the global burden of disease. They may cross the blood–brain barrier (BBB) and cause impairment in human body physiology. Climatic factors like global warming not only ruin natural ecology but also are directly or indirectly related to neurodegenerative diseases. There could be an additional 250,000 deaths per year recorded during the years 2030 and 2050 as per the climate change report. Overall, 38,000 deaths were also reported due to heat exposure and related neurological disorders among older populations. Additionally, genetic variants have been involved in the progressive development of neurodegeneration (Ayeni et al., 2022). One's lifestyle is also directly associated with stress levels which further leads to the overeating of junk food as well as lack of exercise. Altogether it creates a negative impact on quality of life and may play an important role in neurodegeneration (Kip & Parr-Brownlie, 2023). Figure 1.1 summarizes the genetic, environmental, and lifestyle influences on neurodegeneration and approaches to manage them (Ayeni et al., 2022). This chapter aims to provide insight into genetic, environmental, and lifestyle factors associated with the brain health and pathophysiology of neurodegeneration to further find novel targeted therapy to control disease progression.

1.2 GENETIC FACTORS IN BRAIN HEALTH AND NEURODEGENERATION

Genes are highly associated with the origin of neurodegenerative diseases. The genetic mutation leads to oxidative stress, mitochondrial damage, and synaptic loss-like deleterious effects. The role of genetic factors is discussed further in different neurodegenerative diseases.

1.2.1 Genetic Factors Associated with Alzheimer's Disease

Cognitive decline is a major clinical symptom of AD; the pathology includes extracellular deposition of amyloid β plaques. Amyloid β aggregates lead to the activation of microglia which has proved to be dangerous for neurons by causing oxidative stress, mitochondrial dysfunction, and synaptic loss. Triggering Receptor Expressed on Myeloid Cells 2 (TREM2), expressed in microglia,

TABLE 1.1
List of Major Neurotransmitters Related to Neurodegenerative Diseases (Ayeni et al., 2022)

S. No.	Neurotransmitters	Neurodegenerative Diseases	Agonists	Antagonists	Precursors	Receptors	Modulators
1	Acetylcholine	Alzheimer's Parkinson's Dementia	Nicotine Muscarine cholinesterase inhibitors	Tubocurarine Atropine	Choline Acetyl-CoA	Nicotinic Muscarinic	Galantamine Rivastigmine Donepezil
2	Dopamine		Apomorphine Amphetamine	Neuroleptics Domperidone	Phenylalanine Tyrosine L-dopa	G-protein coupled receptors	Bromocriptine Pergolide Cabergoline Ropinirole Pramipexole
3	Serotonin		Azapirones Triptans	Trazodone Nefazodone	5-HTP cation channel	G-protein coupled receptors 5-HT 1-7	Citalopram Escitalopram Fluoxetine Peroxitine Sertraline
4	Histamine		Histamine L-histidine Betahistine	Alimemazine Antazoline Azatadine	Histidine	H1–H4 receptors	Cetirizine Cyproheptadine Vegetables Fexofenadine
5	Epinephrine		Amidephrine Anisodamine	Propranolol Phentolamine	Dopamine	Adrenergic receptors	Micronefrin Vaponefrin
6	Norepinephrine		Sympathomimetic drugs Clonidine Isoprenaline antagonist	Tricyclic antidepressants Beta-blockers Antipsychotics	Dopamine	Adrenergic receptors	Atomoxetine Reboxetine Viloxazine

(Continued)

TABLE 1.1 (*Continued*)
List of Major Neurotransmitters Related to Neurodegenerative Diseases (Ayeni et al., 2022)

S. No.	Neurotransmitters	Neurodegenerative Diseases	Agonists	Antagonists	Precursors	Receptors	Modulators
7	Glutamate	Encephalopathy Parkinson's Huntington's	Kainic acid	N-methyl-d-aspartate (NMDA)	N-acetyl-L-glutamate δ-1-Pyrroline-5-carboxylate β-citryl glutamate L-γ-glutamyl-L-cysteine	α-amino-3-hydroxy-5-methyl-4-isoxazole propionic acid receptor (AMPA) Kainic acid receptors NMDA receptors	GLYX-13, NRX-1074, esketamine, lanicemine
8	Gamma-aminobutyric acid		Muscimol Baclofen	Bicuculline Gabazine Flumazenil	Glucose to glutamate via TCA enzyme pyruvate glutamine	GABA (A–C)	Diazepam Alprazolam
9	Glycine		β-alanine L-serine D-serine L-proline Taurine	Brucine Caffeine Picrotoxin Strychnine	Glutathione Porphyrins Purines Creatine	Glycine receptors and a single β-subunit (GLRB)	D-cycloserine (R,S)-ketamine
10	Nitric Oxide	Alzheimer's Parkinson's Ischemia-reperfusion	S-Nitroso-N-acetyl-DL-penicillamine	3-bromo-7-nitroimidazole N-N-dimethylarginine	L-arginine L-citrulline	Guanylyl cyclase	INomax
11	Hydrogen Sulfide	Anxiety	Trimebutine	Glibenclamide	Methanethiol Ethanethiol Thioglycolic acid	NMDA receptors	

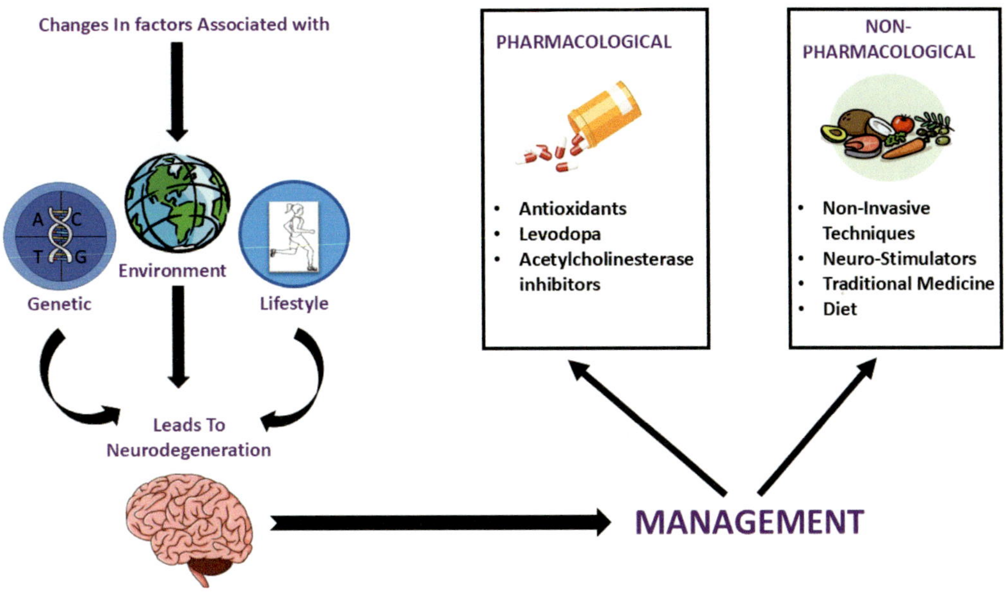

FIGURE 1.1 Summary of influences and approaches to manage neurodegenerative diseases.

has been involved in late-onset sporadic AD. Microglial activation of TREM2 restricts amyloid β-mediated injury by compacting and clearing the plaque. Allele 4 of apolipoprotein E (APOE4) is the other risk factor of AD which is associated with a high concentration of amyloid β in the brain (Shi & Holtzman, 2018). This is in line with the clinical reports showing APOE4 carrier women face high risk for AD (Riedel, Thompson & Brinton, 2016). Mice expressing human APP (hAPP) with XY-sex chromosome complement showed more deleterious effects on cognition and higher mortality in the AD model compared to XX. Lentivirus vector-induced deletion of KDM6a in the neurons of XX aggravated amyloid beta-induced neuronal toxicity *in vitro* whereas its overexpression in XY-neurons decreased toxic effects. Protection was observed *in vivo* also in XY-hAPP mice in the form of low levels of cognitive defects when KDM6a was overexpressed through lentivirus injection into the dentate gyrus (Davis et al., 2020).

1.2.2 Genetic Factors Associated with Parkinson's Disease

Only a few cases of PD are monogenic; the rest are observed to be polygenic and genetically complex. Hypotheses involved are the common disease common variant (CDCV) and common disease rare variant (CDRV) hypothesis. CDCV hypothesis accepts the genetic basis of PD which is a consequence of a large quantity of variants exerting small effects on disease risk. However, their cumulative effects cause substantial risk. CDRV hypothesis postulates that rare genetic variants of little or high effects are risk components for PD with highly functional and deleterious alleles. Genetic mutations in SNCA, LRRK2, microtubule-associated protein tau (MAPT), and GBA play a major role in the pathophysiology of PD (Billingsley et al., 2018).

Aggregates of α-synuclein protein, a major component of Lewy bodies, are considered to be the characteristic feature of PD. Point mutations in the SNCA gene as well as its multiplications take part in severe forms of early-onset PD. LRRK2 with genetic variants is the major pathophysiological factor involved in hereditary PD. Two variants of polymorphic LRRK2 namely p.G2385R and p.R1628P have shown a strong association with two-times increased risk of PD in Asians. p.G2019S, the most ordinary disease-causing variant of LRRK2, has been the reason for 1% of patients with genetic complex PD and 4% of patients with hereditary

PD. People inheriting the LRRK2 p.G2019S variant are observed with increased risk of PD, varying from 28% at 59 years to 51% at 60 years which goes up to 75% at 80 years of age. Ethnicity is one of the factors that play an important role in the frequency of the disease this variant with highest among North African Arab and Jewish people, followed by Middle Eastern and Southern Europe. Caucasians (0.71%), Asians (0.07%), and 30.2% of Arab origin have been reported with PD.

Glucocerebrosidase enzyme encoding gene, GBA with homozygous mutations, gives rise to Gaucher's disease, a lysosomal storage disorder. GBA mutation has been observed in 5–10% of PD patients. Patients suffering from Gaucher's disease have been observed with a high occurrence of PD, indicating GBA as a predisposing factor for PD. Additionally, heterozygous mutations in GBA have been reported as a topmost genetic predisposing factor for PD which enhances the risk of developing PD five times in individuals. GBA p.E326K, p.N370S, and p.L444P are the major variants that reduce glucocerebrosidase enzymatic activity. Decrease in enzymatic activity and low GBA expression levels are shown in PD patients who carry heterozygous GBA mutations. The substantia nigra part of the brain in PD patients has been reported with low levels and activity of glucocerebrosidase enzyme. The risk of developing PD in GBA carriers has been reported to increase from 20% at 70 years to 30% at 80 years. GBA mutations are most frequent among the Ashkenazi Jewish population. GBA p.N370S variant has been only found in the Middle East, European, or American region but not in Chinese or Japanese subjects (Billingsley et al., 2018).

Genetic mutations in MAPT are the other predisposing factor for PD. This along with tau pathology has also been linked to dementia. Haplotypes H1 and H2 are present at the MAPT locus, out of which H2 is found in 20% of the European population. MAPT locus harbors several genes; MAPT H1 and its subtype H1c are linked to the high incidence of PD (Billingsley et al., 2018). Genetic mutations in PD affect autophagy, mitochondrial function, lysosomal storage, and endocytosis, as indicated in Figure 1.2.

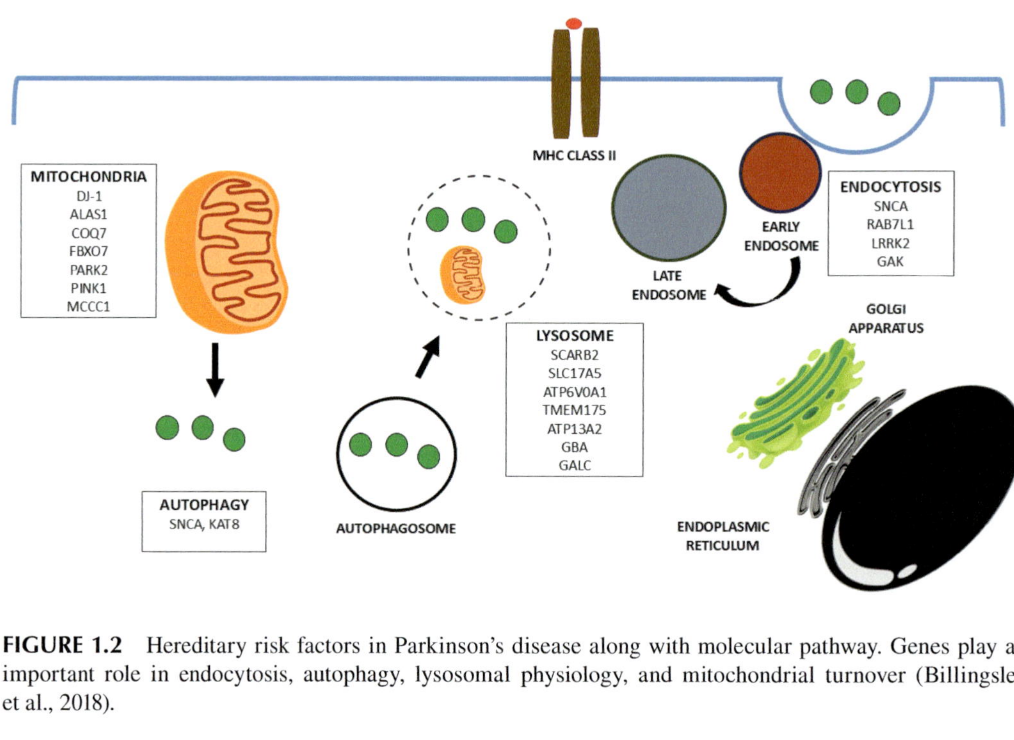

FIGURE 1.2 Hereditary risk factors in Parkinson's disease along with molecular pathway. Genes play an important role in endocytosis, autophagy, lysosomal physiology, and mitochondrial turnover (Billingsley et al., 2018).

1.2.3 Genetic Factors Associated with Huntington's Disease

An inherited neurodegenerative disorder, HD, is associated with the expanded CAG repeat in the huntingtin (HTT) gene. The clinical symptoms include psychiatric and movement disorders as well as cognitive decline (Ellis et al., 2020).

The length of this expanded CAG repeat explains the age variation in motor onset. The pathogenesis of HD is not based on HTT dosage but shows the dominant effects of a single mutant allele. CAG repeat is responsible only for 56% of the differences in age at onset. HD age of onset has been reported to show many variations that may not be explained by expanded CAG repeat only. Additional genetic variations are also involved for the different ages of onset in genes, including GRIN2A, E2F2, GRIN2B, TCERG1, ADORA2A, ATG7, GRIK2, PPARGC1A, and HAP. Subjects having the E2F2 rs2742976 T allele are observed with a mean three-year delay in the earliest age of onset. GRIN2A and GRIN2B genes which encode NR2A and NR2B subunits of N-methyl-D-aspartic acid (NMDA) receptors have also been linked to the age of onset in HD. The diagnostic motor signs are the result of any of the two types of DNA sequence variants including individual sequence variation at the HTT locus that may cause changes in anatomy, physiology, or expression of HTT protein. The second type is the sequence variation in the genome that may change the anatomy, physiology, or expression of genetic factors whose physiology is linked with the pathological process of HD (Valcárcel-Ocete et al., 2015).

1.2.4 Genetic Factors Associated with Amyotrophic Lateral Sclerosis (ALS)

A neurodegenerative disease with the characteristic clinical feature of progressive paralysis, ALS, occurs due to spinal cord and cortical motor neurons. ALS, also known as motor neuron disease, involves the impairment in non-motor neurons which control fronto-executive cognitive function. Overall, 7–10% of cases of ALS are observed with a family history of the disease. The remaining 85–90% of cases are, although sporadic cases, still influenced by genes. This altogether indicates the complex genetic structure of ALS. Genes associated with ALS include TBK1, KIF5A, NEK1, GLT8D1, TUBA4A, CCNF, LGALSL, TIA1, C21orf2, ANXA11, and DNAJC7. Functions that are greatly affected by genetic factors include but are not limited to vesicle trafficking, proteostasis, axonal transport, and neuroinflammation. FUS encodes "prion-like domain" (PLD) and TARDBP (Tar DNA binding protein 43 kDA which encodes TDP-43 protein) containing proteins are associated with RNA processing which is greatly compromised in ALS. TANK-binding kinase (TBK1) plays an important role in autophagy and inflammation against pathogens. However, mutations are observed in TBK1 in ALS. TUBA4A mutations cause dysfunctions in microtubule assembly and negatively affect its stability. KIF5A, a microtubule-associated motor protein named kinesin takes part in organelle transport. Therefore, KIF5A mutations result in impaired axonal transport. Other genes involved in ALS pathology include SOD1, SETX, C9orf72, VAPB, and UBQLN2 (Gregory et al., 2020). Table 1.2 provides a summary of recently discovered ALS genes.

1.2.5 Genetic Factors Associated with Progressive Supranuclear Palsy

PSP, an atypical Parkinsonian syndrome, is a rare neurodegenerative disorder. It is characterized by language, behavior, and motor abnormalities. The occurrence of PSP is linked to the mutation of many genes. However, the generally recognized form is sporadic syndrome. MAPT gene encodes tau protein; 60 mutations of MAPT have been identified till now. An LRRK2 gene mutation is also involved in the genesis of PSP. However, cases are rare. Dynactin subunit 1 (DCTN1), the largest subunit of the dynactin complex, plays an important role in molecular transport. Mutations in other

TABLE 1.2
Recently Discovered ALS Genes (Gregory et al., 2020)

S. No.	Gene	Physiological Role	Involvement in ALS Pathology
1	TBK1 (TANK binding kinase 1)	Autophagy, activation of innate immunity, activator of optineurin gene	Loss-of-function mutations, single amino acid deletions, segregation in familial disease due to a single missense mutation (p.Arg573Gly)
2	TUBA4A (tubulin alpha-4A)	Major protein of microtubules, maintain cytoskeletal integrity and axonal transport	Putative mutations, spinal onset with typical upper and lower motor neuron involvement
3	KIF5A (kinesin family member 5A)	Microtubule-associated motor protein, involve in organelle transport	Missense mutations in N-terminal motor domain of KIF5A, C-terminal loss-of-function mutation
4	ANXA11 (annexin A11)	Responsible for calcium-dependent formation of vesicles, mRNA tethering for transport along axons	Mutation disrupts calcyclin binding; impairs the ability of ANXA11 to tether RNA granules to actively transported lysosomes
5	TIA1	RNA-binding protein, involve in the formation of Stress granules	Mutations are associated with distal myopathy with vacuoles
6	CCNF (cyclin F)	Binds substrates and directs their ubiquitylation for subsequent degradation by ubiquitin proteasomal pathway	Impairs ubiquitin-mediated proteasomal degradation; disrupts ubiquitylation and impairs autophagy
7	DNAJC7 (DNAJ homolog subfamily C member 7)	Regulates key protein-folding chaperones Hsp70 and Hsp90, involved in neuroprotection	Suppresses cellular responses to innate immunity
8	NEK1 (NIMA-related kinase 1)	Serine/threonine kinase, cell cycle regulation, axonal development, DNA damage repair	Loss-of-function mutation or missense mutations lead to DNA damage, and cell death in neurons; causes protein misfolding
9	C21orf2 (chromosome 21 open reading frame 2)	DNA damage control and repair, cilia formation	May disrupt protein structure
10	LGALSL	Galectin-related protein, physiology is largely unknown	Loss-of-function mutations

genes including NPC1, C9orf72, PARK2, TARDBP, GRN, TBK1, and BSN are also observed in PSP. The genetic involvement in the pathology of PSP is getting attention continuously; the mutations directly cause PSP. However, studies using larger cohorts are required to verify the genetic associations (Wen, Zhou, Jiao & Shen, 2021).

1.3 ENVIRONMENTAL FACTORS IN BRAIN HEALTH AND NEURODEGENERATION

Environmental exposures and adverse climatic conditions greatly influence neurodegenerative diseases as they may begin inflammation and other processes during which there may be a deficiency of several neurotransmitters and hormones. The environment has a positive or negative impact on mental health which further takes part in the growth and maturity of the human brain and its functions (Rauh & Margolis, 2016). Midbrain DA concentration in adult substantia nigra was found to be imbalanced due to the effect of environment and activity on the plasticity of DA neurotransmitters. It further leads to bringing out symptoms related to behavioral disorders such

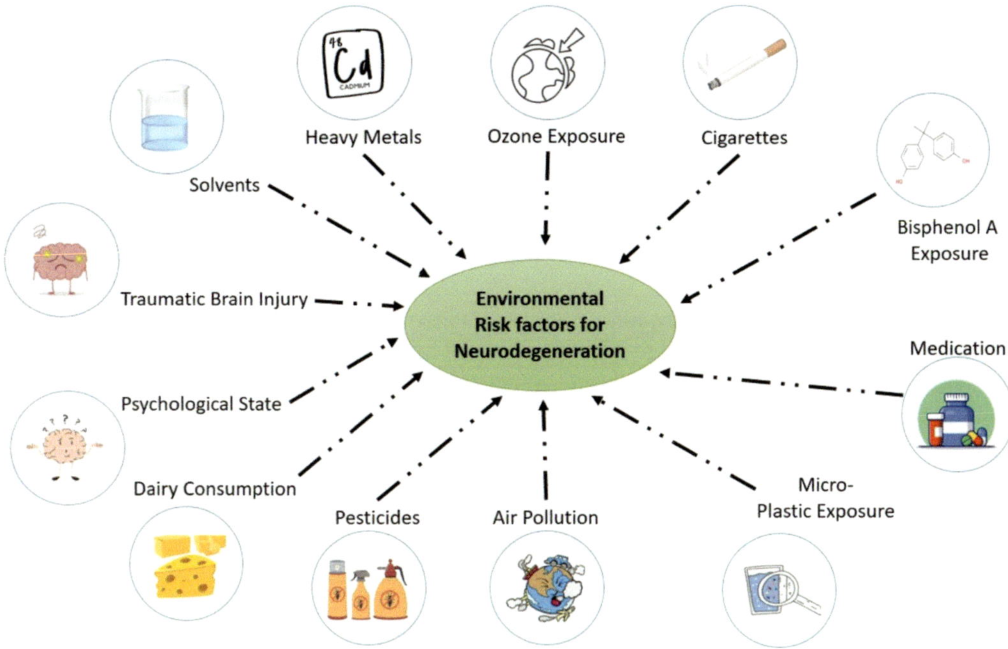

FIGURE 1.3 Environmental risk factors for neurodegeneration (Tsalenchuk, Gentleman & Marzi, 2023).

as addiction, depression, schizophrenia, attention deficit, obsessive–compulsive, and hyperactivity disorder. The underlying mechanism involved may be the role of midbrain DA neurotransmitter plasticity in the etiology of these symptoms (Aumann et al., 2016). Figure 1.3 depicts the various environmental risk factors responsible for neurodegenerative diseases (Tsalenchuk, Gentleman & Marzi, 2023).

1.3.1 Environmental Pollutants

Depletion of dopamine and acetylcholine occurs in PD and AD respectively by exposure to environmental pollutants, oxidative stress, and heavy metals. Additionally, hormones and neurotransmitters including GABA, NE, 5-HT, and tryptophan are also reported to be deficient (Alisky, 2006). Mercury, manganese, pesticides, arsenic, lead, and biotoxins may also cause a deficiency of neural hormones. Mixture of air pollutants which include ozone, lead, and polycyclic aromatic hydrocarbons (PAHs) observed in urban areas also create a negative influence on neurons and hormones of central nervous system (CNS). That's why natives living in these areas are prone to brain shrinkage and stroke (Bellinger, 2018). The reports of deficient volume of brain white matter, intellectual disability, and symptoms of attention deficit disorders on prenatal exposure to PAHs are also available. Air pollutants can cross BBB and cause immunological responses as well as neuroinflammation (Mortamais et al., 2017). Additionally, there are some metals including zinc and copper which cross BBB and are also involved in the biosynthesis of various neurotransmitters, therefore can be further investigated to explore novel pathways to find the effective treatment of neurodegenerative diseases (Zatta & Frank, 2007). Invisible neurotoxins are found in the atmosphere due to pollution and accumulate on edible items as well as drinking water (Kinney, 2018), continuous exposure of which further leads to slow cognitive development in children, and neurodegeneration in adults (Perera, 2018). Plantation in urban areas and the availability of green spaces have the potential to improve air quality and negatively affect toxic pollutants either directly or indirectly.

1.3.2 Metals and Pesticides

Poisonous impurities exist in large amounts in the environment. Popularly called heavy metals, these accumulated contaminants are toxic to natural habitats. Once reach the systemic circulation, they further collect near the BBB. However, BBB is strong enough to halt its pathway toward the brain. Iron, aluminum, copper, lead, zinc, and mercury are found in water and soil which are correlated to the pathogenesis of NDD. However, research findings are highly inconsistent in terms of the repercussions of these metals on health. For example, a high risk of AD has been associated with excessive amounts of aluminum in drinking water whereas other studies did not report any such correlation between them. Many pesticides and radioactive materials take entry into the body through the ocular route. Cyanide, a highly dangerous natural gas, and a neurotoxicant, is synthesized by algae, bacteria, and fungi. It impairs the mitochondrial function by inhibiting mitochondrial complex IV which further creates a negative impact on cellular respiration and neuronal immunity. Water sources, gastrointestinal and respiratory tracts, and cutaneous contact are the other sources of contamination through which insecticides like carbamates and organochlorines as well as other agrochemicals may gain access to the human body. This is directly correlated to a high risk of contamination in farmers working in agricultural areas. Others using those products are also exposed to food contamination, indicating the principal role of pesticides which affects the regulatory system and initiates the disease progression (Ayeni et al., 2022).

1.3.3 Chemotransmission

Release, reuptake, and metabolism of neurotransmitters are highly regulated processes of a living body. Any disturbances in this system may further lead to neuronal dysfunction and neurodegeneration. Sometimes, the toxicants share so many similarities with the neurotransmitters that they use the same transporter as a neurotransmitter to gain entry inside the neurons. For example, toxicant 1-methyl-4-phenyl pyridinium uses the DA transporter of the neurotransmitter dopamine to impair the functioning of DA neurons. Loss in neuronal volume is the principal etiological factor involved in cognitive disabilities and motor impairment. Chemicals and solvents used in laboratories may be one of the probable sources for the deposition of neurotoxicants in the body (Ayeni et al., 2022).

1.3.4 Climatic Conditions

First ever evidence of the relationship between environmental stimuli and the quantity of human midbrain DA neurons was reported in 2016 where lower DA levels and signaling were found in the winter season than in summer. Anti-neuronal hormone deficits and/or pro-neurotransmitter benefits were observed due to deficiency in neuronal hormones through variations in climatic conditions in various seasons (Aumann et al., 2016). In addition to DA deficiency, the occurrence of depression-related symptoms was also observed in winter seasons (Dunlop & Nemeroff, 2007). To get a clearer picture of the consequences of climatic factors on the physiology of the human body should be further investigated. Any alterations in weather compositions including but not limited to floods, hurricanes, fires, glaciers, heat, tornadoes as well as desertification play an important role in gradual changes in climatic conditions which are strong enough to affect the ecosystem and biological cycles of living bodies. Alterations in climatic conditions negatively affect basic life processes including sensation and responsiveness and further lead to cognitive decline. Pressing issues like global warming cause abnormal increases in temperature and give rise to NDDs (Sun et al., 2013). Oxidative stress and excitotoxicity are the effects of heat stress in older patients whose body regulatory processes are thermo compromised (Lim, 2019). Negative effects of heat stroke have been reported in experimental animals of mild traumatic brain injury also in the form of low integrity of BBB and intellectual deficits (Titus, Furones, Atkins & Dietrich, 2015). Other diseases including chikungunya, Zika, yellow fever, and dengue which are directly linked with changes in climatic conditions have shown neurological implications, such as encephalopathy, neuroinflammation, encephalitis, neuronal damage, and cortical thinness (Brooks, 2013).

1.3.5 In Utero Conditions

Heavy metals, hormonal disturbances, hypoxia, and chronic placental insufficiency may affect the human fetus at the time of pregnancy. A fetus affected by neurotoxins in the uterus may further turn into an individual with delayed growth after delivery (Saidj et al., 2020). Stress in expecting mothers may further lead to increased concentration of glucocorticoid in the placenta which alters the quantity of glucocorticoid receptors in the fetus and negatively affects its growth (Kapoor, Petropoulos & Matthews, 2008). Studies related to radioactive materials reported dysfunctional visual processing due to prenatal exposure to large amounts of mercury and lead in the umbilical cord. (García-Esquinas et al., 2013). Hormonal irregularities may also produce harmful effects on the psychological growth and development of the fetus (Miranda & Sousa, 2018). However, the human-based research on in utero exposure related to neurodegeneration is still not sufficient enough to completely understand the exact period of pregnancy during which a fetus is prone to such disorders.

1.4 LIFESTYLE FACTORS IN BRAIN HEALTH AND NEURODEGENERATION

Modernization brings lifestyle changes which can further lead to brain and heart-related disorders. Subjects living disciplined lifestyles, avoiding smoking, having a modest amount of alcohol consumption with moderate body weight in their midforties are less prone to neurodegenerative diseases compared to individuals living high-risk lifestyles (no physical exercises, imbalanced diet, smoking, alcoholics) (Popa-Wagner et al., 2020). Some common lifestyle factors related to brain health and neurodegeneration are discussed here.

1.4.1 Nutrition

Iron-rich diet is an important source of overall nutrition. The role of iron is well known in the pathophysiology of various conditions present in the elderly population. Low hemoglobin concentration with a high amount of ferritin is reported to be involved in the pathology of many diseases. Inflammation is one of the major pathways taking part in the pathophysiology of neurodegenerative diseases. Inflammation affects iron overload whereas iron homeostasis is important for normal mitochondrial physiology. An overabundance of iron is directly responsible for the overflow of hydroxyl radicals which further create oxidative stress. However, iron deficiency obstructs the various physiological processes using iron as a cofactor (Popa-Wagner et al., 2020). It was reported in patients with idiopathic PD that plants and fish-based diets significantly reduced PD severity. Fish oil and coenzyme Q10 also decreased the rate of PD progression. Nutritional behavior of patients also affected severity scores in PD. Patients having home-cooked foods had significant protection against PD compared to those eating foods from local markets (Mischley, Lau & Bennett, 2017). Higher coffee consumption was found to be protective against AD occurrence whereas higher concentrations of 25(OH) vitamin D were reported to cause AD (Larsson et al., 2017). A multicenter case control retrospective study performed among 151 patients with frontotemporal dementia rejected any direct correlation between tobacco-coffee use and frontotemporal dementia but reported a low risk of this disease in alcohol users. However, the amount of alcohol consumed was not defined in the study (Tremolizzo et al., 2017).

1.4.2 Smoking

Smoking may affect AD occurrence directly indicating that the more a person smokes, the higher the probability of suffering from AD (Larsson et al., 2017). Oxidative stress is one of the factors affecting aging and neurodegeneration. Ascorbic acid, a strong antioxidant proved to be effective against amyloid-beta peptide deposition. However, even the protective role of ascorbic acid was observed only in non-smokers (Kim et al., 2015).

1.4.3 Caloric Restriction

The incidence of various age-related diseases in primates has been lowered by the method of caloric restriction, which also decreased the incidence of AD and PD. The underlying mechanism involved is its autophagy-promoting effects which may also play an important role in lifespan extension in organisms (Bishop & Guarente, 2007; Colman et al., 2009).

1.4.4 Physical Activity

Earlier studies performed showed no or little benefit of physical activity on the occurrence of AD (Stephen et al., 2017). However, recent studies reported that six weeks of aerobic exercises improved functional ability in AD subjects probably due to the changes in peak VO_2 and bilateral hippocampal volume (Morris et al., 2017). Nature is the best source of rejuvenation which plays an important role in overall health and wellbeing. The more exposure to natural sceneries including animals and plants, the more positive effects on one's mental, physical, and emotional health and resistance to disease (Buttazzoni, Parker & Minaker, 2021). Outdoor physical activities are proven to be beneficial in terms of gaining energy and concentration which further decrease anxiety, anger, and sadness like distressing emotions (Lahart et al., 2019).

1.4.5 Economic Factors

A social disadvantage including poverty is associated with brain physiology and stress levels due to peer pressure (Izett et al., 2020). It is one of the main reasons behind insufficient social support, dietary requirements, and skills. Infants with childhood poverty when monitored up to three months since birth showed high symptoms of behavioral disorders during their teenage years (Johnson, Riis & Noble, 2016).

The anatomical changes were observed in these people in terms of the low volume of the orbitofrontal cortex, an area involved in the regulation of emotional transmission. It further leads to impairment in motivational and emotional control and the occurrence of psychopathology (Holz et al., 2015; Izett et al., 2020).

1.4.6 Sleeping Pattern

Circadian rhythms and sleep are highly controlled physiology of the living body which mainly involves CNS areas. Figure 1.4 shows the two-way relationship between sleep and neurodegeneration (Ibrahim, Högl & Stefani, 2024). AD patients have reported insomnia, daytime sleepiness, and sleep apnea (Holth, Patel & Holtzman, 2017) due to neuronal loss of the suprachiasmatic nucleus and affected synchronization of circadian oscillators in the brain. Sleep disturbances are also common in patients of PD which include circadian rhythm disorders, rapid eye movement sleep behavior disorder, excessive daytime sleepiness, and insomnia (Stefani & Högl, 2020). Chronic loss of sleep accentuates proinflammatory markers and induces inflammation which further leads to impaired physiology of glia and loss of synapse (Zielinski & Gibbons, 2022). Alterations in sleep patterns also lead to the accumulation of misfolded proteins including phosphorylated tau, alpha synuclein,

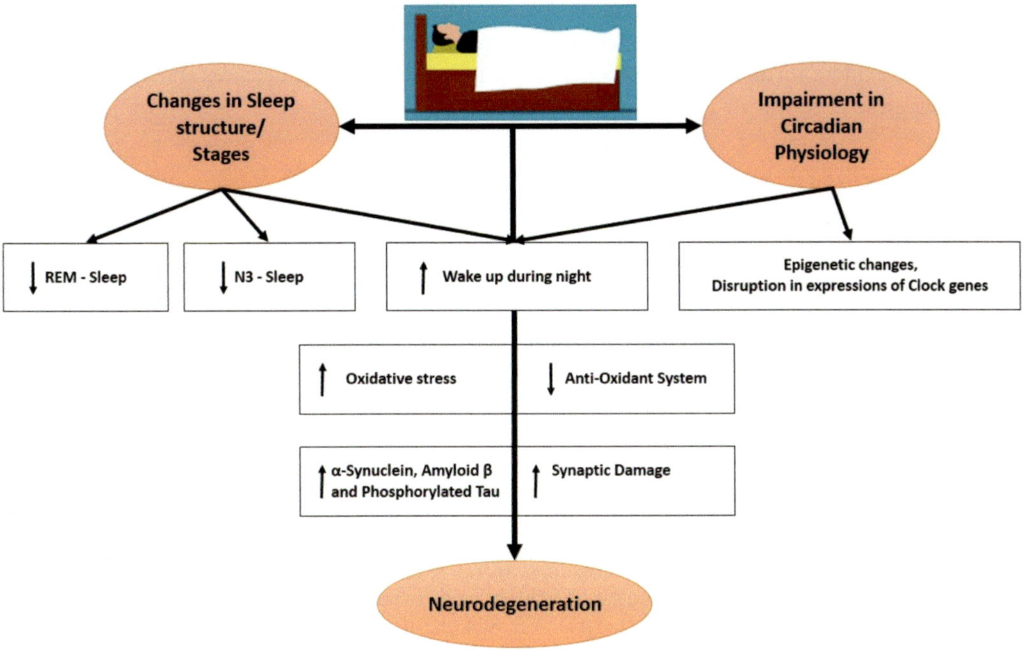

FIGURE 1.4 Two-way relationship between sleep and neurodegeneration (Ibrahim, Högl & Stefani, 2024).

and amyloid beta which altogether bring about impairment in synaptic physiology and give rise to neurodegeneration (Ibrahim, Högl & Stefani, 2024).

1.5 CONCLUSION AND FUTURE PROSPECTS

Genetic, environmental, and lifestyle factors are linked with the etiopathogenesis of neurodegenerative diseases as they produce direct or indirect effects on various neuronal circuits in the brain. However, many factors are yet to be investigated further which include emotional stability, environmental exposure, involved neurotransmitters, and early childhood lifestyle. The interconnection between brain-related hormones and gender, age, and ethnicity may further open the pathway to newer perspectives to find the disease-modifying treatment of neurodegeneration.

In this chapter, effects of genetic factors like TREM2, APOE4, sex chromosome, SNCA, LRRK2, MAPT, GBA, GRIN2A, E2F2, GRIN2B, TCERG1, ADORA2A, ATG7, GRIK2, PPARGC1A and HAP, ALS including TBK1, KIF5A, NEK1, GLT8D1, TUBA4A, CCNF, LGALSL, TIA1, C21orf2, ANXA11, DNAJC7, DCTN1, NPC1, C9orf72, PARK2, TARDBP, GRN, TBK1, BSN on the pathogenesis of neurodegenerative diseases such as AD, PD, HD, ALS, PSP were discussed. Furthermore, effects of environmental factors like environmental pollutants, oxidative stress and heavy metals, mercury, manganese, pesticides, arsenic, lead, biotoxins, cyanides, insecticides like carbamates and organochlorines as well as other agrochemicals, chemotransmission, climatic conditions as well as in utero conditions on the occurrence of neurodegenerative diseases have also been described. Additionally, lifestyle factors are also involved in the etiopathogenesis of these diseases. The factors include but are not limited to diet, nutrition, smoking, caloric restriction, physical activity, economic factors, and sleeping patterns. Studies are further needed to get a complete picture of the mechanism and involved neuronal brain circuits and the extent to which it affects the neuronal degeneration. *In vitro* studies involving cell culture together with the combination of *in vivo* and other neurodegenerative models should be utilized to further investigate the intracellular signaling pathway involved in the process. Additionally, the pathogenesis of neurodegeneration should be linked with other brain-related disorders including anxiety, and depression which can be highly important to explore further the etiopathogenesis of neurodegenerative diseases and find novel treatment strategies.

REFERENCES

Alisky, J. M. (2006). Neurotransmitter depletion may be a cause of dementia pathology rather than an effect. *Medical Hypotheses*, *67*(3), 556–560. doi:10.1016/j.mehy.2006.02.043.

Aumann, T. D., Raabus, M., Tomas, D., Prijanto, A., Churilov, L., Spitzer, N. C., & Horne, M. K. (2016). Differences in number of midbrain dopamine neurons associated with summer and winter photoperiods in humans. *PLoS One*, *11*(7), e0158847. doi:10.1371/journal.pone.0158847.

Ayeni, E. A., Aldossary, A. M., Ayejoto, D. A., Gbadegesin, L. A., Alshehri, A. A., Alfassam, H. A., ... Tawfik, E. A. (2022). Neurodegenerative diseases: Implications of environmental and climatic influences on neurotransmitters and neuronal hormones activities. *International Journal of Environmental Research and Public Health*, *19*(19). doi:10.3390/ijerph191912495.

Bellinger, D. C. (2018). An overview of environmental chemical exposures and neurodevelopmental impairments in children. *Pediatric Medicine*, *1*. doi:10.21037/pm.2018.11.03.

Billingsley, K. J., Bandres-Ciga, S., Saez-Atienzar, S., & Singleton, A. B. (2018). Genetic risk factors in Parkinson's disease. *Cell and Tissue Research*, *373*(1), 9–20. doi:10.1007/s00441-018-2817-y.

Bishop, N. A., & Guarente, L. (2007). Genetic links between diet and lifespan: Shared mechanisms from yeast to humans. *Nature Reviews Genetics*, *8*(11), 835–844. doi:10.1038/nrg2188.

Brooks, H. E. (2013). Severe thunderstorms and climate change. *Atmospheric Research*, *123*, 129–138. doi:10.1016/j.atmosres.2012.04.002.

Buttazzoni, A., Parker, A., & Minaker, L. (2021). Investigating the mental health implications of urban environments with neuroscientific methods and mobile technologies: A systematic literature review. *Health & Place*, *70*, 102597. doi:10.1016/j.healthplace.2021.102597.

Colman, R. J., Anderson, R. M., Johnson, S. C., Kastman, E. K., Kosmatka, K. J., Beasley, T. M., ... Weindruch, R. (2009). Caloric restriction delays disease onset and mortality in rhesus monkeys. *Science*, *325*(5937), 201–204. doi:10.1126/science.1173635.

Davis, E. J., Broestl, L., Abdulai-Saiku, S., Worden, K., Bonham, L. W., Miñones-Moyano, E., ... Dubal, D. B. (2020). A second X chromosome contributes to resilience in a mouse model of Alzheimer's disease. *Science Translational Medicine*, *12*(558). doi:10.1126/scitranslmed.aaz5677.

Dunlop, B. W., & Nemeroff, C. B (2007). The role of dopamine in the pathophysiology of depression. *Archives of General Psychiatry*, *64*(3), 327–337. doi:10.1001/archpsyc.64.3.327.

Ellis, N., Tee, A., McAllister, B., Massey, T., McLauchlan, D., Stone, T., ... Holmans, P. (2020). Genetic risk underlying psychiatric and cognitive symptoms in Huntington's disease. *Biological Psychiatry*, *87*(9), 857–865. doi:10.1016/j.biopsych.2019.12.010.

García-Esquinas, E., Pérez-Gómez, B., Fernández-Navarro, P., Fernández, M. A., De Paz, C., Pérez-Meixeira, A. M., ... Astray, J. (2013). Lead, mercury and cadmium in umbilical cord blood and its association with parental epidemiological variables and birth factors. *BMC Public Health*, *13*, 1–11. doi:10.1186/1471-2458-13-841.

Gregory, J. M., Fagegaltier, D., Phatnani, H., & Harms, M. B. (2020). Genetics of amyotrophic lateral sclerosis. *Current Genetic Medicine Reports*, *8*(4), 121–131. doi:10.1007/s40142-020-00194-8.

Holth, J., Patel, T., & Holtzman, D. M. (2017). Sleep in Alzheimer's disease – Beyond amyloid. *Neurobiology of Sleep and Circadian Rhythms*, *2*, 4–14. doi:10.1016/j.nbscr.2016.08.002.

Holz, N. E., Boecker, R., Hohm, E., Zohsel, K., Buchmann, A. F., Blomeyer, D., ... Laucht, M. (2015). The long-term impact of early life poverty on orbitofrontal cortex volume in adulthood: Results from a prospective study over 25 years. *Neuropsychopharmacology*, *40*(4), 996–1004. doi:10.1038/npp.2014.277.

Ibrahim, A., Högl, B., & Stefani, A. (2024). The bidirectional relationship between sleep and neurodegeneration: Actionability to improve brain health. *Clinical and Translational Neuroscience*, *8*(1), 11. doi:10.3390/ctn8010011.

Izett, E., Rooney, R., Prescott, S. L., De Palma, M., & McDevitt, M. (2020). Prevention of mental health difficulties for children aged 0-3 years: A review. *Frontiers in Psychology*, *11*, 500361. doi:10.3389/fpsyg.2020.500361.

Johnson, S. B., Riis, J. L., & Noble, K. G. (2016). State of the art review: Poverty and the developing brain. *Pediatrics*, *137*(4). doi:10.1542/peds.2015-3075.

Kapoor, A., Petropoulos, S., & Matthews, S. G (2008). Fetal programming of hypothalamic–pituitary–adrenal (HPA) axis function and behavior by synthetic glucocorticoids. *Brain Research Reviews*, *57*(2), 586–595. doi:10.1016/j.brainresrev.2007.06.013.

Kim, S. M., Lim, S. M., Yoo, J. A., Woo, M. J., & Cho, K. H. (2015). Consumption of high-dose vitamin C (1250 mg per day) enhances functional and structural properties of serum lipoprotein to improve anti-oxidant, anti-atherosclerotic, and anti-aging effects via regulation of anti-inflammatory microRNA. *Food & Function*, *6*(11), 3604–3612. doi:10.1039/c5fo00738k.

Kinney, P. L. (2018). Interactions of climate change, air pollution, and human health. *Current Environmental Health Reports*, *5*(1), 179–186. doi:10.1007/s40572-018-0188-x.

Kip, E., & Parr-Brownlie, L. C. (2023). Healthy lifestyles and wellbeing reduce neuroinflammation and prevent neurodegenerative and psychiatric disorders. *Frontiers in Neuroscience*, *17*, 1092537. doi:10.3389/fnins.2023.1092537.

Lahart, I., Darcy, P., Gidlow, C., & Calogiuri, G. (2019). The effects of green exercise on physical and mental wellbeing: A systematic review. *International Journal of Environmental Research and Public Health*, *16*(8). doi:10.3390/ijerph16081352.

Larsson, S. C., Traylor, M., Malik, R., Dichgans, M., Burgess, S., & Markus, H. S. (2017). Modifiable pathways in Alzheimer's disease: Mendelian randomisation analysis. *BMJ*, *359*, j5375. doi:10.1136/bmj.j5375.

Lim, K. H. (2019). Diverse misfolded conformational strains and cross-seeding of misfolded proteins implicated in neurodegenerative diseases. *Frontiers in Molecular Neuroscience*, *12*, 158. doi:10.3389/fnmol.2019.00158.

Miranda, A., & Sousa, N. (2018). Maternal hormonal milieu influence on fetal brain development. *Brain and Behavior*, *8*(2), e00920. doi:10.1002/brb3.920.

Mischley, L. K., Lau, R. C., & Bennett, R. D. (2017). Role of diet and nutritional supplements in Parkinson's disease progression. *Oxidative Medicine and Cellular Longevity*, *2017*, 6405278. doi:10.1155/2017/6405278.

Morris, J. K., Vidoni, E. D., Johnson, D. K., Van Sciver, A., Mahnken, J. D., Honea, R. A., ... Burns, J. M. (2017). Aerobic exercise for Alzheimer's disease: A randomized controlled pilot trial. *PLoS One*, *12*(2), e0170547. doi:10.1371/journal.pone.0170547.

Mortamais, M., Pujol, J., van Drooge, B. L., Macià, D., Martínez-Vilavella, G., Reynes, C., ... Sunyer, J. (2017). Effect of exposure to polycyclic aromatic hydrocarbons on basal ganglia and attention-deficit hyperactivity disorder symptoms in primary school children. *Environment International*, *105*, 12–19. doi:10.1016/j.envint.2017.04.011.

Perera, F. (2018). Pollution from fossil-fuel combustion is the leading environmental threat to global pediatric health and equity: Solutions exist. *International Journal of Environmental Research and Public Health*, *15*(1), 16. doi:10.3390/ijerph15010016.

Petersen, R. B., Lissemore, F. M., Appleby, B., Aggarwal, N., Boyatzis, R., Casadesus, G., ... Lerner, A. (2015). From neurodegeneration to brain health: An Integrated approach. *Journal of Alzheimer's Disease*, *46*, 271–283. doi:10.3233/JAD-150043.

Popa-Wagner, A., Dumitrascu, D. I., Capitanescu, B., Petcu, E. B., Surugiu, R., Fang, W. H., & Dumbrava, D. A. (2020). Dietary habits, lifestyle factors and neurodegenerative diseases. *Neural Regeneration Research*, *15*(3), 394–400. doi:10.4103/1673-5374.266045.

Ramakrishna, K., Asokwah, S. O., Njuguna, G. M., Mohammed, H., Dumala, N., Viswanadh, M. K., ... Anusha, V. L. (2025). Relevance of genetic model and linked therapeutics in Parkinson's disease. *Neurodegenerative Diseases*, 42–52. doi:10.1201/9781032620442-4

Ramakrishna, K., Mohammed, H., Njuguna, G. M., Asokwah, S. O., Singh, V., Dumala, N., ... Nalluri, B. N. (2025). Importance of newer 3D organoids model-based therapeutic approach in Parkinson's disease. *Neurodegenerative Diseases*, 53–59. doi:10.1201/9781032620442-5

Rauh, V. A., & Margolis, A. E (2016). Research review: Environmental exposures, neurodevelopment, and child mental health – New paradigms for the study of brain and behavioral effects. *Journal of Child Psychology and Psychiatry*, *57*(7), 775–793. doi:10.1111/jcpp.12537.

Riedel, B. C., Thompson, P. M., & Brinton, R. D. (2016). Age, APOE and sex: Triad of risk of Alzheimer's disease. *Journal of Steroid Biochemistry and Molecular Biology*, *160*, 134–147. doi:10.1016/j.jsbmb.2016.03.012.

Saidj, S., Henderson, M., Ruchat, S.-M., Paradis, G., Hulst, A. V., Zappitelli, M., & Mathieu, M.-E. (2020). Are suboptimal in utero conditions associated with obesity and cardiometabolic risk factors in Caucasian children? *medRxiv*. doi:10.1101/2020.06.18.20132837

Shi, Y., & Holtzman, D. M (2018). Interplay between innate immunity and Alzheimer disease: APOE and TREM2 in the spotlight. *Nature Reviews Immunology*, *18*(12), 759–772. doi:10.1038/s41577-018-0051-1.

Stefani, A., & Högl, B. (2020). Sleep in Parkinson's disease. *Neuropsychopharmacology*, *45*(1), 121–128. doi:10.1038/s41386-019-0448-y.

Stephen, R., Hongisto, K., Solomon, A., & Lönnroos, E (2017). Physical activity and Alzheimer's disease: A systematic review. *Journals of Gerontology. Series A, Biological Sciences and Medical Sciences*, *72*(6), 733–739. doi:10.1093/gerona/glw251.

Sun, G., Qian, S., Jiang, Q., Liu, K., Li, B., Li, M., … Liu, Y. (2013). Hyperthermia-induced disruption of functional connectivity in the human brain network. *PloS One*, *8*(4), e61157. doi:10.1371/journal.pone.0061157.

Titus, D. J., Furones, C., Atkins, C. M., & Dietrich, W. D. (2015). Emergence of cognitive deficits after mild traumatic brain injury due to hyperthermia. *Experimental Neurology*, *263*, 254–262. doi:10.1016/j.expneurol.2014.10.020.

Tremolizzo, L., Bianchi, E., Susani, E., Pupillo, E., Messina, P., Aliprandi, A., … Ferrarese, C. (2017). Voluptuary habits and risk of frontotemporal dementia: A case control retrospective study. *Journal of Alzheimer's Disease*, *60*, 1–6. doi:10.3233/JAD-170260.

Tsalenchuk, M., Gentleman, S. M., & Marzi, S. J. (2023). Linking environmental risk factors with epigenetic mechanisms in Parkinson's disease. *npj Parkinson's Disease*, *9*(1), 123. doi:10.1038/s41531-023-00568-z.

Valcárcel-Ocete, L., Alkorta-Aranburu, G., Iriondo, M., Fullaondo, A., García-Barcina, M., Fernández-García, J. M., … Aguirre, A. (2015). Exploring genetic factors involved in Huntington disease age of onset: E2F2 as a new potential modifier Gene. *PLoS One*, *10*(7), e0131573. doi:10.1371/journal.pone.0131573.

Wen, Y., Zhou, Y., Jiao, B., & Shen, L. (2021). Genetics of progressive supranuclear palsy: A review. *Journal of Parkinson's Disease*, *11*(1), 93–105. doi:10.3233/jpd-202302.

Zatta, P., & Frank, A. (2007). Copper deficiency and neurological disorders in man and animals. *Brain Research Reviews*, *54*(1), 19–33. doi:10.1016/j.brainresrev.2006.10.001.

Zielinski, M. R., & Gibbons, A. J. (2022). Neuroinflammation, sleep, and circadian rhythms. *Frontiers in Cellular and Infection Microbiology*, *12*, 853096. doi:10.3389/fcimb.2022.853096.

2 Blood–Brain Barrier in Brain Health, Aging, and Neurodegenerative Diseases

Swati Prakash and Vishakha Jaiswal*
Amity Institute of Pharmacy, Amity University,
Lucknow, Uttar Pradesh, India

Shradha Bisht
Uttaranchal Institute of Pharmaceutical Sciences,
Uttaranchal University, Dehradun, Uttarakhand, India

Mousumi Besan
Department of Pharmacy, Gayatri College of Pharmacy,
Jamadarpali, Sambalpur, Odisha, India

2.1 INTRODUCTION

The brain and spinal cord together govern all the vital activities and make up most important system known as the central nervous system (CNS). It has unique physiological and anatomical structures that provide the CNS an immune-privileged status. Therefore, preserving the brain's health is essential for it to function in optimal capacity. This can be ensured by checking the movement of the vital molecules inside the brain and not the harmful substances. The brain being a delicate organ, a highly specialized environment has to be maintained inside it for the maintenance of ionic homeostasis of the interstitial fluid inside the brain throughout life (Yu et al., 2024). The mobility of the ions, molecules, and cells is regulated by a dynamic and semi-permeable multi-functional natural barrier, called as blood–brain barrier (BBB), existing between the brain and its external environment. It is a crucial physiological structure that is vital in preserving the homeostasis of the brain (Lacoste et al., 2025). BBB acts as a selective semi-permeable membrane and is made up of endothelial cells (ECs) connected by tight junctions, along with a supporting network of astrocytes, pericytes, and the extracellular matrix that protects the brain from toxins, pathogens, and fluctuations in blood-borne substances while allowing essential nutrients to pass through it. The receptors, efflux pumps, and transporters also aid the facilitation of the movement of the molecules inside the brain. The primary function of BBB is to prevent the entry of harmful substances and only allow the movement of vital molecules inside it (Andjelkovic et al., 2023). The transport of essential nutrients, metabolites, and oxygen to the brain and the removal of carbon dioxide from the brain take place through the circulatory system, which is aided by the BBB.

Although the integrity and functions of the BBB are crucial for regulating brain health, they are also quite vulnerable to alterations with advancing age, because, aging is found to have structural and functional changes in the BBB, leading to increased permeability and impaired protective functions (Andjelkovic et al., 2023). Reduced cognitive function and heightened susceptibility to

*Corresponding Author. E-mail: sprakash@lko.amity.edu, swati.prakash.0206@gmail.com

DOI: 10.1201/9781003530510-2

neurodegenerative diseases including Alzheimer's, Parkinson's, and multiple sclerosis (MS) occur because of age-related changes in the BBB (Hang et al., 2023). In the context of neurodegenerative disorders, the deterioration of BBB is not only a result but frequently an early event that exacerbates the course of the neurological disease. The disrupted BBB integrity permits toxic chemicals to enter the brain, sets off inflammatory responses, and contributes to neuronal cell death (Hang et al., 2023).

In the present chapter, we will be discussing the overview of BBB and its role in brain health with respect to aging and neurodegenerative diseases. We will explore the mechanisms behind dysfunction in the BBB, and its effect on the development of neurodegenerative disorders, along with the possible therapeutic approaches toward the restoration of BBB integrity in the aging population. For the development of effective therapies in the management of neurological disorders and aging, understanding the intricacies of BBB becomes vital and crucial for improving brain health.

2.2 STRUCTURE AND FUNCTION OF THE BBB

A comprehensive and integrated neurovascular unit present in the CNS responds to the energy needs and regulates the blood flow in the brain. It is the structural and functional unit composed of

TABLE 2.1
The Core Cellular and Structural Elements of the BBB (Wu et al., 2023)

1. Cellular Components and Their Description	
Endothelial cells (EC)	Cerebral endothelial cells are specialized to form the blood–brain barrier, featuring tight junctions, high mitochondrial density, and polarized membranes. These adaptations regulate molecular entry, maintain brain homeostasis, and distinguish them from other endothelial cells.
Glycocalyx	Vascular endothelial cells secrete a glycocalyx layer, composed of proteoglycans and glycosaminoglycans like hyaluronic acid and heparan sulfate. This layer helps maintain blood flow and inhibits coagulation and inflammation.
Astrocytes	Astrocytes are abundant, star-shaped glial cells that support brain homeostasis by regulating immune responses, neuronal activity, and the blood–brain barrier (BBB). They manage metabolic functions such as pH balance, potassium buffering, and neurotransmitter uptake. Astrocytic end feet tightly cover cerebral capillaries, contributing to BBB integrity. Astrocytes are classified as protoplasmic in gray matter and fibrous in white matter, and their end feet attach to the basement membrane via aquaporin-4 and the dystroglycan–dystrophin complex.
Pericytes	Pericytes are contractile cells located between endothelial cells and astrocyte end feet, embedded in the basal lamina around brain microvessels. Abundant in the CNS, they play key roles in blood–brain barrier (BBB) formation, brain homeostasis, and regulation of capillary diameter and blood flow. Pericytes also interact with other cells to control angiogenesis, phagocytosis, and inflammation, helping remove harmful substances from the brain.
2. Structural Components	
Tight junctions (TJs)	Junctional proteins between adjacent endothelial cells form tight seals that regulate paracellular permeability and maintain the blood–brain barrier by preventing harmful substances and hydrophilic molecules from entering the brain. Key tight junction proteins include claudins, occludins, junctional adhesion molecules, F-actin, and zonula occludens proteins. These junctions can be bicellular or tricellular, depending on the cells they connect.
Basal lamina	The basal lamina, or basement membrane, is a key component of the blood–brain barrier, providing structural support, cell anchorage, immune regulation, permeability control, and signal transduction. It is mainly composed of collagen IV, heparan sulfate, nidogen, laminin, and perlecan.
Perivascular space	The perivascular space is the area between glial cells, especially astrocytes, and the endothelial cells of blood vessels, acting as a channel for fluid, solute, and immune cell exchange between blood vessels and the brain's interstitial fluid.

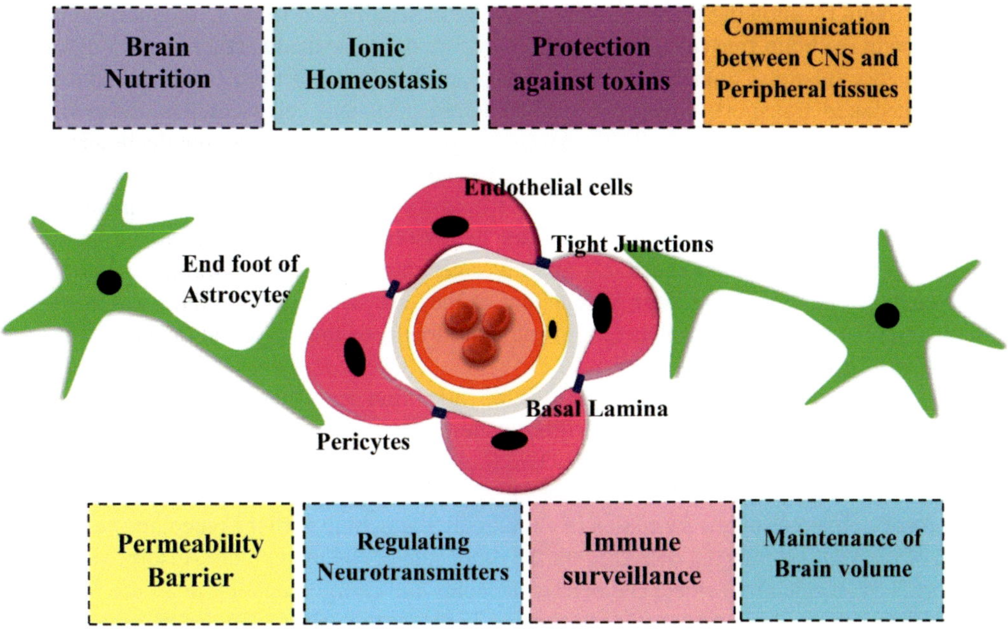

FIGURE 2.1 Cellular and structural components of the BBB and an overview of its functions. This figure shows the main structural components of the BBB; endothelial cells, tight junctions, pericytes, basal lamina, and astrocyte end feet, along with its key functions, such as nutrient supply, ionic balance, protection from toxins, selective permeability, neurotransmitter regulation, immune defense, and brain volume maintenance.

multiple cells, including ECs, astrocytes, mural cells, neurons, microglia, and extracellular matrix components, working in a synchronized manner to maintain the nutrient delivery, waste clearance, and homeostasis in the brain (Segarra, Aburto & Acker-Palmer, 2021). Being the central component of the whole system, a semi-permeable is present in the brain that prevents the entry of harmful substances and allows only required movements. The microvascular ECs of the NeuroVascular Unit (NVU) form this selective and semi-permeable membrane called BBB. This metabolic and physical barrier is crucial for maintaining homeostasis in the brain by preventing the movement of potentially hazardous substances in the bloodstream. The conception of BBB was laid by the idea of Ridley (around 1653), after observing the differences in the permeability of beeswax and mercury in the brain tissues. Following this, through his experiments, phenomenal research was conducted by the Nobel laureate Paul Ehrlich and later by Edwin Goldman on the existence of BBB. BBB is a complex structure primarily composed of several key cellular components that work together to maintain the integrity and functionality of the CNS. These structural components provide a supportive framework for regulating the functions of this integral homeostatic barrier. It is composed of ECs, tight junctions, basement membrane, and the end feet of the astrocytes. The different cellular and structural components of the BBB are included in Table 2.1.

As discussed above, the BBB is a selective transport system regulating the exchange of molecules from the blood to the brain for the maintenance of homeostasis. An overview of the functions of the BBB is represented in Figure 2.1. During aging and neurological disorders, the integrity and the functions of the BBB get compromised.

2.3 TRANSPORT MECHANISMS OF THE BBB

The BBB executes numerous significant functions as listed in Figure 2.1. To perform these functions, it needs transport mechanisms that can help it to maintain its permeability between the brain and the bloodstream by allowing the movement of useful molecules and limiting the harmful ones, so as to maintain the brain free from the alterations occurring in the blood. The transport mechanism includes paracellular and transcellular diffusional pathways, transport-protein-mediated transport, receptor-mediated transport, adsorptive-mediated transport, and cell-mediated transport (Knox et al., 2022). The movement of small, water-soluble molecules occurring without the requirement of energy in a concentration-dependent manner across the membrane is paracellular diffusion, while transcellular diffusion involves the migration of lipid-soluble molecules (gaseous molecules like O_2, CO_2, and anesthetics) through the membrane by dissolving in the lipid membrane of the ECs. Besides this transport, if any undesirable molecules get entry inside the brain, it is either degraded by the enzymes present in the brain vasculature or they are sent back into the blood through the efflux pumps, such as P-glycoprotein (P-gp) and multidrug resistance proteins (MRPs) (Ding et al., 2020) (Figure 2.2).

The transport of essential nutrients like glucose and amino acids occurs through the specialized transport proteins such as Adenosine TriPhosphate (ATP)-binding cassette (ABC) transporters, against the concentration gradient by the active transport channel. The drugs can also get access to the brain through this route (Cockerill et al., 2018). Receptor-mediated transport involves the migration of a molecule by the formation of vesicles after attaching with the receptors present on the surface of the ECs, which gets delivered on the other side of the membrane. Adsorptive-mediated transport is the mechanism where the macromolecules and charged nanoparticles are transported through the BBB by the electrostatic interactions between the said molecules and the negatively

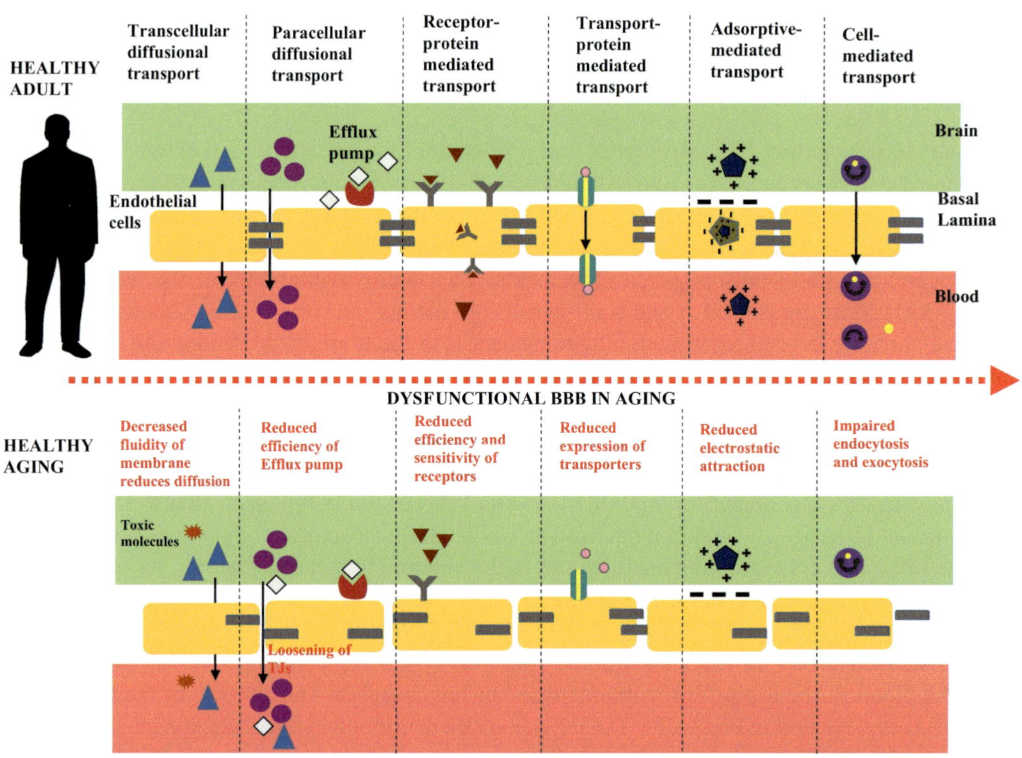

FIGURE 2.2 The transport mechanisms of BBB in normal adult vs aging. The figure compares the BBB in healthy adults and during aging. In healthy adults, various transport mechanisms work efficiently to protect the brain. With aging, these processes become impaired; membrane diffusion, efflux pumps, receptors, and transporters are less effective, tight junctions loosen, and endocytosis/exocytosis is reduced, leading to a more permeable and dysfunctional BBB.

charged endothelial membrane. Cell-mediated transport involves the migration of leukocytes across the BBB (Terstappen et al., 2021). The above-discussed transport mechanisms work in a synchronized manner maintaining the permeability of the BBB effectively. Also, a deeper understanding of these mechanisms is essential for enhanced drug delivery and therapeutic interventions.

2.4 AGING AND THE BBB

Studies have defined aging as the deterioration of the normal physiological process with age, which is considered around 30 years. Also, the inability to do physical activities in daily life can be related to the advancing age which reflects the physical and pathological changes and is regarded as aging. BBB is a highly regulated interface whose morphology changes with healthy aging contributing to the occurrence and progression of neurodegenerative diseases and also cognitive decline, which is triggered by inflammation (Banks et al., 2021). The permeability changes in BBB with the advancing age have garnered special attention in recent times. Various factors including oxidative stress, inflammation, genomic and epigenetic mutations, and disrupted cell signaling are influenced in a time-dependent manner during healthy aging. The time-dependent cellular damage factors, such as oxidative stress, epigenetic modifications, genomic instability, telomere attrition, dysregulation of cell signaling, and inflammatory responses, accumulate during healthy aging. When these natural disruptions are further exacerbated, they may become harmful (Knox et al., 2022). The alterations in the BBB during aging can be adaptive contributing to neurodegenerative diseases. The cellular structures including ECs, glycocalyx, astrocytes, and pericytes are altered with healthy aging along with the structural components, such as the TJs, basal lamina, and perivascular space. The loss of the integrity and the regularity of the BBB in the brain vasculature in the hippocampus is the first incident in aging as observed in the magnetic resonance imaging (MRI) scans. The hallmarks of the broken BBB-reduced tight junctional proteins, and reduced pericyte expression are observed in the aged mice, increasing the permeability and the cerebral blood flow (CBF). The property of neovascularization is reduced, and the density of the brain vasculature is reduced (Tarumi & Zhang, 2018). Degenerative changes decreased mitochondrial content, and decreased density of the endothelium make it progressively dysfunctional. An increase in the thickness and fragmentation of the extracellular matrix of the basal lamina is observed. The glycocalyx layer serves as the first point of defense between blood and the brain. Its composition changes with aging resulting in a thinner glycocalyx layer reducing the density of BBB (Banks et al., 2021). The density of pericytes is reduced, and a few ultrastructure changes are observed such as increased mitochondria size, foamy transformations, and increased platelet-derived growth factor-β (PDGFβ) receptors (Lacoste et al., 2025). Aging reduces the aquaporin-4 (AQP4) expression and increases the astrocyte's end feet size followed by astrogliosis. The expression of TJ proteins Ceroid Lipofuscinosis Neuronal- 5 (CLN-5), OC, and ZO-1 is reduced increasing the permeability of the BBB. Due to the structural changes with aging, the functions of the BBB including permeability, transport, and clearance are compromised. Advancing age challenges the BBB in numerous ways which can be adaptive but understanding the changes in the cellular and structural components of the BBB is significant for the development of therapeutic strategies for maintaining the integrity of BBB in neurodegenerative diseases.

2.5 BBB IN NEURODEGENERATIVE DISEASES

2.5.1 Alzheimer's Disease (AD)

Alzheimer's disease (AD), the most common cause of dementia, is primarily a cognition-associated neurodegenerative disorder that affects memory, thinking, and reasoning in daily life. The symptoms of AD usually start subtly and worsen over time. Early signs often include mild forgetfulness. With due course of the disease individuals experience memory loss, confusion about time and place, difficulty in speaking, and changes in personality and behavior. In the later stages, they may struggle with basic tasks and become dependent on others for care. The exact cause of AD is not fully understood, but it is believed to involve a combination of age-related changes in the brain, along with genetic, health, and lifestyle factors (Schachter & Davis, 2000). Age is the biggest risk factor, with most cases occurring in individuals over 65. Other risk factors include family history, certain genetic markers [such as the Apolipoprotein E ε4, (APOE) allele], cardiovascular health, and a history of head injuries (Raulin et al., 2022).

AD is characterized by the accumulation of amyloid-beta (Aβ) plaques, i.e., clumps of protein fragments, and tau tangles, i.e., twisted tau protein fibers of a protein in the brain. These abnormalities disrupt communication between neurons and lead to cell death, ultimately resulting in brain shrinkage seen in advanced stages of the disease (Moujalled, Strasser & Liddell, 2021).

2.5.2 BBB Dysfunction in AD

BBB dysfunction plays a critical role in the development and progression of AD. Being a highly selective barrier, BBB regulates the movement of molecules between the bloodstream and the brain and maintains a stable environment for optimal brain function. In AD, several mechanisms contribute to the disruption of this barrier, exacerbating neurodegeneration.

2.5.3 Mechanisms of BBB Dysfunction in AD

ECs Dysfunction: ECs line the blood vessels and form tight junctions that control the passage of substances. In AD, these cells become damaged due to oxidative stress, inflammation, and Aβ toxicity (Tarafdar et al., 2022). The tight junctions between ECs are compromised, leading to increased permeability of the BBB. This allows harmful substances, such as circulating Aβ, inflammatory cytokines, and neurotoxic molecules, to enter the brain and cause neuronal damage.

Pericyte Loss and Dysfunction: Pericytes are contractile cells that wrap around ECs and play a crucial role in maintaining BBB integrity. In AD, pericytes are lost or become dysfunctional due to exposure to Aβ and other toxic factors. The loss of pericytes leads to a breakdown of the BBB, resulting in leakage of blood-derived products like fibrinogen and iron into the brain parenchyma. These substances can promote neuroinflammation, oxidative stress, and further Aβ deposition (Montagne et al., 2020).

Astrocyte Activation: In AD, astrocytes become reactive in response to Aβ and other pathological signals. Reactive astrocytes release inflammatory cytokines and metalloproteinases that degrade the basement membrane and tight junction proteins, further weakening the BBB (Price, Johnson & Norris, 2021). Additionally, reactive astrocytes contribute to the impaired clearance of Aβ, promoting its accumulation in the brain.

Aβ Accumulation: Aβ is a toxic protein that accumulates in the brains of Alzheimer's patients, forming plaques. Aβ can directly damage ECs, such as the receptor for advanced glycation end-products (RAGE), leading to increased transport of Aβ across the BBB into the brain (Petrushanko, Mitkevich & Makarov, 2023). Furthermore, Aβ can impair the function of transporters responsible for clearing it from the brain, such as the low-density lipoprotein receptor-related protein 1 (LRP1), leading to its accumulation and further damaging the BBB.

Inflammatory Processes: Chronic inflammation is a hallmark of AD and significantly contributes to BBB dysfunction. Microglia, the brain's resident immune cells, become activated in response to Aβ and other pathological stimuli. Activated microglia release pro-inflammatory cytokines like TNF-α, IL-1β, and IL-6, which can disrupt endothelial tight junctions and increase BBB permeability. Inflammation also promotes the recruitment of peripheral immune cells to the brain, which can exacerbate neuroinflammation and contribute to neuronal damage.

2.5.4 Parkinson's Disease

Parkinson's disease (PD) is a progressive motor activity-associated neurological disorder affecting other bodily functions over time. It is characterized by the gradual loss of dopaminergic neurons in the substantia nigra in the brain. The neurotransmitter dopamine plays a key role in sending messages to the part of the brain that controls movement and coordination. The reduced dopamine levels in PD produce motor activity-associated symptoms such as bradykinesia, tremors, stiffness, and balance issues. In prolonged PD, non-motor symptoms like sleep disturbances, cognitive changes, and mood disorders may also develop.

2.5.5 BBB Alterations in PD

In PD, growing evidence suggests that this barrier becomes compromised, contributing to the disease's progression. The alterations in the BBB during PD involve multiple mechanisms that can exacerbate neuronal damage and accelerate the disease.

2.5.6 ECs Dysfunction

In PD, these cells often become dysfunctional, resulting in a breakdown of tight junctions, which are critical for preventing unwanted substances from crossing into the brain such as toxins, inflammatory cytokines, and pathogens, to enter the brain. These substances can exacerbate neuronal damage, particularly in the dopaminergic neurons of the substantia nigra, which are already vulnerable to PD (Cossu, Hatano & Hattori, 2023).

2.5.7 Neuroinflammation

PD is associated with chronic neuroinflammation, which can further compromise BBB integrity. Activated microglia, the brain's immune cells, release pro-inflammatory cytokines that can damage the BBB by increasing its permeability. This allows peripheral immune cells and additional inflammatory molecules to enter the brain, creating a vicious cycle of inflammation and neurodegeneration. The ongoing neuroinflammation in PD not only damages neurons directly but also weakens the BBB, exacerbating the entry of harmful substances into the brain (Pathak & Sriram, 2023).

2.5.8 Oxidative Stress

The brain's environment in Parkinson's is characterized by high levels of oxidative stress, which results from an imbalance between the production of free radicals and the brain's ability to detoxify them. This oxidative stress can damage ECs and the tight junctions they form, leading to a compromised BBB. The entry of oxidative agents and other harmful molecules into the brain can then cause further neuronal injury, particularly to the dopaminergic neurons, thereby accelerating disease progression.

2.5.9 α-Synuclein Pathology

α-synuclein, a protein that aggregates to form Lewy bodies in PD, is also implicated in BBB dysfunction. Research suggests that α-synuclein can disrupt BBB integrity either directly, by affecting the ECs, or indirectly, by promoting neuroinflammation. The accumulation of α-synuclein in the brain can increase BBB permeability, allowing toxic proteins and immune cells to infiltrate the brain, which may worsen neuronal damage and contribute to the spread of the disease.

2.5.10 Vascular Dysfunction

Vascular changes are increasingly recognized as contributors to BBB dysfunction in PD. Reduced CBF, commonly observed in Parkinson's, can lead to hypoxia, which stresses the BBB and makes it more permeable. Vascular dysfunction can also impair the clearance of waste products from the brain, leading to the accumulation of toxic substances that can further damage neurons and the BBB itself (Rajeev et al., 2022).

2.5.11 Multiple Sclerosis

MS is an autoimmune disease affecting the brain and spinal cord. It is characterized by the loss of the myelin sheath due to the failure of the immune system to differentiate between the self and non-self, eventually destroying the insulating myelin sheath. This disturbs and hampers the signaling of the electrical impulses along the nerves producing a wide range of neurological symptoms. MS is often described as a disease with a highly variable course, where symptoms can vary greatly from one person to another, and the disease can fluctuate between periods of relapse and remission.

2.5.12 BBB Disruption in MS

In MS, the BBB becomes compromised, leading to a cascade of pathological events that contribute to the disease's progression. BBB disruption in MS is a multifaceted process that plays a significant role in the autoimmune attack on the CNS.

EC Dysfunction: The ECs that line the blood vessels in the brain lose their tight junctions due to inflammation and oxidative stress. This loosening of tight junctions increases BBB permeability, allowing immune cells, particularly T-cells, to cross into the CNS (Lubetzki & Stankoff, 2014). These T-cells play a key role in attacking the myelin sheath, leading to the demyelination characteristic of MS.

Increased Expression of Adhesion Molecules: In MS, ECs express higher levels of adhesion molecules, such as VCAM-1 and ICAM-1. These molecules act like "sticky" signals that help immune cells adhere to the blood vessel walls and migrate across the BBB into the CNS (Engelhardt & Ransohoff, 2012). This migration is a critical step in the initiation of the autoimmune response in MS.

Matrix Metalloproteinases (MMPs): MMPs are enzymes that degrade the extracellular matrix, including the proteins that form tight junctions between ECs. In MS, the production of MMPs, especially MMP-9, is upregulated. These enzymes break down the BBB's structural components, further increasing its permeability and facilitating the infiltration of immune cells into the CNS (Cabral-Pacheco et al., 2020).

2.5.13 Amyotrophic Lateral Sclerosis (ALS)

The second name of amyotrophic lateral sclerosis (ALS) is Lou Gehrig's disease, which is a progressive neurodegenerative disorder affecting the motor neurons in the CNS. The gradual loss of the motor neurons causes muscle weakness, atrophy, and eventually paralysis. ALS is characterized by its relentless progression and lack of effective treatments, making it a particularly challenging condition for both patients and healthcare providers.

2.5.14 BBB Role in ALS Pathogenesis

In ALS, BBB dysfunction has been increasingly recognized as a significant factor in the disease's pathogenesis. The disruption of the BBB contributes to neuroinflammation, oxidative stress, and neurodegeneration, all of which are central to ALS pathology.

2.6 MECHANISMS OF BBB DYSFUNCTION IN ALS

2.6.1 Endothelial Cell Dysfunction

In ALS, the ECs that line the blood vessels in the brain and spinal cord become dysfunctional. This dysfunction is characterized by the disruption of tight junctions, which are crucial for maintaining the integrity of the BBB. The breakdown of these junctions leads to increased BBB permeability, allowing potentially harmful substances, such as immune cells and pro-inflammatory cytokines, to infiltrate the CNS. This infiltration exacerbates the neuroinflammatory response and contributes to motor neuron damage (Takata et al., 2021).

2.6.2 Inflammatory Mediators

The presence of inflammatory mediators in ALS further compromises BBB integrity. Activated microglia and astrocytes in the CNS release cytokines such as TNF-α, IL-1β, and IL-6, which can induce the expression of adhesion molecules on ECs. These adhesion molecules facilitate the migration of immune cells across the BBB. The increased presence of inflammatory cells and cytokines within the CNS contributes to a chronic inflammatory environment, which accelerates motor neuron degeneration (Zhang et al., 2023).

2.6.3 Oxidative Stress

Oxidative stress is a significant factor in ALS and contributes to BBB dysfunction (Hemerková & Vališ, 2021). The accumulation of reactive oxygen species (ROS) and reactive nitrogen species (RNS) can damage ECs and disrupt tight junctions. This oxidative damage not only weakens the BBB but also exacerbates neuronal injury. In ALS, increased oxidative stress leads to the activation of MMPs, enzymes that degrade extracellular matrix components and further compromise BBB integrity (Cabral-Pacheco et al., 2020).

2.6.4 Protein Aggregation and Toxicity

Abnormal protein aggregation, a hallmark of ALS, also impacts the BBB. Proteins such as mutant TAR DNA-binding protein 43 (TDP-43) and neuronal superoxide dismutase-1 (SOD1), which aggregate in motor neurons, can affect ECs function and contribute to BBB breakdown. These aggregated proteins can indirectly influence BBB permeability by inducing inflammatory responses and oxidative stress. The presence of these toxic proteins in the brain can lead to dysfunction of the ECs and disruption of the BBB (Trist et al., 2022).

2.6.5 Genetic Factors

Genetic mutations associated with familial ALS, such as those in the SOD1, C9orf72, and TDP-43 genes, may also impact BBB function. For instance, mutations in SOD1 are known to affect EC function and increase oxidative stress, leading to BBB disruption. The genetic factors underlying familial ALS can contribute to the development of BBB dysfunction by altering cellular stress responses and promoting neuroinflammation (Song et al., 2020).

BBB alterations in neurodegenerative diseases are complex and involve a combination of multiple mechanisms including EC dysfunction, loosening of TJs, neuroinflammation, oxidative stress, and vascular changes. These alterations not only compromise the barrier's ability to protect the brain but also contribute to the progression of neurodegenerative processes. Understanding these mechanisms is crucial for developing therapeutic strategies aimed at preserving BBB integrity and slowing the progression of the devastating conditions. Table 2.2 provides an outcome of research related to the role of the BBB in various neurodegenerative diseases. It summarizes the

TABLE 2.2
Studies Identifying the Involvement of BBB in Neurodegenerative Diseases (Yuan et al., 2023)

Disease	Type of Study Conducted	Outcomes
AD	Review/mechanistic study	BBB dysfunction is central to AD pathology; impaired clearance of amyloid-beta.
	Experimental/preclinical study	Angiogenesis and BBB breakdown contribute to cognitive decline in AD.
PD	Clinical/experimental study	BBB disruption and neuroinflammation play significant roles in PD pathology.
	Experimental/preclinical study	Alterations in BBB permeability are associated with dopaminergic neuron loss.
MS	Review/mechanistic study	BBB disruption is a key factor in MS pathogenesis; immune cell infiltration.
	Review/mechanistic study	Increased expression of adhesion molecules leads to immune cell migration.
ALS	Experimental/preclinical study	BBB dysfunction contributes to neuroinflammation and motor neuron degeneration.
	Experimental/preclinical study	TDP-43-mediated BBB permeability and leukocyte infiltration promote neurodegeneration and contributes to ALS pathology.

type of study and the key outcomes, reflecting the current understanding and ongoing research in these areas.

2.7 GUT–BRAIN AXIS AND PROBIOTICS

The gut and the brain communicate with each other in a two-way manner, both influencing each other's functions, also called as gut–brain axis. The brain drives the functions of the gut through the hypothalamic–pituitary–adrenal (HPA) axis and the autonomic nervous system, and in turn gut talks to the brain through the multiple mediators that includes microbiota-derived metabolites, neuroactive substances, and the gut hormones. It is an intricate communication, where the signals are communicated through the enteric nervous system (ENS), also referred to as second brain which is deeply innervated in the gut wall, sympathetic and parasympathetic nervous system, immune system, gut microbiota metabolic systems, and also the neuroendocrine signaling pathways (Ashique et al., 2024). ENS is present within the gastrointestinal tract and is an extensive reflex control system that works in association with the CNS. It can operate independently if separated from the CNS; however, it is not autonomous. The ENS is composed of myenteric (auerbach's) and submucosal (meissner's) plexus, having sensory and motor neurons. They regulate the gastrointestinal motility, gastric acid secretion, blood flow, nutrient absorption, and the maintenance of gut health by interacting with the immune system. The key connecting link between the gut microbes and the brain is the vagus nerve; others include the metabolites of the microbial colonies, i.e., Short Chain Fatty Acids (SCFAs) and indole derivatives, secondary bile acids, and polysaccharides (Liu et al., 2022). The microbes influence the healthy and diseased states of the host by controlling the neurons of the ENS, synthesis of neurotransmitters, SCFAs, and gut hormones. The synthesis of more than 90% of serotonin occurs in gut (*Enterococcus*, *Streptococcus*), other NTs include dopamine (*Bacillus*), adrenaline (*Saccharomyces*, *Escherichia*), Gamma-Amino Butryic Acid (GABA) (*Bifidobacterium*, *Lactobacillus*), glutamine. These NTs cannot cross the BBB except GABA but indirectly affect the brain via interacting with the CNS. They also produce the pro- and anti-inflammatory signaling molecules. The mucosal epithelium in the gut has abundant immune cells, which contributes to the defense system. Its extensive involvement has shed light on the pathophysiology of several neurodegenerative diseases. Prebiotics, probiotics, and synbiotics supplementation have shown to modulate the gut microbiota for the treating the neurodegenerative disorders in addition to the fecal microbiota transplantation (FMT) technique (Matheson & Holsinger, 2023). Table 2.3 summarizes the probiotic suppliments and their role in various neurological disorders.

TABLE 2.3
The Effect of Probiotic Intervention on Neurological Disorders

Probiotic Supplementation	Disease
Lactobacillus species	AD
Bifidobacterium, *Lactobacillus*, and *Streptococcus* species	Cognitive dysfunction
Lactobacillus helveticus R0052 and *Bifidobacterium longum* R0175	AD
Lactobacillus rhamnosus HA-114 (1015 CFU) or *B. longum R0175*	AD
Lactobacillus and *Bifidobacterium* species	Memory deficits in AD
Bifidobacterium and *Lactobacillus* species	Improved motor functions in PD
Lactobacillus acidophilus, *Bifidobacterium bifidum*, *Lactobacillus reuteri*, and *Lactobacillus fermentum*	Neuroprotective effects in PD
Bacillus subtilis	Inhibits α-synuclein aggregation in PD
Fecal microbiota transplantation	Improved motor symptoms in PD
Probiotic capsule	Improved MS symptoms
Probiotic supplementation	Improved health in RRMS patients
Probiotic supplementation	Improved stroke treatment

2.8 DIAGNOSTIC TOOLS AND BIOMARKERS FOR BBB DYSFUNCTION

2.8.1 Imaging Techniques

To assess BBB integrity for research purposes, MRI techniques that call for administration are frequently used as contrast medium (CA). The detecting mechanism is the passive diffusion of a tiny molecule CA via the compromised BBB. Dynamic susceptibility contrast (DSC) MRI and dynamic contrast-enhanced (DCE) MRI are two methods that take advantage of a gadolinium-based CA (GBCA) (Heye et al., 2014).

2.8.1.1 DCE-MRI

T1-weighted DCE-MRI is commonly used for MRI-based assessments of BBB integrity. This technique involves injecting an intravenous contrast agent, which is usually a paramagnetic gadolinium-based contrast agent (GBCA) and detecting any leakage of the agent across the BBB as a change in T1 of brain tissue. The aim of DCE-MRI is the estimation of the transfer constant, K^{trans} of the contrast agent. It is a quantitative analysis of the rate of indicator transfer from the area between arteries and extravasation. People and animals can have region-specific K^{trans} values determined thanks to DCE-MRI's 3D macroscopic resolution. This method has been used to demonstrate BBB leakage in cerebral small vessel disease, AD, and MS, as well as in conditions with more severe leakage such as stroke and brain tumors (Thrippleton et al., 2019).

2.8.1.2 Dynamic Susceptibility Contrast Perfusion Imaging

Furthermore, the administration of a paramagnetic CA is necessary for DSC-MRI. However, this technique uses T2-weighted or largely T2*-weighted scans, which are vulnerable to reductions in signal strength caused by the effects of magnetic susceptibility caused by relatively high bloodstream concentrations of the CA. Conversely, the blooming effect describes how the changes in the T2* signal caused by the paramagnetic CA particles inside the blood vessel extend beyond the blood space and slightly expand into the surrounding area outside the blood vessels. Rapid imaging sequences measure the first pass of the contrast bolus to get brain perfusion measurements. One of the primary benefits of DSC-MRI is its brief scan length. Hemodynamic modeling uses arterial blood concentration as the input and applies computational deconvolution techniques to the tissue enhancement curve. This allows the blood perfusion parameters cerebral blood volume (CBV), cerebral blood flow (CBF), and mean transit time (MTT) to be calculated from the DSC-MRI data. The extravasation of CA has been demonstrated to impact DSC perfusion measurements. The amount of gadolinium that typically leaks into the parenchyma is very small and does not contribute to T2 or T2* signal decreases; instead, it causes the signal to reverse to an increase (the T1 shine-through effect) and contaminates the intended perfusion signal, particularly later in time after the first-pass peak. These effects cause an overestimation of CBF and CBV, interfere with the linear behavior of CA in relation to the relaxation rate, and affect the temporal dynamics of the DSC-MRI signal (Blockley et al., 2008).

2.8.1.3 Glucose Chemical Exchange Saturation Transfer Imaging/ Dynamic Glucose-Enhanced Imaging

The application of D-glucose as CA for MRI is a more recent development. Since glucose is an endogenous metabolite, it can be broken down naturally. Since glucose serves as the brain's primary energy source, studying glucose metabolism can provide insight into cerebral energy metabolism. Dynamic glucose-enhanced (DGE) MRI can be used to measure the D-glucose signal over time following the delivery of a D-glucose bolus. Chemical exchange saturation transfer (CEST) MRI is used to detect the signal; as a result, glucose CEST (glucoCEST) is another name for the method. Utilizing a frequency-selective radiofrequency pulse, the method saturates the protons in a target molecule. The CEST signal can be derived by examining the exchange of protons from water molecules with saturated protons. Low amounts of particular chemicals that are undetectable by traditional MRI methods can be quantified with this technology. The D-glucose molecule has five hydroxyl groups, making it an excellent candidate for CEST-MRI excitation and measurement. By filling the hydroxyl groups' protons to a resonance, the exchange with bulk water protons may be measured at a frequency of 1.2 ppm, which can be used to indirectly measure the metabolism of glucose.

2.8.1.4 Arterial Spin Labeling

Across the BBB, water is exchanged through a significant number of transport routes. Water exchange can be measured via arterial spin labeling (ASL), which may provide an intriguing signal for BBB impairment identification. Given that water is transported across the BBB by both passive and active diffusion mechanisms, and because water molecules have a much smaller molecular weight and size than gadolinium CAs, the technique may be sensitive to a wide range of BBB pathologies (Dickie, Parker & Parkes, 2020). Magnetic labeling, or inverting the magnetization of the incoming blood, is used in ASL. Following magnetic labeling, magnetic spins enter the imaging plane of the brain for a duration determined by the post-labeling delay time. The ASL signal is represented by the signal difference between the two images taken prior to and following the labeling of the blood inflow.

2.8.1.5 Intravoxel Incoherent Motion Imaging

Since this technique is based on diffusion-weighted imaging methods, no contrast administration is required. Water molecules move randomly in free and constrained ways, which is exploited by diffusion-weighted imaging. Thermal motion is the term used to describe the movement of free water in a fluid under the influence of temperature, fluid viscosity, and molecular diffusivity. The vessel wall limits the mobility of water molecules inside capillaries. It's called pseudo diffusion when these water molecules move through the capillary network with the bloodstream. These two distinct kinds of circulating water molecules are distinguished in IntraVoxel Incoherent Motion (IVIM) imaging by using a biexponential model to describe the signal. In this manner, the parameters for blood volume perfusion fraction (f), intravascular pseudo diffusivity (D*), and parenchymal diffusivity (D) can be determined (Paschoal et al., 2018). These parameters provide us with information about BBB integrity.

2.8.1.6 Molecular and Cellular Biomarkers of BBB Dysfunction

When determining the BBB permeability, physicochemical characteristics like size and polarity should generally be considered. One potential method to assess the permeability and integrity of the BBB is to look at the likelihood of big molecular weight markers like dextran entering the brain when the BBB is damaged and loses its integrity. Small molecular weight markers, on the other hand (MW of less than 400 Da), are ideal for precisely analyzing minute changes in BBB permeability that arise in numerous neurological disorders. Small molecular weight markers, like sucrose, should be employed to quantify slight BBB impairment since even mild barrier degradation is expected to have a detectable impact on their permeability, which bigger molecular weight markers cannot measure. Evans blue is a high-molecular-weight permeability marker that is commonly used nowadays to evaluate the permeability of cellular and capillary membranes. As soon as serum albumin enters the circulatory system, Evans blue binds to it substantially. When the BBB is disrupted, the dye seeps in and stains the area of the brain injury where the BBB is compromised. One benefit of employing the Evans blue tracer is its capacity to make the zone of changed permeability easily visible. This is highly helpful in animal research where lesions causing BBB disruption in certain brain regions are examined. The introduction of HRP (Horseradish peroxidase) is significant because the peroxidase's reaction product generates an electron-dense material that is visible under an electron microscope. The location and characteristics of the BBB, as well as the crucial role the brain endothelium played in its development, are revealed by the application of horseradish peroxidase (Brightman & Reese, 1969). HRP can traverse the BBB in a variety of pathological and experimental settings, which offers morphological support for a barrier opening sodium fluorescein, a 376 Da molecule, was the first visualizable small molecular-sized marker to be introduced into the BBB. To help with its detection in BBB permeability tests, it was proposed that sodium fluorescein may be evaluated spectro photo – fluorometrically (excitation at 440 nm and emission at 525 nm). Sodium fluorescein, however, appears to be far less hazardous than HRP or Evans blue. Its low binding to plasma proteins, in contrast to Evans blue dye, supports its application as a tiny molecular size indicator for BBB integrity. In general, labeled dextran are valuable markers of BBB integrity, which can be used safely in small concentrations. The use of biotin-labeled dextrans and fluorophore helps to clarify that, contrary to what is commonly thought, the pathway of the entrance from the blood to the CSF is intracellular via the plexus epithelial cells, not intercellularly via the tight junctions. Molecules labeled with biotin can be observed using electron and light microscopy. Its permeability via the blood–CSF barrier is comparable to that of sucrose, inulin, and more conventional permeability-tagged indicators (Ek et al., 2006).

2.8.1.7 Clinical Relevance of BBB Biomarkers

For several neurological disorders, biomarkers are crucial diagnostic and prognostic instruments. The studies have highlighted the significance of various markers in BBB breakdown. serum visinin-like protein 1 (sVILIP-1) and serum caveolin 1 (sCAV-1) can predict the breakdown of BBB which increases in epileptic patients. Another protein biomarker that is generated by astrocyte end feet is called s100β, and it is released into the peripheral circulation as soon as the BBB is damaged (106). Additionally, a study found that A-kinase anchoring protein 7 (AKAP7) was highly expressed in lymphocytes in peripheral blood (PB), suggesting that it could be used to detect BBB collapse for stroke. Also, neuron-specific enolase (NSE) and Glial Fibrillary Acidic Protein (GFAP) are the promising biomarkers for the detection of BBB breakdown. According to a study, a higher level of sPDGFRβ is linked to pericyte injury and BBB disruption, which can impair cognition. Vascular Dementia (VaD) was also found to have soluble PDGFRβ as a biomarker. The expression of the soluble adhesion molecules PECAM-1, P-selectin, and E-selectin are upregulated in individuals with compromised BBB and hence can be used as biomarkers for evaluation of the irregular BBB (D'Ambrosio et al., 2015). The medical ailments that will affect the BBB will also affect the vascular tone and blood flow to the brain, which will further affect BBB transport. Therefore, a deeper understanding of the vascular system will aid in the development of new, effective treatment approaches for cognitive impairment.

2.9 THERAPEUTIC APPROACHES TO PROTECT AND RESTORE BBB FUNCTION

Multiple therapeutic approaches are significant in the protection and restoration of BBB, improving the treatment outcomes in neurological disorders. Some of the emerging strategies include the modulation of TJs, stem cell-based approaches, nanotechnology, and microRNA (miRNAs)-based therapies are listed below (Figure 2.3).

2.9.1 Target Specific Tight Junction Modulators

2.9.1.1 CLN-5

CLN-5, a key TJ protein, plays a crucial role in regulating the integrity and permeability of the BBB. Several upstream signaling pathways influence CLN-5 levels, and targeting these pathways to modulate its expression in response to disease presents a promising therapeutic approach. Elevated CLN-5 expression has a neuroprotective effect in neurological disorders by selectively opening the BBB to small molecules of <800 Da (Greene, Hanley & Campbell, 2019).

Perampanel, a novel non-competitive antagonist of the α-amino-3-hydroxy-5-methyl-4-isoxazole propionate receptor (AMPAR), may regulate CLN-5 expression at the protein level. The antibiotic doxycycline and progesterone significantly increase CLN-5 expression, reducing BBB leakage and cerebral infarct size in rat models. Additionally, progesterone improves thrombin-induced BBB

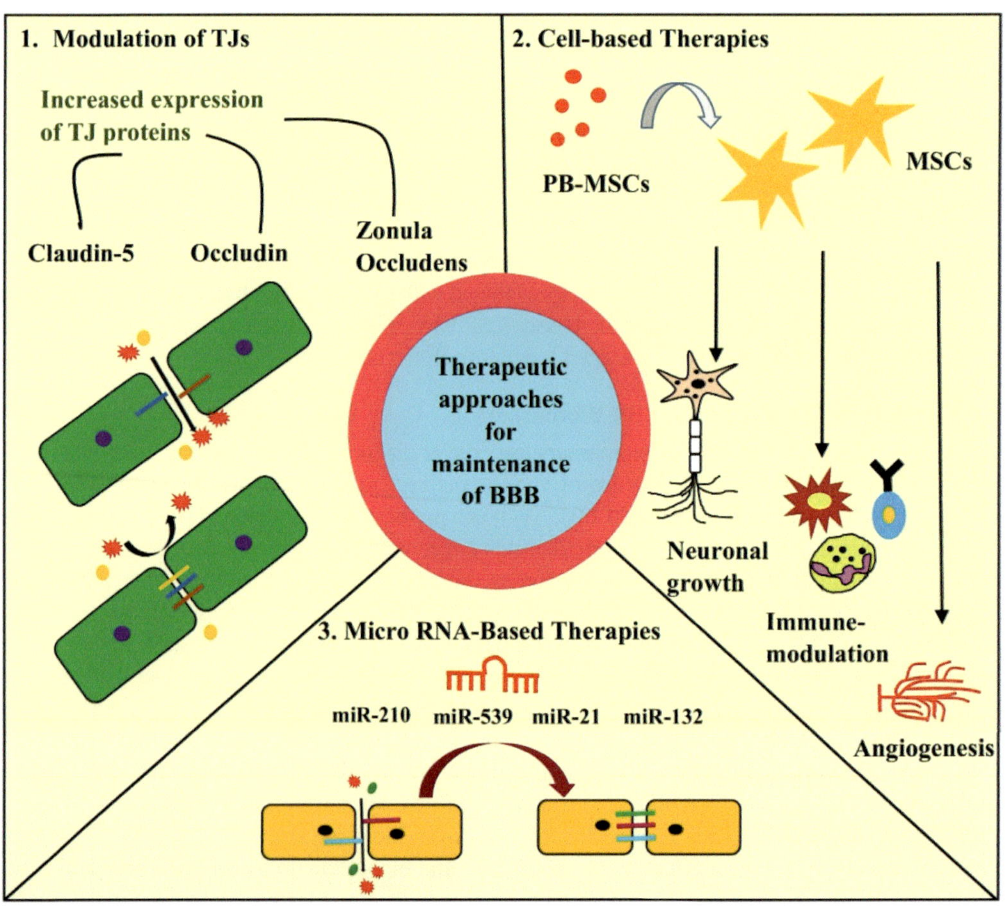

FIGURE 2.3 Therapeutic approaches to restore BBB.

dysfunction. Activation of estrogen receptor-β (ER-β) helps preserve CLN-5 protein levels by specifically suppressing HIF-1α/VEGF pathways. The G protein-coupled estrogen receptor 1 (GPER-1) also plays a key role in estrogen-mediated neuroprotection. Agonists of GPER-1 significantly reduce immunoglobulin G (IgG) leakage and enhance CLN-5 levels (Lv et al., 2018).

2.9.1.2 Occludin (OC)

OC, a 65-kDa tetraspan integral membrane protein, plays a crucial role in stabilizing TJs and ensuring optimal barrier function. One of its transmembrane domains is involved in regulating epithelial integrity and paracellular permeability. Preclinical studies are essential to further evaluate and validate OC as a potential therapeutic target (Sugiyama et al., 2023).

2.9.1.3 Zona Occludens

Zona occludens (ZO) is the third major protein involved in TJs and consists of three subtypes: ZO-1, ZO-2, and ZO-3. The depletion of ZO proteins has been found to cause delayed formation of TJs and lead to apical reorganization of actin and myosin. In embryos, this disruption is lethal. ZO-1 expression can be modulated by interferon-γ, and other research has indicated that interleukins also play a role in the localization and expression of TJ proteins (Brunner, Ragupathy & Borchard, 2021).

2.9.2 Stem Cell-Based Approaches

2.9.2.1 For Damaged ECs

Mesenchymal stem cells (MSCs) can be isolated from various sources, including adult and perinatal tissues. Their impact on neural tissue includes differentiating into neuron-like cells and neuroglia, immunomodulation, and promoting neovascularization. Due to their high neurogenic potential and the non-invasive isolation procedure, peripheral blood MSCs (PB-MSCs) have been proposed as a promising alternative for neural tissue regeneration in animal models. When administered systemically, MSCs have the capacity to home in on infarcted regions, facilitating neural repair through paracrine signaling and differentiating into neuron-like cells or astrocytes to support tissue regeneration. Owing to their remarkable neuroprotective properties, MSCs are increasingly being investigated as therapeutic vectors for targeted interventions in neurodegenerative conditions such as Parkinson and Alzheimer (Do et al., 2021).

2.9.2.2 For Pericyte Transplantation

Pericytes can be derived from embryonic mouse MSCs as well as from human pluripotent stem cells (hPSCs) or human induced pluripotent stem cells (iPSCs) following neural crest fate induction from human embryonic fibroblasts. The induction of pericytes from these neural crest-like cells is facilitated by the addition of basic fibroblast growth factor (bFGF) and PDGF-BB (PDGF-BB). The therapeutic potential of pericyte transplantation for addressing neurovascular damage has garnered increasing interest due to its regenerative capabilities and role in restoring the integrity of the BBB. Pericyte transplantation could potentially enhance pericyte coverage in patients with AD and PD, contributing to the restoration of the BBB, reducing serum extravasation, promoting Aβ clearance, and normalizing efflux mechanisms (Kim et al., 2024).

2.9.3 Nanoparticle-Based Drug Delivery

Liposomes are a promising treatment strategy for brain-related injuries by enhancing drug delivery and therapeutic efficacy. Long et al. developed brain-targeted borneol–baicalin liposome (BO-BA-LP) to improve BBB integrity after injury, which significantly increased the concentration of BA in the brain and improved therapeutic outcomes by protecting the BBB (Long et al., 2023).

2.9.4 MicroRNA-Based Therapies

miRNAs, a class of small non-coding RNAs (20–25 nucleotides), are gaining recognition as key regulators of BBB integrity. They directly target TJ proteins, such as ZO, OCN, and CLN-5, as well as adherens junction proteins like VE-cadherin, resulting in altered BBB integrity. For instance, inhibiting miR-210 has been shown to regulate the tight junction protein OC and the adherens junction protein β-catenin strengthening the BBB.

2.9.5 Natural Products in Neuroprotection

Restoring the integrity of BBB integrity is a significant area of research. The diverse chemical structures and sustainable nature of natural products have made them promising candidates for neuroprotection. Consequently, several natural agents have been studied for their potential to support or restore BBB integrity (Table 2.4).

This figure summarizes therapeutic approaches for maintaining the BBB. It highlights three main strategies: (1) modulation of TJ proteins like claudin-5, occludin, and zonula occludens to strengthen the BBB; (2) cell-based therapies using MSCs and PB-MSCs to promote neuronal growth, immune modulation, and angiogenesis; and (3) miRNA-based therapies (e.g., miR-210, miR-539, miR-21, miR-132) to regulate BBB integrity. The central circle emphasizes that these methods collectively support BBB maintenance.

TABLE 2.4
Natural Agents for the Maintenance of BBB Integrity (Isabel, Belén & Elena, 2024)

Natural Product	Chemical Category	Mechanism for Protecting BBB Integrity	Impact on BBB Health
Carvacrol	Monoterpenoid phenol	Reduction in matrix metalloproteinase-9 (MMP-9) expression and elevation of tight junction protein (ZO-1, occludin, and claudin-5 levels)	Maintaining BBB integrity
Rhein	Anthracene carboxylic acid	Inhibition of the gp91phox subunit of NADPH oxidase, leading to reduced MMP-9 expression and enhanced ZO-1 levels	BBB protection
Resveratrol	3,5,4′-trihydroxystilbene	Lowering the half of the concentration of matrix metalloproteinase-9 (MMP-9) in cerebrospinal fluid (CSF) can facilitate clearance of amyloid-beta (Aβ)	Restore BBB integrity and beneficial in AD
Ginkgolides and bilobalide (*Ginkgo biloba*)	Terpene trilactone	Increased the expression of tight junction proteins, including ZO-1 and occludin	Restore BBB integrity
Omega-3 polyunsaturated fatty acids (Omega-3 PUFAs)	Polyunsaturated fatty acids (PUFAs)	Enhances the expression of the efflux transporter LRP-1 and tight junction proteins, improves glymphatic function, and facilitates clearance	Promotes neuroprotection and boosts brain health
Sulforaphane	1-isothiocyanato-4-methylsulfinylbutane	Increases the expression of nuclear factor erythroid 2-related factor 2 (Nrf2) and HO-1 in perivascular astrocytes	Prevents the breakdown of the blood–brain barrier (BBB) integrity

2.10 CONCLUSION

The intricate and dynamic structure of the BBB plays a critical role in maintaining the delicate balance of the CNS. Through its cellular components, including ECs and tight junctions, and its various transport mechanisms, the BBB regulates the exchange of substances between the bloodstream and the brain, ensuring neural homeostasis and protecting the brain from harmful agents. However, this barrier's functionality is not static, as it is influenced by a myriad of factors such as age, stress, and diseases.

Understanding the molecular mechanisms underlying BBB dysfunction, particularly the roles of oxidative stress, inflammation, and environmental factors, has shed light on how the integrity of the BBB can be compromised. Age-related changes, characterized by increased permeability and impaired transport mechanisms, further exacerbate BBB dysfunction, contributing to reduced cognition and the progression of neurological diseases such as AD, PD, MS, and ALS. The interplay between the gut–brain axis and the BBB, along with the emerging role of probiotics, highlights the complex and interconnected nature of CNS health. Diagnostic tools and biomarkers have become invaluable in detecting BBB dysfunction, offering insights into disease progression and potential therapeutic targets. Therapeutic strategies aimed at protecting and restoring BBB function, including drug delivery systems, lifestyle interventions, and cutting-edge therapies like stem cells, represent promising avenues for treatment. Ongoing research and clinical trials continue to advance our understanding of BBB dynamics, paving the way for innovative approaches to address the challenges posed by neurodegenerative diseases.

In conclusion, the BBB is a fundamental component of brain health, and its dysfunction is intricately linked to the pathogenesis of various neurological disorders. Continued research into the molecular mechanisms of BBB regulation and dysfunction, along with the development of novel therapeutic strategies, holds the potential to significantly impact the management and treatment of CNS diseases, ultimately improving patient outcomes.

ABBREVIATIONS

BBB	blood–brain barrier
EC	endothelial cells
CA	contrast medium
DSC MRI	dynamic susceptibility contrast MRI
DCE MRI	dynamic contrast-enhanced (DCE) MRI
CNS	central nervous system
Aβ *plaques*	amyloid-beta plaques
MMM	matrix metalloproteinases
ALS	amyotrophic lateral sclerosis
Da	Dalton
PD	Parkinson's disease
AD	Alzheimer's disease
MS	multiple sclerosis
MSCs	mesenchymal stem cells

REFERENCES

Andjelkovic, A. V., Situ, M., Citalan-Madrid, A. F., Stamatovic, S. M., Xiang, J., & Keep, R. F. (2023). Blood-brain barrier dysfunction in normal aging and neurodegeneration: Mechanisms, impact, and treatments. *Stroke*, *54*(3), 661–672.

Ashique, S., Mohanto, S., Ahmed, M. G., Mishra, N., Garg, A., Chellappan, D. K., …, & Kahwa, I. (2024). Gut-brain axis: A cutting-edge approach to target neurological disorders and potential synbiotic application. *Heliyon*, *10*(13).

Banks, W. A., Reed, M. J., Logsdon, A. F., Rhea, E. M., & Erickson, M. A. (2021). Healthy aging and the blood–brain barrier. *Nature Aging*, *1*(3), 243–254.

Blockley, N., Jiang, L., Gardener, A., Ludman, C., Francis, S., & Gowland, P. (2008). Field strength dependence of R1 and R relaxivities of human whole blood to prohance, vasovist, and deoxyhemoglobin. *Magnetic Resonance in Medicine: An Official Journal of the International Society for Magnetic Resonance in Medicine*, *60*(6), 1313–1320.

Brightman, M., & Reese, T. (1969). Junctions between intimately apposed cell membranes in the vertebrate brain. *The Journal of Cell Biology*, *40*(3), 648–677.

Brunner, J., Ragupathy, S., & Borchard, G. (2021). Target specific tight junction modulators. *Advanced Drug Delivery Reviews*, *171*, 266–288.

Cabral-Pacheco, G. A., Garza-Veloz, I., Castruita-De la Rosa, C., Ramirez-Acuña, J. M., Perez-Romero, B. A., Guerrero-Rodriguez, J. F., …, & Martinez-Fierro, M. L. (2020). The roles of matrix metalloproteinases and their inhibitors in human diseases. *International Journal of Molecular Sciences*, *21*(24), 9739.

Cockerill, I., Oliver, J.-A., Xu, H., Fu, B. M., & Zhu, D. (2018). Blood-brain barrier integrity and clearance of amyloid-β from the BBB. *Molecular, Cellular, and Tissue Engineering of the Vascular System*, 261–278.

Cossu, D., Hatano, T., & Hattori, N. (2023). The role of immune dysfunction in Parkinson's disease development. *International Journal of Molecular Sciences*, *24*(23), 16766.

D'Ambrosio, A., Pontecorvo, S., Colasanti, T., Zamboni, S., Francia, A., & Margutti, P. (2015). Peripheral blood biomarkers in multiple sclerosis. *Autoimmunity Reviews*, *14*(12), 1097–1110.

Dickie, B. R., Parker, G. J., & Parkes, L. M. (2020). Measuring water exchange across the blood-brain barrier using MRI. *Progress in Nuclear Magnetic Resonance Spectroscopy*, *116*, 19–39.

Ding, S., Khan, A. I., Cai, X., Song, Y., Lyu, Z., Du, D., …, & Lin, Y. (2020). Overcoming blood–brain barrier transport: Advances in nanoparticle-based drug delivery strategies. *Materials Today*, *37*, 112–125.

Do, P. T., Wu, C.-C., Chiang, Y.-H., Hu, C.-J., & Chen, K.-Y. (2021). Mesenchymal stem/stromal cell therapy in blood–brain barrier preservation following ischemia: Molecular mechanisms and prospects. *International Journal of Molecular Sciences*, *22*(18), 10045.

Ek, C. J., Dziegielewska, K. M., Stolp, H., & Saunders, N. R. (2006). Functional effectiveness of the blood-brain barrier to small water-soluble molecules in developing and adult opossum (*Monodelphis domestica*). *Journal of Comparative Neurology*, *496*(1), 13–26.

Engelhardt, B., & Ransohoff, R. M. (2012). Capture, crawl, cross: The T cell code to breach the blood–brain barriers. *Trends in Immunology*, *33*(12), 579–589.

Greene, C., Hanley, N., & Campbell, M. (2019). Claudin-5: Gatekeeper of neurological function. *Fluids and Barriers of the CNS*, *16*, 1–15.

Hang, Z., Zhou, L., Xing, C., Wen, Y., & Du, H. (2023). The blood-brain barrier, a key bridge to treat neurodegenerative diseases. *Ageing Research Reviews*, *91*, 102070.

Hemerková, P., & Vališ, M. (2021). Role of oxidative stress in the pathogenesis of amyotrophic lateral sclerosis: Antioxidant metalloenzymes and therapeutic strategies. *Biomolecules*, *11*(3), 437.

Heye, A. K., Culling, R. D., Hernández, M. d. C. V., Thrippleton, M. J., & Wardlaw, J. M. (2014). Assessment of blood–brain barrier disruption using dynamic contrast-enhanced MRI. A systematic review. *NeuroImage: Clinical*, *6*, 262–274.

Isabel, U.-V., Belén, A. d. l. R. M., & Elena, G.-B. (2024). A new frontier in neuropharmacology: Recent progress in natural products research for blood–brain barrier crossing. *Current Research in Biotechnology*, *8*, 100235.

Kim, H. W., Wu, K. L. K., Tam, K.-W., Chan, Y.-S., & Shum, D. K.-Y. (2024). Pericyte derivation and transplantation for blood-CNS barrier reconstitution in CNS disorders. *IBRO Neuroscience Reports*, *16*, 147–154.

Knox, E. G., Aburto, M. R., Clarke, G., Cryan, J. F., & O'Driscoll, C. M. (2022). The blood-brain barrier in aging and neurodegeneration. *Molecular Psychiatry*, *27*(6), 2659–2673.

Lacoste, B., Prat, A., Freitas-Andrade, M., & Gu, C. (2025). The blood–brain barrier: Composition, properties, and roles in brain health. *Cold Spring Harbor Perspectives in Biology*, *17*(5), a041422.

Liu, J., Tan, Y., Cheng, H., Zhang, D., Feng, W., & Peng, C. (2022). Functions of gut microbiota metabolites, current status and future perspectives. *Aging and Disease*, *13*(4), 1106.

Long, Y., Liu, S., Wan, J., Zhang, Y., Li, D., Yu, S., …, & He, F. (2023). Brain targeted borneol-baicalin liposome improves blood-brain barrier integrity after cerebral ischemia-reperfusion injury via inhibiting HIF-1α/VEGF/eNOS/NO signal pathway. *Biomedicine & Pharmacotherapy*, *160*, 114240.

Lubetzki, C., & Stankoff, B. (2014). Demyelination in multiple sclerosis. *Handbook of Clinical Neurology*, *122*, 89–99.

Lv, J., Hu, W., Yang, Z., Li, T., Jiang, S., Ma, Z., …, & Yang, Y. (2018). Focusing on claudin-5: A promising candidate in the regulation of BBB to treat ischemic stroke. *Progress in Neurobiology, 161*, 79–96.

Matheson, J.-A. T., & Holsinger, R. D. (2023). The role of fecal microbiota transplantation in the treatment of neurodegenerative diseases: A review. *International Journal of Molecular Sciences, 24*(2), 1001.

Montagne, A., Nation, D. A., Sagare, A. P., Barisano, G., Sweeney, M. D., Chakhoyan, A., …, & D'Orazio, L. M. (2020). APOE4 leads to blood–brain barrier dysfunction predicting cognitive decline. *Nature, 581*(7806), 71–76.

Moujalled, D., Strasser, A., & Liddell, J. R. (2021). Molecular mechanisms of cell death in neurological diseases. *Cell Death & Differentiation, 28*(7), 2029–2044.

Paschoal, A. M., Leoni, R. F., Dos Santos, A. C., & Paiva, F. F. (2018). Intravoxel incoherent motion MRI in neurological and cerebrovascular diseases. *NeuroImage: Clinical, 20*, 705–714.

Pathak, D., & Sriram, K. (2023). Molecular mechanisms underlying neuroinflammation elicited by occupational injuries and toxicants. *International Journal of Molecular Sciences, 24*(3), 2272.

Petrushanko, I. Y., Mitkevich, V. A., & Makarov, A. A. (2023). Effect of β-amyloid on blood-brain barrier properties and function. *Biophysical Reviews, 15*(2), 183–197.

Price, B. R., Johnson, L. A., & Norris, C. M. (2021). Reactive astrocytes: The nexus of pathological and clinical hallmarks of Alzheimer's disease. *Ageing Research Reviews, 68*, 101335.

Rajeev, V., Fann, D. Y., Dinh, Q. N., Kim, H. A., De Silva, T. M., Lai, M. K., …, & Arumugam, T. V. (2022). Pathophysiology of blood brain barrier dysfunction during chronic cerebral hypoperfusion in vascular cognitive impairment. *Theranostics, 12*(4), 1639.

Raulin, A.-C., Doss, S. V., Trottier, Z. A., Ikezu, T. C., Bu, G., & Liu, C.-C. (2022). ApoE in Alzheimer's disease: Pathophysiology and therapeutic strategies. *Molecular Neurodegeneration, 17*(1), 72.

Schachter, A. S., & Davis, K. L. (2000). Alzheimer's disease. *Dialogues in Clinical Neuroscience, 2*(2), 91–100.

Segarra, M., Aburto, M. R., & Acker-Palmer, A. (2021). Blood–brain barrier dynamics to maintain brain homeostasis. *Trends in Neurosciences, 44*(5), 393–405.

Song, K., Li, Y., Zhang, H., An, N., Wei, Y., Wang, L., …, & Xing, Y.. (2020). Oxidative stress-mediated blood-brain barrier (BBB) disruption in neurological diseases. *Oxidative Medicine and Cellular Longevity, 2020*(1), 4356386.

Sugiyama, S., Sasaki, T., Tanaka, H., Yan, H., Ikegami, T., Kanki, H., …, & Okazaki, S. (2023). The tight junction protein occludin modulates blood–brain barrier integrity and neurological function after ischemic stroke in mice. *Scientific Reports, 13*(1), 2892.

Takata, F., Nakagawa, S., Matsumoto, J., & Dohgu, S. (2021). Blood-brain barrier dysfunction amplifies the development of neuroinflammation: Understanding of cellular events in brain microvascular endothelial cells for prevention and treatment of BBB dysfunction. *Frontiers in Cellular Neuroscience, 15*, 661838.

Tarafdar, A., Wolska, N., Krisp, C., Schlüter, H., & Pula, G. (2022). The amyloid peptide β disrupts intercellular junctions and increases endothelial permeability in a NADPH oxidase 1-dependent manner. *Redox Biology, 52*, 102287.

Tarumi, T., & Zhang, R. (2018). Cerebral blood flow in normal aging adults: Cardiovascular determinants, clinical implications, and aerobic fitness. *Journal of Neurochemistry, 144*(5), 595–608.

Terstappen, G. C., Meyer, A. H., Bell, R. D., & Zhang, W. (2021). Strategies for delivering therapeutics across the blood–brain barrier. *Nature Reviews Drug Discovery, 20*(5), 362–383.

Thrippleton, M. J., Backes, W. H., Sourbron, S., Ingrisch, M., van Osch, M. J., Dichgans, M., …, & van Oostenbrugge, R. J. (2019). Quantifying blood-brain barrier leakage in small vessel disease: Review and consensus recommendations. *Alzheimer's & Dementia, 15*(6), 840–858.

Trist, B. G., Fifita, J. A., Hogan, A., Grima, N., Smith, B., Troakes, C., …, & Blair, I. P. (2022). Co-deposition of SOD1, TDP-43 and p62 proteinopathies in ALS: Evidence for multifaceted pathways underlying neurodegeneration. *Acta Neuropathologica Communications, 10*(1), 122.

Wu, D., Chen, Q., Chen, X., Han, F., Chen, Z., & Wang, Y. (2023). The blood–brain barrier: Structure, regulation and drug delivery. *Signal Transduction and Targeted Therapy, 8*(1), 217.

Yu, S., Chen, X., Yang, T., Cheng, J., Liu, E., Jiang, L., …, & Ma, Y. (2024). Revealing the mechanisms of blood–brain barrier in chronic neurodegenerative disease: An opportunity for therapeutic intervention. *Reviews in the Neurosciences, 35*(8), 895–916.

Yuan, Y., Sun, J., Dong, Q., & Cui, M. (2023). Blood–brain barrier endothelial cells in neurodegenerative diseases: signals from the "barrier". *Frontiers in Neuroscience, 17*, 1047778.

Zhang, W., Xiao, D., Mao, Q., & Xia, H. (2023). Role of neuroinflammation in neurodegeneration development. *Signal Transduction and Targeted Therapy, 8*(1), 267.

3 Mechanism of Neurodegeneration in Alzheimer's Disease

Pranshul Sethi
Department of Pharmacology, College of Pharmacy,
Shri Venkateshwara University, Gajraula, Uttar-Pradesh, India
Chitkara College of Pharmacy, Chitkara University, Rajpura, Punjab, India

Megha Bahuguna
University Institute of Biotechnology (UIBT), Chandigarh University,
Gharuan, Mohali, Punjab, India

Ramandeep Saini
Chandigarh College of Technology, Chandigarh Group of Colleges,
Landran, Greater Mohali, Punjab, India

Sachchida Nand Rai
University Centre for Research & Development (UCRD),
Chandigarh University, Mohali, 140413, Punjab, India

*Ayush Madan**
Department of Biotechnology, School of Research & Technology,
People's University, Bhanpur, Bhopal, Madhya Pradesh, India
Institute of Tropical Aquaculture and Fisheries (AKUATROP),
Universiti Malaysia Terengganu, Kuala Nerus, Terengganu, Malaysia

3.1 INTRODUCTION

3.1.1 Definition and Overview of Alzheimer's Disease

Dementia is a medical condition marked by a gradual deterioration in two or more cognitive areas, such as memory, language, problem-solving, spatial awareness, personality, and behavior. This decline leads to a loss of ability to carry out everyday tasks, both fundamental and more complex. Alzheimer's disease (AD) is the predominant cause of dementia, responsible for approximately 80% of all dementia diagnoses (Tahami Monfared et al., 2022). While the total mortality rate in the United States due to stroke and cardiovascular disease is declining, there has been a significant increase in the percentage of deaths attributed to AD, which increases by 89% between 2000 and 2014. The yearly costs associated with healthcare for AD, including both direct and indirect

*Corresponding Author. E-mail: madanayushmadan@gmail.com

DOI: 10.1201/9781003530510-3

expenses, are anticipated to be close to $500 billion. Post-mortem assessment of brain tissue is necessary for the conclusive diagnosis of AD. However, the use of cerebrospinal fluid (CSF) and positron emission tomography (PET) biomarkers, along with novel clinical criteria, can assist in diagnosing living individuals with AD (Leuzy et al., 2021). Presently, the available treatments consist of cholinesterase inhibitors for those at any stage of AD dementia, and memantine for those with moderate-to-severe AD dementia. When administered at the appropriate time during the illness, these medications have been proven to improve the quality of life for both the patient and caregiver.

3.1.2 Prevalence and Impact on Society

AD is a growing public health concern in India, with an incidence that is increasing in line with global patterns. India's population surpassed 1.4 billion people in 2023, while the elderly population is experiencing significant growth. AD is anticipated to become more widespread in India as the population of aged adults continues to grow. Studies suggest that there are more than 4 million people in India who have dementia, with AD being the primary cause, responsible for around 60–70% of these instances. India's healthcare system is already facing resource limits, and the increasing burden of AD poses considerable issues (Ravindranath & Sundarakumar, 2021). The societal ramifications of AD in India are significant, impacting not just individuals who have been diagnosed but also their families and those responsible for their care. Due to the traditional structure of Indian society, which prioritizes family care, the majority of individuals with AD are taken care of at home by their family members. The demanding nature of the caregiving position typically leads to caregivers experiencing high levels of stress, anxiety, and depression. This has important repercussions for their emotional, physical, and economic well-being. The financial impact of AD on Indian families is significant, encompassing expenses related to providing care, medical interventions, and reduced income resulting from the incapacity to work on a full-time basis.

Furthermore, there is a lack of knowledge and comprehension regarding AD in India, which leads to late identification and inadequate treatment of the condition. The limited knowledge about mental illness in India is further exacerbated by the negative perception and discrimination that are linked to it. This can result in patients and their families experiencing social exclusion and a decline in their overall well-being. The Indian healthcare system encounters difficulties in delivering sufficient care for individuals with AD, due to a scarcity of well-trained healthcare workers, restricted availability of diagnostic equipment, and inadequate infrastructure for long-term care.

3.2 PATHOLOGICAL HALLMARKS

3.2.1 Amyloid-beta (Aβ) Plaques

AD is the predominant kind of dementia, marked by a gradual deterioration in cognitive abilities and degenerative alterations in the brain. One of the most notable and extensively studied characteristics of AD is the build-up of amyloid-beta (Aβ) plaques, which is among its multiple pathological markers. Aβ plaques are deposits seen outside of cells that are mainly made up of Aβ peptides. These peptides are produced from the amyloid precursor protein (APP) by a series of cleavage reactions involving β-secretase and γ-secretase enzymes (Roda et al., 2022).

The amyloid cascade hypothesis, which has been the primary focus of research on AD for many years, suggests that the build-up of Aβ peptides, specifically the Aβ42 isoform, is the initial trigger for the development of AD. Aβ42 exhibits a higher propensity for aggregation compared to other isoforms, resulting in the creation of oligomers, fibrils, and ultimately insoluble plaques, as mentioned in Table 3.1. The plaques are primarily located in the cerebral cortex, hippocampus, and other brain regions that are linked to memory and cognitive functioning. While it was once believed that Aβ plaques were the main cause of neuronal death, recent studies indicate that soluble oligomers, rather than the plaques themselves, are the most harmful type of Aβ. The oligomers cause disruption

TABLE 3.1
Characteristics of Amyloid-Beta (Aβ) Plaques and Neurofibrillary Tangles (NFTs)

Pathological Hallmarks	Composition	Location in the Brain	Effect on Neurons	Genetic Link	References
Amyloid-beta (Aβ) plaques	Aggregates of amyloid-beta peptides (mainly Aβ42)	Cerebral cortex, hippocampus, and other regions linked to memory and cognition (including the prefrontal cortex and posterior cingulate)	Aβ oligomers, the toxic form of Aβ, disrupt synaptic communication, impair long-term potentiation (LTP) and hinder synaptic plasticity. This causes oxidative stress through the generation of reactive oxygen species (ROS), disrupts mitochondrial function, and leads to synapse loss, inflammation, and eventual neuronal death.	Mutations in APP, PSEN1, PSEN2 genes increase Aβ42 production, leading to plaque formation. The APOE ε4 allele is strongly linked to late-onset Alzheimer's disease, affecting Aβ clearance and increasing the risk of plaque deposition.	Azargoonjahromi (2024) and Roda et al. (2022)
Neurofibrillary tangles (NFTs)	Aggregates of hyperphosphorylated tau protein	Hippocampus, entorhinal cortex, and brain regions involved in cognitive function, including the temporal and parietal lobes	Hyperphosphorylated tau detaches from microtubules, destabilizing their structure. This disrupts axonal transport, leading to impaired synaptic signaling and eventual synaptic degradation. NFTs also lead to neuron loss and synaptic dysfunction, particularly in areas associated with memory, such as the hippocampus.	Mutations in the MAPT gene (which encodes the tau protein) are linked to tauopathies, including Alzheimer's. Tau pathology is also exacerbated by Aβ presence, as Aβ accelerates tau phosphorylation.	Plascencia-Villa and Perry (2022) and Kotarba et al. (2024)

of synaptic function, hinder long-term potentiation (LTP), and trigger oxidative stress. These factors contribute to the reported synapse loss and neuronal death in AD (Azargoonjahromi, 2024).

The presence of Aβ plaques initiates a series of neuroinflammatory reactions, which worsen the damage to neurons. Microglia, the immune cells that dwell in the central nervous system, become activated when there is a deposit of Aβ. While microglia first try to remove Aβ by phagocytosis, prolonged activation results in the secretion of pro-inflammatory cytokines, chemokines, and reactive oxygen species (ROS). This can harm neighboring neurons and worsen the advancement of AD. The persistent inflammation related to Aβ plaques has been connected to changes in tau protein, which is another significant pathogenic feature of AD. Hyperphosphorylated tau undergoes aggregation, resulting in the formation of neurofibrillary tangles (NFTs). These tangles interfere with the movement of substances within cells and have a role in causing malfunction and death of neurons.

Genetic studies have provided additional clarity on the role of Aβ in AD. Genetic mutations in genes responsible for the production and removal of Aβ, such as APP, PSEN1, and PSEN2, are linked to early-onset familial AD.

3.2.2 Neurofibrillary Tangles (NFTs)

NFTs are prominent pathological features of AD, playing a fundamental role in the neurodegenerative mechanisms that define the disease. NFTs are accumulations of hyperphosphorylated tau protein, which is a microtubule-associated protein responsible for stabilizing microtubules in neurons, as mentioned in Table 3.1. During the development of AD, tau proteins undergo aberrant phosphorylation, causing them to separate from microtubules. As a result, the tau proteins self-assemble into paired helical filaments (PHFs), which then aggregate to form NFTs (Plascencia-Villa & Perry, 2022).

NFTs are formed in close correlation with the impairment of neuronal activity and the demise of cells. When tau separates from microtubules, it causes the microtubule network to become unstable, which hampers the movement of substances within cells and results in problems with communication between neurons. The interruption of axonal transport is a contributing factor to the deterioration of neurons, namely in brain regions that are essential for memory and cognition, such as the hippocampus and entorhinal cortex. The dissemination of tau pathology across the brain is associated with the advancement of AD, as the presence of NFTs is observed in greater quantities and in a wider range of brain locations as the illness progresses.

Both genetic and neuropathological investigations have provided support for the involvement of NFTs in the development of AD. Genetic alterations in the MAPT gene, responsible for producing the tau protein, are associated with tauopathies, a collection of neurodegenerative disorders that encompass AD. These mutations result in atypical phosphorylation and accumulation of tau, which is comparable to what is seen in AD. Furthermore, examinations conducted after death on brains affected by AD regularly show a significant presence of NFTs. The degree of NFT accumulation is closely linked to the severity of cognitive loss and neurodegeneration, surpassing the impact of Aβ plaque load (Matsubara et al., 2024).

Tau pathology not only directly affects neuronal function but also interacts with other pathologies associated with AD, specifically Aβ plaques. Research indicates that the existence of Aβ speeds up the process of tau phosphorylation and the creation of NFTs, demonstrating a mutually reinforcing connection between these two significant characteristics of AD. This interaction emphasizes the intricacy of the development of AD, where various pathogenic processes come together to cause the disease (Kotarba et al., 2024).

Table 3.1 compares the two primary pathological hallmarks of AD, Aβ plaques and NFTs, based on their composition, brain localization, neuronal effects, and genetic associations. Aβ plaques mainly result from APP processing, while NFTs are formed by hyperphosphorylated tau. Both contribute significantly to synaptic dysfunction and neuronal degeneration in Alzheimer's pathology.

3.2.3 Tau Protein Pathology

The presence of tau protein pathology is a characteristic that clearly identifies AD and is strongly associated with the degenerative processes that lead to cognitive loss. Tau is a protein that is mostly present in neurons and has a crucial function in maintaining microtubules and aiding in intracellular transport. Tau experiences aberrant hyperphosphorylation in AD, resulting in its separation from microtubules and subsequent formation of insoluble PHFs. The PHFs combine together to create NFTs, which are a characteristic feature of AD and other conditions involving abnormal tau protein accumulation (Thal and Tomé, 2022).

NFTs production exhibits a robust correlation with neuronal dysfunction and demise. When tau separates from microtubules, it weakens the structural stability of the cytoskeleton, which hinders the movement of important chemicals and organelles in neurons. The disturbance is especially harmful in the brain regions that are accountable for memory and learning, such as the hippocampus and entorhinal cortex. These areas are among the first regions to be impacted in AD. The presence of NFTs in these areas is directly linked to the extent of cognitive decline, highlighting the crucial role of tau pathology in the evolution of AD.

Tau pathology does not arise independently but rather interacts with other processes relevant to AD, including Aβ pathology. Research indicates that Aβ has the ability to expedite the process of tau hyperphosphorylation and aggregation, resulting in a combined action that worsens neurodegeneration. The interaction between Aβ and tau underscores the intricate and multifaceted nature of AD, where various pathogenic pathways come together to facilitate the advancement of the disease.

Due to its significant involvement in AD, tau has been a focal point for therapeutic strategies targeted at decreasing tau phosphorylation, impeding aggregation, and halting the propagation of tau pathology throughout the brain (Khan et al., 2025).

3.3 MECHANISMS OF NEURODEGENERATION

3.3.1 Amyloid-beta Pathology

The presence of Aβ pathology is a key factor in the neurodegeneration seen in AD. Aβ is a peptide that is produced from the APPAPP through a series of cleavage steps by the enzymes β-secretase and γ-secretase. The Aβ peptides that are produced as a result, especially Aβ42, have a tendency to come together and form aggregates, including oligomers and fibrils. Eventually, these aggregates develop into extracellular amyloid plaques, which are a characteristic feature of AD.

Aβ oligomers are thought to be a crucial initial occurrence in the development of AD, as these soluble versions of Aβ are more harmful to neurons compared to the insoluble plaques. Aβ oligomers affect synaptic function by interfering with synaptic plasticity, hindering LTP, and causing synapse loss, which is associated with cognitive decline. The oligomers also cause oxidative stress by producing ROS, which harm neurons by interfering with mitochondrial activity and causing lipid peroxidation, DNA damage, and protein oxidation (Ichimata et al., 2024).

Aβ pathology induces a persistent inflammatory reaction in the brain. Microglia, the immune cells that dwell in the brain, become activated in response to the presence of Aβ deposits and try to eliminate them by engulfing them by phagocytosis (Sehar et al., 2022). Nevertheless, prolonged activation of microglia results in the secretion of pro-inflammatory cytokines and chemokines, which worsen damage to neurons and facilitate the propagation of Aβ pathology. The interaction between Aβ pathology and other key features of AD, such as tau pathology, intensifies the progression of neurodegenerative processes.

3.3.2 Aβ Production and Aggregation

Aβ production and aggregation are central to the neurodegenerative mechanisms underlying AD. Aβ is generated from the APP, a transmembrane protein expressed in many tissues, particularly in the brain. The production of Aβ involves two proteolytic cleavages: the first by β-secretase (BACE1) and the second by γ-secretase. This sequential cleavage results in the formation of Aβ peptides, primarily Aβ40 and Aβ42, with the latter being more prone to aggregation.

The aggregation of Aβ begins with the formation of soluble oligomers, which are considered the most neurotoxic species. These oligomers can interfere with synaptic function, leading to impaired synaptic plasticity, reduced LTP, and eventual synaptic loss, which correlates with cognitive decline in AD patients. Over time, oligomers aggregate into insoluble fibrils and eventually form amyloid plaques, the hallmark extracellular deposits observed in the brains of individuals with AD (Yu & Wu, 2021).

The process of Aβ aggregation is influenced by several factors, including genetic mutations and post-translational modifications. For instance, mutations in the APP gene, or in genes encoding presenilin-1 and presenilin-2 (components of the γ-secretase complex), can increase the production of the more aggregation-prone Aβ42, leading to early-onset familial Alzheimer's disease (FAD).

3.3.3 Synaptic Dysfunction and Memory Impairment

Synaptic dysfunction is a critical mechanism underlying neurodegeneration and memory impairment in AD. The synapse is the fundamental unit of communication between neurons, essential for cognitive functions such as learning and memory. In AD, the integrity of synaptic connections is compromised early in the disease process, even before significant neuronal loss occurs. This synaptic damage is strongly correlated with cognitive decline, making it a central feature of AD pathogenesis (Pelucchi et al., 2022).

One of the primary drivers of synaptic dysfunction in AD is the accumulation of Aβ oligomers, which are considered the most neurotoxic form of Aβ. These soluble oligomers disrupt synaptic plasticity by impairing LTP, a process critical for memory formation and storage. Aβ oligomers interfere with the function of neurotransmitter receptors, particularly the N-methyl-D-aspartate (NMDA) receptors, leading to excessive calcium influx, synaptic weakening, and eventual synapse loss. This disruption of synaptic signaling contributes to the characteristic memory deficits seen in AD patients.

Additionally, synaptic dysfunction in AD is exacerbated by tau pathology. Hyperphosphorylated tau, which accumulates in neurons as NFTs, can disrupt axonal transport, leading to synaptic degradation. The interaction between Aβ and tau pathology creates a vicious cycle, where Aβ-induced synaptic damage promotes tau hyperphosphorylation, further impairing synaptic function.

3.3.4 Tau Hyperphosphorylation

Tau hyperphosphorylation is a crucial mechanism driving neurodegeneration in AD. Tau is a microtubule-associated protein that plays a vital role in stabilizing microtubules and facilitating intracellular transport within neurons. In the healthy brain, tau is phosphorylated at specific sites to regulate its function. However, in AD, tau undergoes abnormal hyperphosphorylation at multiple sites, leading to its detachment from microtubules and subsequent aggregation into PHFs and NFTs, which are hallmark features of the disease.

Hyperphosphorylated tau is toxic to neurons for several reasons. First, the loss of tau's normal binding to microtubules destabilizes the microtubule network, disrupting axonal transport and leading to synaptic dysfunction. This disruption is particularly detrimental to neurons in brain regions critical for memory and cognition, such as the hippocampus and entorhinal cortex, which are among the first areas affected in AD. As tau continues to aggregate, it forms NFTs, which interfere with neuronal function and ultimately lead to cell death (Moore et al., 2023).

The hyperphosphorylation of tau is influenced by several kinases, such as glycogen synthase kinase-3β (GSK-3β) and cyclin-dependent kinase 5 (CDK5), which are found to be dysregulated in AD. Additionally, evidence suggests that Aβ pathology may exacerbate tau hyperphosphorylation, creating a feedback loop that accelerates neurodegeneration. The accumulation of hyperphosphorylated tau and NFTs in the brain correlates strongly with the severity of cognitive decline in AD, making tau pathology a critical target for therapeutic intervention. Targeting tau hyperphosphorylation and preventing its aggregation are promising strategies for slowing the progression of AD and mitigating its devastating effects on cognition (Rawat et al., 2022).

3.3.5 Axonal Transport Disruption

Disruption of axonal transport is a crucial process that contributes to the neurodegeneration observed in AD. Axonal transport is crucial for the optimal operation of neurons, as it enables the transportation of organelles, proteins, and other vital substances between the cell body and synaptic terminals. The transportation mechanism depends on the microtubule network, which acts as pathways for motor proteins like kinesin and dynein to carry goods to their intended locations. During the progression of AD, the transportation of axons is significantly hindered, mostly as a result of tau pathology.

Tau is a protein that binds to microtubules and enhances their stability specifically in neurons. Nevertheless, in AD, tau experiences anomalous hyperphosphorylation, resulting in its separation from microtubules. The separation of the microtubule network disrupts its stability, hindering the motor proteins' efficiency in transporting cargoes along axons. Axonal transport disruption causes an accumulation of organelles, proteins, and other substances in the axons, leading to the formation of swollen areas and ultimately, the degeneration of the axons.

The disruption of axonal transport has a particularly harmful effect on neurons in areas of the brain that are crucial for memory and cognition, such as the hippocampus and cortex (Dan & Zhang, 2023).

3.3.6 Oxidative Stress

Oxidative stress is a significant mechanism driving neurodegeneration in AD. It occurs when there is an imbalance between the production of ROS and the brain's ability to detoxify these reactive intermediates or repair the resultant damage. In AD, oxidative stress is triggered by several factors, including Aβ accumulation, mitochondrial dysfunction, and neuroinflammation, all of which contribute to neuronal injury and death (Bai et al., 2022).

Aβ oligomers, which are toxic aggregates of the Aβ peptide, are key contributors to oxidative stress in AD. These oligomers can induce the production of ROS, including superoxide radicals, hydrogen peroxide, and hydroxyl radicals. The generation of ROS leads to oxidative damage to lipids, proteins, and nucleic acids, which impairs cellular function and promotes neuronal death. Lipid peroxidation, for example, results in the formation of toxic by-products like 4-hydroxynonenal (HNE), which can further exacerbate oxidative damage and disrupt cellular homeostasis.

3.3.7 Sources and Impact of Reactive Oxygen Species (ROS)

ROS are highly reactive molecules that play a crucial role in the neurodegenerative processes observed in AD. ROS are primarily generated as by-products of cellular metabolism, particularly in the mitochondria during oxidative phosphorylation. In the context of AD, the production of ROS is significantly elevated due to several pathological factors, including Aβ accumulation, mitochondrial dysfunction, and neuroinflammatory processes (Walia et al., 2022).

One of the primary sources of ROS in AD is mitochondrial dysfunction, which is exacerbated by Aβ toxicity. Aβ peptides, particularly in their oligomeric form, are known to localize to mitochondria and disrupt the electron transport chain (ETC). This disruption leads to electron leakage and the subsequent generation of superoxide radicals, a type of ROS, which can cause extensive oxidative damage to mitochondrial DNA, proteins, and lipids, further impairing mitochondrial function and exacerbating ROS production.

In addition to mitochondrial sources, ROS are also produced by activated microglia and astrocytes during neuroinflammatory responses in AD. These immune cells generate ROS as part of their defensive response to clear Aβ plaques; however, chronic activation leads to sustained ROS production, contributing to widespread oxidative damage (Yeung et al., 2021).

3.3.8 Cellular Damage and Impairment

Cellular damage and dysfunction play a crucial role in the progression of neurodegeneration in AD. The clinical characteristics of AD, such as the presence of Aβ plaques, NFTs made up of hyperphosphorylated tau, and persistent neuroinflammation, play a role in causing extensive cellular dysfunction and neuronal death. Aβ plaques, namely the soluble oligomeric variants of Aβ, possess significant neurotoxicity and interfere with cellular function at various levels. The oligomers hinder the ability of synapses to change and adapt by interfering with the functioning of neurotransmitter receptors, namely those involved in glutamate signaling. This disruption results in the loss of synapses and cognitive impairments. Moreover, persistent neuroinflammation in AD worsens the harm to cells (Monteiro et al., 2023).

3.3.9 Neuroinflammation

Neuroinflammation is a key mechanism contributing to neurodegeneration in AD. It is primarily driven by the chronic activation of glial cells, including microglia and astrocytes, in response to the accumulation of Aβ plaques and tau pathology. While acute inflammation is a protective response to injury, chronic neuroinflammation in AD leads to sustained release of pro-inflammatory cytokines, chemokines, and ROS, which exacerbate neuronal damage and accelerate disease progression. Microglia, the resident immune cells of the brain, play a central role in neuroinflammation (Heneka et al., 2025).

3.3.10 Role of Microglia and Astrocytes

AD is a degenerative ailment of the nervous system that worsens with time and is marked by a deterioration in cognitive function and loss of memory. The main clinical characteristics of AD consist of the presence of Aβ plaques outside the cells, NFTs inside the cells made up of tau protein that is excessively phosphorylated, and significant neuroinflammation. Neuronal degeneration plays a fundamental role in AD, but there is growing evidence that glial cells, particularly microglia and astrocytes, may have important functions in the development of the illness. Microglia, which are the immune cells that dwell in the brain, have a dual function in AD. Under normal biological settings, microglia remove Aβ through the process of phagocytosis, which stops the build-up of plaques. Nevertheless, prolonged exposure to elevated amounts of Aβ stimulates the activation of microglia, resulting in the secretion of pro-inflammatory cytokines, ROS, and complement proteins. The neuroinflammatory condition worsens damage to neurons and promotes the accumulation of Aβ, leading to a harmful loop of neurodegeneration. Astrocytes, responsible for preserving synaptic equilibrium and assisting in neural operations, are also associated with AD. Astrocytes undergo reactive changes in response to AD pathology, which are characterized by an increase in size (hypertrophy) and an elevation of glial fibrillary acidic protein (GFAP). Reactive astrocytes generate substances that cause inflammation and enhance the harmful effects of glutamate, leading to impaired communication between neurons and the death of nerve cells. Astrocytes also have a role in regulating microglial activity, which affects the balance between protecting and harming neurons. Malfunctioning astrocytic processes, such as poor clearance of Aβ and disrupted calcium signaling, also play a role in the advancement of the disease.

The interaction among microglia, astrocytes, and neurons plays a vital role in AD. Persistent stimulation of glial cells leads to long-lasting inflammation, which in turn causes damage to synapse and the degeneration of neurons (Deng et al., 2024).

3.3.11 Pro-inflammatory Cytokines and Neuronal Injury

Chronic neuroinflammation, caused in part by the continuous release of pro-inflammatory cytokines, is a major factor in the development of this neurodegenerative condition. Within the context of AD, microglia, which are the immune cells that reside in the central nervous system, undergo persistent activation in response to the accumulation of Aβ deposits. This stimulation leads to the continuous release of pro-inflammatory cytokines, including interleukin-1β (IL-1β), tumor necrosis factor-α (TNF-α), and interleukin-6 (IL-6). These cytokines enhance the inflammatory response and provide a harmful environment that contributes to damage to neurons (Florescu et al., 2023).

Pro-inflammatory cytokines exert neurotoxic effects through multiple routes. Increased concentrations of TNF-α and IL-1β stimulate the generation of ROS and nitric oxide (NO), resulting in oxidative stress and impairment of mitochondrial function.

3.3.12 Mitochondrial Dysfunction

AD is an intricate neurodegenerative condition characterized by gradual deterioration of cognitive function and loss of memory. Within the several pathways that contribute to AD, mitochondrial dysfunction is recognized as a crucial factor in the process of neurodegeneration. Mitochondria are vital organelles that play a crucial role in energy generation, maintaining calcium balance, and regulating apoptosis. In the context of AD, these functions are impaired because of various pathological changes, as mentioned in Table 3.2. An important factor contributing to mitochondrial dysfunction in AD is the impairment of oxidative phosphorylation, a process essential for the generation of adenosine triphosphate (ATP). Neurons, being reliant on a significant amount of energy, experience a decrease in ATP levels, resulting in impairments in synaptic transmission and plasticity, crucial processes for the acquisition of knowledge.

AD is marked by an abnormality in the functioning of mitochondria, which results in an overproduction of ROS (Ke et al., 2021). The heightened oxidative stress leads to harm to cellular constituents, including lipids, proteins, and DNA, which in turn contributes to neuroinflammation and the demise of neurons (Ashleigh et al., 2023).

3.3.13 Energy Metabolism Impairment

One critical mechanism underlying AD pathogenesis is the disruption of brain energy metabolism. Neurons are highly energy-demanding cells, relying on glucose metabolism for the generation of ATP, which is essential for synaptic activity, neurotransmission, and cellular maintenance. In AD, the impairment of glucose metabolism is one of the earliest detectable abnormalities, even before the onset of significant cognitive symptoms. This metabolic dysfunction is evident as reduced glucose uptake and utilization in affected brain regions, particularly the hippocampus and cerebral cortex.

The energy metabolism impairment in AD is driven by several interrelated factors, including mitochondrial dysfunction, insulin resistance, and impaired glycolysis. Mitochondria, which are crucial for ATP production through oxidative phosphorylation, show marked dysfunction in AD. This includes defects in the ETC, leading to reduced ATP synthesis and increased production of ROS, which in turn cause oxidative damage to neuronal components (Yu et al., 2022).

3.3.14 Increased Apoptotic Susceptibility

AD is a progressive neurological ailment characterized by many pathogenic causes, such as heightened vulnerability to apoptosis, a type of programmed cell death. Apoptosis is a highly controlled process that is crucial for maintaining the balance of cells in the body. However, when this process is disrupted, it can result in an excessive loss of neurons, which is a contributing factor to the cognitive decline observed in AD. During AD, neurons become more susceptible to apoptotic signals because of various factors, including the toxicity of Aβ, excessive phosphorylation of tau proteins, and dysfunction of mitochondria.

Aβ, a characteristic feature of AD, has a vital function in making neurons more susceptible to death. The build-up of Aβ oligomers initiates oxidative stress and disturbs calcium balance, leading

TABLE 3.2
Mechanisms Contributing to Neurodegeneration in AD

Mechanisms	Description	Primary Pathological Effects	Related Therapeutic Approaches	References
Aβ production and aggregation	Amyloid-beta (Aβ) is produced when amyloid precursor protein (APP), a transmembrane protein, is cleaved by β-secretase and γ-secretase. The Aβ42 isoform is more prone to aggregation. These peptides aggregate into oligomers, fibrils, and eventually form extracellular amyloid plaques in the brain.	Aβ oligomers, rather than plaques themselves, are highly neurotoxic. They disrupt synaptic communication by impairing long-term potentiation (LTP), necessary for memory formation. Aβ also causes oxidative stress by generating reactive oxygen species (ROS), mitochondrial dysfunction, and chronic neuroinflammation. Together, these processes lead to synaptic dysfunction, loss of neuronal plasticity, and eventual neuronal death. These plaques are primarily found in the hippocampus, cerebral cortex, and other areas important for memory and cognition.	Therapeutic approaches targeting Aβ include monoclonal antibodies (e.g., aducanumab and lecanemab) that aim to clear amyloid plaques from the brain by enhancing immune-mediated clearance. Additionally, β-secretase inhibitors (e.g., verubecestat) attempt to reduce Aβ production by preventing APP cleavage. Small molecules aiming to inhibit Aβ aggregation are also being explored.	Yu & Wu (2021)
Tau hyperphosphorylation	Tau is a microtubule-associated protein that stabilizes the cytoskeleton of neurons. In AD, tau undergoes hyperphosphorylation, leading to its detachment from microtubules and aggregation into paired helical filaments (PHFs), which form neurofibrillary tangles (NFTs). These tangles disrupt the structural integrity and function of neurons.	Hyperphosphorylated tau protein destabilizes the microtubule network, leading to impaired intracellular transport of essential nutrients, proteins, and organelles within neurons. This disruption impairs synaptic signaling and contributes to the death of neurons, particularly in regions crucial for memory and cognition, such as the hippocampus and entorhinal cortex. The presence of NFTs is closely correlated with cognitive decline, and their spread throughout the brain contributes significantly to neurodegeneration. Tau pathology also interacts with Aβ, exacerbating the severity of neuronal dysfunction.	Anti-tau therapies focus on preventing tau phosphorylation and aggregation. Monoclonal antibodies (e.g., gosuranemab) target extracellular tau to prevent its spread between neurons. Vaccines (AADvac1 and ACI-35) aim to stimulate the immune system to clear hyperphosphorylated tau. Small molecules that inhibit tau kinases (e.g., GSK-3β inhibitors) are also being studied. These strategies aim to reduce the toxic effects of tau and slow disease progression.	Moore et al. (2023)
Oxidative stress	Oxidative stress is a state where there is an imbalance between the production of reactive oxygen species (ROS) and the brain's ability to neutralize them with antioxidants. In AD, Aβ accumulation, mitochondrial dysfunction, and neuroinflammation all contribute to excessive ROS production, which damages neurons.	ROS cause widespread damage to cellular components, including lipids, proteins, and nucleic acids. This damage leads to mitochondrial dysfunction, exacerbating energy deficits in neurons, and further increasing ROS production. Oxidative stress also plays a critical role in the pathogenesis of tau hyperphosphorylation, creating a vicious cycle that accelerates neurodegeneration. In addition, ROS contribute to synaptic dysfunction and cell death, worsening cognitive decline. Oxidative stress interacts with neuroinflammation, amplifying both processes and contributing to chronic neuronal injury in AD.	Antioxidant therapies, such as high doses of vitamin E, coenzyme Q10, and other compounds that target mitochondrial dysfunction, aim to reduce oxidative damage in neurons. Several small molecules that target ROS production or enhance the brain's antioxidant defenses are being investigated to mitigate the effects of oxidative stress. These therapies may help reduce cognitive decline by protecting neurons from oxidative damage.	Bai et al. (2022)

(Continued)

TABLE 3.2 (*Continued*)
Mechanisms Contributing to Neurodegeneration in AD

Mechanisms	Description	Primary Pathological Effects	Related Therapeutic Approaches	References
Mitochondrial dysfunction	Mitochondria are responsible for generating the energy needed for neuronal function through oxidative phosphorylation. In AD, mitochondrial function is impaired due to the accumulation of Aβ, which interacts with mitochondrial proteins and disrupts the electron transport chain (ETC). This leads to reduced ATP production and increased ROS generation.	The reduction in ATP production due to mitochondrial dysfunction results in an energy deficit that impairs synaptic transmission and plasticity, critical for learning and memory. Mitochondrial dysfunction also increases ROS production, leading to further oxidative damage and promoting apoptotic pathways, which result in neuronal death. In AD, mitochondria become fragmented due to imbalances in mitochondrial dynamics (fission and fusion), and impaired mitophagy leads to the accumulation of dysfunctional mitochondria. These events contribute to widespread neuronal degeneration, particularly in brain regions such as the hippocampus and cerebral cortex.	Therapeutic approaches include mitochondrial protectants such as coenzyme Q10, which may help improve mitochondrial function and reduce ROS production. Other treatments focus on restoring mitochondrial dynamics, improving ATP production, and enhancing mitophagy to eliminate damaged mitochondria. Antioxidant therapies that target mitochondrial dysfunction are also being explored to mitigate the impact of oxidative stress on neuronal health.	Ashleigh et al. (2023)
Neuroinflammation	Chronic neuroinflammation is driven by prolonged activation of glial cells, including microglia and astrocytes, in response to Aβ plaques and tau tangles. These glial cells produce pro-inflammatory cytokines and reactive oxygen species (ROS), which further damage neurons and exacerbate disease progression.	Activated microglia and astrocytes release pro-inflammatory cytokines such as IL-1β, TNF-α, and IL-6, which disrupt synaptic plasticity and promote tau pathology. Chronic neuroinflammation contributes to synaptic dysfunction, neuronal death, and the spread of tau tangles. The inflammatory response also weakens the blood–brain barrier, allowing peripheral immune cells to infiltrate the brain and worsen inflammation. This creates a vicious cycle where neuroinflammation and oxidative stress reinforce each other, accelerating neurodegeneration in AD.	Anti-inflammatory treatments targeting glial cells are being explored to reduce chronic neuroinflammation in AD. Therapies such as non-steroidal anti-inflammatory drugs (NSAIDs) have shown limited success, but novel approaches aim to modulate glial cell activity, reduce cytokine production, and enhance the clearance of Aβ and tau. Glial-targeted therapies are a promising area of research for slowing disease progression.	Heneka et al. (2025)
Axonal transport disruption	Axonal transport is critical for the movement of organelles, proteins, and other molecules between the neuronal cell body and synaptic terminals. In AD, hyperphosphorylated tau disrupts the microtubule network, which impairs the transport of essential cargo along the axons, leading to axonal swelling and degeneration.	The disruption of axonal transport leads to synaptic dysfunction and the accumulation of damaged organelles in neurons, contributing to synaptic loss. This process primarily affects neurons in memory-related brain regions, such as the hippocampus and cortex, contributing to the early cognitive impairments observed in AD. The disruption of axonal transport also impairs neuronal communication and weakens the structural integrity of the neuron, hastening neuronal death.	Therapies aimed at stabilizing the microtubule network and preventing tau hyperphosphorylation are being investigated to restore axonal transport. Kinase inhibitors that target enzymes involved in tau phosphorylation (e.g., GSK-3β inhibitors) are of interest. Other approaches aim to improve axonal transport by enhancing microtubule stability and preventing tau from aggregating into NFTs.	Dan & Zhang (2023)

Synaptic dysfunction	Synaptic dysfunction is a hallmark of AD and occurs early in the disease process. Aβ oligomers impair synaptic plasticity by interfering with neurotransmitter receptors, such as NMDA receptors. This disruption leads to excessive calcium influx, synaptic weakening, and eventual loss of synapses, critical for memory and learning.	Impaired synaptic plasticity, particularly long-term potentiation (LTP), is closely associated with memory loss and cognitive decline in AD. Synaptic dysfunction occurs before significant neuronal loss and is an early feature of the disease. The loss of synapses reduces neuronal communication and disrupts the neural circuits responsible for cognition. Aβ oligomers and tau hyperphosphorylation work synergistically to worsen synaptic dysfunction, leading to rapid cognitive deterioration. This process underlies the characteristic memory deficits seen in AD patients.	Therapeutic approaches that target synaptic dysfunction focus on clearing Aβ oligomers and preventing tau hyperphosphorylation. Monoclonal antibodies against Aβ and tau, as well as drugs that target neurotransmitter receptors (e.g., NMDA receptor antagonists like memantine), are being developed to prevent synapse loss and improve cognitive function. Early intervention targeting synaptic dysfunction may help slow the progression of AD.	Pelucchi et al. (2022)
Increased apoptotic susceptibility	Apoptosis, or programmed cell death, is heightened in neurons affected by AD. Aβ oligomers induce oxidative stress and disrupt calcium homeostasis, activating pro-apoptotic pathways. Mitochondrial dysfunction also contributes to apoptosis by releasing cytochrome c, which activates caspase enzymes that mediate cell death.	Increased susceptibility to apoptosis leads to widespread neuronal death, particularly in brain regions involved in memory and cognition, such as the hippocampus. This accelerates cognitive decline, as neurons critical for learning and memory are lost. Apoptosis contributes to the neurodegenerative process in AD, as damaged neurons undergo programmed cell death in response to oxidative stress, mitochondrial dysfunction, and tau pathology.	Therapeutic approaches that target apoptotic pathways focus on reducing mitochondrial dysfunction and blocking pro-apoptotic signaling. Mitochondrial stabilizers and caspase inhibitors are being investigated to prevent apoptosis in neurons and reduce cognitive decline. Enhancing mitochondrial function and protecting against oxidative stress are key strategies for reducing apoptosis in AD.	Kumari et al. (2023)

to the activation of pro-apoptotic pathways. Aβ can also bind to death receptors, including Fas and TNF receptors, on the surface of neurons, triggering a series of caspase cascades that ultimately lead to apoptosis. In addition, the presence of hyperphosphorylated tau, which is another defining characteristic of AD, causes destabilization of the cytoskeleton and disruption of axonal transport, resulting in impaired neuronal function. Additionally, this tau pathology increases the likelihood of undergoing programmed cell death by activating stress-related kinases and intensifying mitochondrial dysfunction (Kumari et al., 2023).

AD is characterized by an increased vulnerability to apoptosis, which is primarily caused by malfunction in the mitochondria. When mitochondria are damaged, they release cytochrome c into the cytosol. This cytochrome c then attaches to apoptotic protease-activating factor-1 (Apaf-1) and triggers the activation of caspase-9. This activation ultimately leads to the initiation of apoptosis.

Table 3.2 outlines key molecular and cellular mechanisms driving neurodegeneration in AD, including Aβ accumulation, tau hyperphosphorylation, oxidative stress, and synaptic dysfunction. Each mechanism is linked to specific pathological effects and potential therapeutic strategies, reflecting the complex and multifactorial nature of the disease.

3.4 GENETIC FACTORS

AD is a type of dementia that affects memory, cognition, and behavior. It is characterized by the way the illness progresses. Early-onset AD is predominantly linked to three essential genes: PSEN 1 (presenilin 1), PSEN 2 (presenilin 2), and APP genes. APP, PSEN 1, and PSEN 2 mutations cause early-onset AD (Kabir et al., 2020). In most cases, these mutations lead to increased production or aggregation of beta-amyloid peptides, which form plaques in the brain. These mutations are autosomal dominant, meaning that an individual must inherit only one mutated gene from either side of the family to be affected.

3.4.1 APP, PSEN1, AND PSEN2 MUTATIONS

AD is caused by mutations in the APP gene. These mutations affect the synthesis, deposition, and degradation of amyloid-β (Aβ) peptides, which are central to the pathogenesis of AD. FAD is a subtype of AD that is caused by a mutation in the APP gene. APP is a transmembrane protein expressed in many tissues, including the brain (produced in large quantities in neurons and is metabolized very rapidly). This transmembrane protein is involved in cell signaling, neuronal proliferation, and neural repair (Marsh et al., 2025).

3.4.1.1 Amyloid Precursor Protein (APP)

The APP gene is localized on chromosome 21 and codes for a protein that undergoes multiple processing pathways. These pathways include the amyloidogenic and non-amyloidogenic pathways. In the amyloidogenic pathway APP is fragmented to Aβ peptides through enzymatic digestion by β-secretase and γ-secretase. These Aβ peptides especially Aβ42 tend to aggregate and deposit beneath the cortex forming amyloid plaques, a hallmark of AD while APP is digested by α-secretase within the Aβ domain in the non-amyloidogenic pathway, preventing the development of Aβ peptides, as illustrated in Figure 3.1. This cleavage results in the creation of the soluble component known as sAPPα, which is believed to promote synaptic activity and have neuroprotective properties. Any alterations or mutations in the APP gene can interfere with its regular processing, which cause an increase in the synthesis of harmful Aβ peptides and hasten the development of plaque which results in the early onset of the AD (Yang et al., 2022).

3.4.1.2 Presenilin 1 (PSEN 1)

A gene that is important in the development of early-onset familial AD is PSEN 1 (presenilin 1). It synthesizes a protein that is a component of gamma-secretase complex which is an enzyme

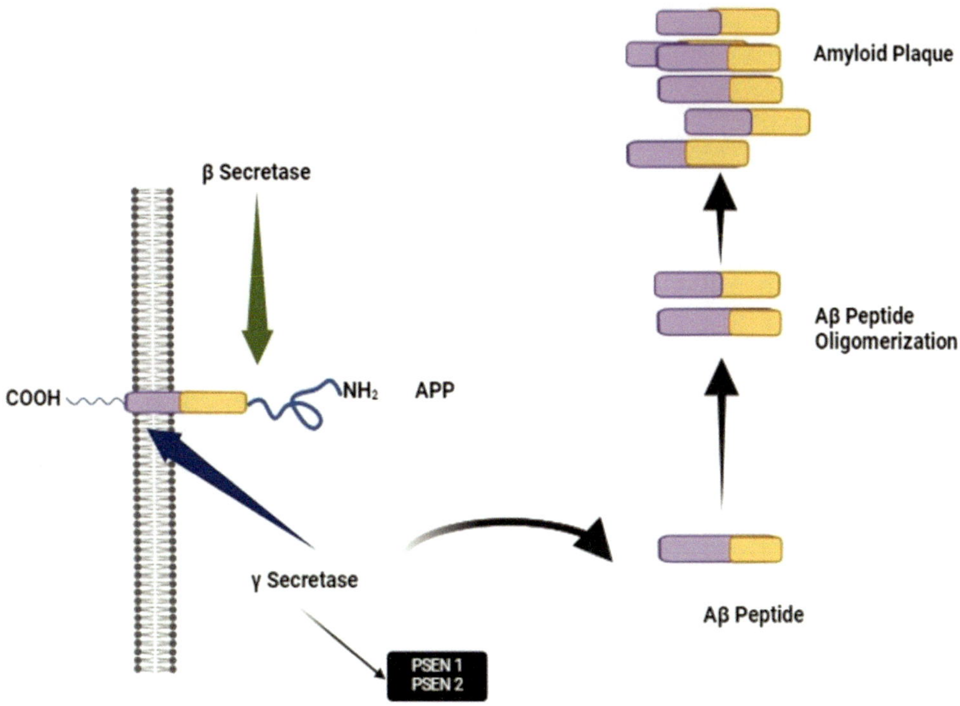

FIGURE 3.1 Structure of amyloid precursor protein (APP) and synthesis of amyloid-beta (Aβ) peptides. (Source: Biorender.)

responsible for cleavage of APP-generating Aβ peptides. PSEN1 mutations lead to the production of more Aβ42 peptide, which then forms amyloid plaques. The hallmark signs of AD, neuronal damage and cognitive impairment, are caused by the production of amyloid plaques (Yang et al., 2023).

3.4.1.3 Presenilin 2 (PSEN 2)

Similar to PSEN1, less common mutations in PSEN2 are known to affect the gamma-secretase complex, which is responsible for cleaving the APP. This indicates that they can promote the aggregating peptide known as Aβ 42 synthesis. Thus, by experiencing a conformational change hallmark of AD, the condensation of Aβ42 results in the production of amyloid plaques and sharpens dementia by causing a rupture in neuronal transmission (Sun et al., 2024).

Figure 3.1 illustrates the proteolytic processing of APP by β-secretase and γ-secretase (involving PSEN1 and PSEN2), resulting in the generation of Aβ peptides. These peptides undergo oligomerization and aggregate into extracellular amyloid plaques, a hallmark of Alzheimer's disease pathology.

3.4.2 APOE ε4 Allele

Many genetic risk factors are recognized with AD, and some of the most critical of them are the gene apolipoprotein E (APOE) and its ε4 allele. The APOE gene is located on chromosome 19 and codes for APOE. This protein is essential for lipid metabolism and the transport of lipoproteins and cholesterol in the brain and body tissues. APOE exists in three main allelic forms: ε2, ε3, and ε4. These alleles are defined by SNPs, which are single nucleotide polymorphisms that cause differences in the amino acid sequence and the structure of the APOE protein (Gal et al., 2022). Among these three allelic forms, ε4 is a major contributor to late-onset AD. Any alteration or mutation in the APOE ε4 variant results in an altered APOE protein that is less effective at lipid transport and neuronal repair, as mentioned in Table 3.3. This altered or mutated APOE ε4 is less efficient

TABLE 3.3
Genetic Mutations Associated with Alzheimer's Disease

Gene	Mutation Type	Associated Onset (Early or Late)	Effect on Aβ or Tau Pathways	References
APP (amyloid precursor protein)	Mutations such as V717I (London mutation), A673T (protective), E693Q (arctic mutation), and gene duplications lead to increased or altered cleavage of APP by β-secretase and γ-secretase.	Early-onset (familial Alzheimer's disease)	APP mutations result in either overproduction of Aβ42 or increased aggregation propensity. Aβ42 aggregates faster than Aβ40, leading to early amyloid plaque formation. Mutations like the London mutation enhance the γ-secretase cleavage, resulting in more Aβ42, while others like the arctic mutation promote Aβ oligomerization. These mutations cause accelerated synaptic dysfunction, neuroinflammation, and cognitive decline. Some APP mutations (like A673T) may provide protection by reducing β-secretase cleavage.	Yang et al. (2022)
PSEN1 (presenilin 1)	More than 300 mutations were identified, including A246E, L286P, and ΔE9, affecting the γ-secretase complex responsible for cleaving APP.	Early-onset (familial Alzheimer's disease)	PSEN1 mutations alter the activity of γ-secretase, increasing the ratio of Aβ42 to Aβ40. Aβ42 is more prone to aggregation, leading to amyloid plaque deposition. These mutations cause aggressive, early-onset AD, often before age 50, and are highly penetrant. They also affect mitochondrial function, calcium homeostasis, and autophagy, exacerbating neurodegeneration. PSEN1 mutations are the most common cause of familial AD, leading to more severe and rapid progression compared to sporadic AD.	Yang et al. (2023)
PSEN2 (presenilin 2)	Mutations such as N141I, M239V, and R62H that influence γ-secretase activity. PSEN2 mutations are less frequent and have later onset than PSEN1 mutations.	Early-onset (familial Alzheimer's disease, rarer than PSEN1 mutations)	Similar to PSEN1, PSEN2 mutations affect γ-secretase, leading to increased production of Aβ42, though with a somewhat lower ratio compared to PSEN1 mutations. This results in slower plaque formation and later onset compared to PSEN1-related familial AD. PSEN2 mutations also contribute to mitochondrial dysfunction and oxidative stress, which are implicated in neuronal death. These mutations are associated with autosomal dominant inheritance but are less frequent than PSEN1 mutations.	Sun et al. (2024)
APOE ε4 (apolipoprotein E)	APOE ε4 allele involves two key single nucleotide polymorphisms (SNPs): rs429358 and rs7412. The presence of one ε4 allele increases risk, and two alleles further heighten the risk.	Late-onset (sporadic Alzheimer's disease)	APOE ε4 reduces the efficiency of amyloid-beta clearance from the brain, promoting the accumulation of amyloid plaques. It also enhances the aggregation of Aβ and facilitates tau phosphorylation, leading to neurofibrillary tangle formation. APOE ε4 carriers have a higher risk of developing AD, with those homozygous for ε4 (carrying two copies) at significantly higher risk. APOE ε4 also impairs blood–brain barrier integrity, promoting neuroinflammation. APOE ε4's interaction with both Aβ and tau makes it a critical factor in late-onset AD.	Gal et al. (2022)

at clearing Aβ peptides from the brain, which leads to the accumulation and aggregation of Aβ plaques. The presence of Aβ plaques activates microglia, the brain's immune cells, resulting in chronic inflammation, which is further responsible for neuronal damage.

Table 3.3 summarizes the key genetic mutations implicated in AD, highlighting their mutation types, onset patterns (early or late), and effects on Aβ or tau pathways. Mutations in APP, PSEN1, and PSEN2 are primarily linked to early-onset familial AD, while the APOE ε4 allele is a major genetic risk factor for late-onset sporadic AD. These genetic alterations contribute to disease progression through enhanced Aβ production, aggregation, impaired clearance, and tau pathology.

3.5 INTERPLAY OF MECHANISMS

3.5.1 Interaction between Amyloid-Beta and Tau Pathology

AD is a common type of dementia and it is a chronic, slowly, and progressively worsening neurological disorder that affects cognitive functioning and memory. Two hallmark pathological features define AD: the deposition of Aβ aggregates also called plaques, and the aggregation of NFTs made of phosphorylated tau protein. Aβ peptides are produced from beta cleavage of APP. These peptides deposit as insoluble structures referred to as extracellular plaques which are believed to interfere with signaling between cells and incite immune responses that cause inflammation (Sehar et al., 2022). On the other hand, Tau proteins are mainly involved in the stabilization of microtubules in neurons. In AD, tau acquires abnormally elevated phosphate groups or hyperphosphorylation. It ceases to bind with microtubules forming intracellular NFTs and ultimately making them dysfunctional and leading to the death of neurons. Recent research suggests that Aβ plaque formation and tau pathology are linked in a pathogenic cascade rather than being separate entities (Kapasi et al., 2021). A theory suggests that the build-up of Aβ plaques could initiate a sequence of events resulting in tau hyperphosphorylation and aggregation. It is thought that this interaction may worsen neuronal loss, synaptic dysfunction, and eventually cognitive decline.

3.5.2 Contribution of Oxidative Stress and Inflammation

Oxidative stress is a condition in which the body cannot eliminate ROS or repair the damage they cause. ROS are molecules that can damage cell components such as DNA, proteins, and lipids. Oxidative stress is thought to play a central role in the pathogenesis of neuronal damage and death in AD. Several mechanisms can cause increased oxidative stress in AD like mitochondrial dysfunction, amyloid-beta accumulation etc. In addition to being important for the synthesis of cellular energy, mitochondria are also a key generator of ROS. Oxidative damage is exacerbated in AD by mitochondrial failure, which results in increased ROS production. Aggregating Aβ peptides can cause plaques that disturb cellular homeostasis and directly produce ROS, which intensifies oxidative stress (Buccellato et al., 2021).

Oxidative stress and inflammation are two processes that are not independent of each other in AD. Instead, they form a positive feedback loop that increases the extent of neuronal injury. ROS, which are produced during oxidative stress, can trigger inflammasome activation. Cytokines promote inflammation, which can enhance ROS production in a cycle that tightens neurodegeneration. This proves the interaction between the two factors of AD pathology and underlines the need to target both oxidative stress and inflammation for effective treatment (Ganguly et al., 2021).

3.5.3 Mitochondrial Dysfunction and Neurodegeneration

Mitochondria are the "powerhouses of the cell," and their main function is to generate the energy needed for a variety of cellular processes. Mitochondria are especially important in neurons, which are very metabolically active and require a lot of energy. Growing evidence suggests that mitochondrial dysfunction is a key process in the development of neurodegenerative diseases such as AD which is a progressive neurodegenerative disorder affecting the mind and causing the deterioration of memory and other cognitive abilities (Johnson et al., 2021). Some of the characteristics of AD are Aβ plaques and tau protein tangles. These interfere with nerve cell signaling and cause cell death. Recent research indicates a significant connection between these pathological features and mitochondrial dysfunction concerning dynamism, energy generation, and biosynthesis. Several anomalies are evident in the mitochondria of AD-affected neurons. These anomalies include decreased energy production, elevated oxidative stress, and poor calcium handling. These dysfunctions aid in the disease's overall progression, synapse failure, and neuronal death (Bhatia et al., 2022).

3.6 CHALLENGES AND FUTURE DIRECTIONS

Despite having major advancements in the understanding of AD, many challenges remain in its effective treatment and management. Present treatments, i.e., cholinesterase inhibitors (e.g., donepezil, rivastigmine) and NMDA receptor antagonists (e.g., memantine), offer only improvement in symptoms such as memory and behavior. Alas, they are not capable of arresting or reversing the progression of the disease (Khan et al., 2023). Among the most critical needs is the lack of FDA-approved disease-modifying therapies that significantly halt or reverse the course of AD. This owes to the multifactorial nature of the disease's pathophysiology, which is characterized by several concurrent processes, such as Aβ deposition, tau protein tangles, persistent neuroinflammation, and massive neuronal loss (Musiek & Morris, 2021). Current treatments generally target only one or two of these mechanisms, which limits their overall efficacy.

In future, researchers are increasingly exploring combination therapies that simultaneously target various pathological features, aiming for a more holistic and effective approach (Pardo-Moreno et al., 2022). Precision medicine or personalized therapy according to a person's genetic makeup, biomarkers promises much for enhancing the efficacy of treatment and reducing the risk of side effects (Silva-Spínola et al., 2022). Furthermore, gene therapy, especially with technologies such as Clustered Regularly Interspaced Short Palindromic Repeats (CRISPR), provides a promising avenue for potentially curing the root genetic causes of AD (Doshi et al., 2024). Collectively, these future directions underscore the need for multifaceted and personalized strategies to combat this devastating disease.

3.7 CONCLUSION

The neurodegeneration seen in AD is the result of a complex interplay of molecular and cellular processes. Key pathological features include the presence of Aβ plaques and tau NFTs, which disrupt neuronal communication, leading to synaptic dysfunction and cell death. In addition to these hallmarks, neuroinflammation, oxidative stress, and mitochondrial dysfunction have emerged as significant contributors to the disease's progression. The multifaceted nature of AD suggests that targeting a singular therapeutic approach may be inadequate. Instead, addressing the various underlying mechanisms comprehensively is likely necessary to slow or halt disease progression. Recent advances in understanding the genetic and environmental factors influencing these processes offer new avenues for early diagnosis and personalized treatment strategies. Ultimately, a deeper understanding of these processes will pave the way for innovative therapeutic approaches that can significantly improve patient outcomes and address the growing global burden of AD.

ACKNOWLEDGMENTS

Figure 3.1 was created using BioRender.com. We acknowledge BioRender for providing the platform and tools to design the illustration used in this book chapter.

REFERENCES

Ashleigh, T., Swerdlow, R. H., & Beal, M. F. (2023). The role of mitochondrial dysfunction in Alzheimer's disease pathogenesis. *Alzheimer's & Dementia: The Journal of the Alzheimer's Association, 19*(1), 333–342. https://doi.org/10.1002/alz.12683

Azargoonjahromi A. (2024). The duality of amyloid-β: Its role in normal and Alzheimer's disease states. *Molecular Brain, 17*(1), 44. https://doi.org/10.1186/s13041-024-01118-1

Bai, R., Guo, J., Ye, X. Y., Xie, Y., & Xie, T. (2022). Oxidative stress: The core pathogenesis and mechanism of Alzheimer's disease. *Ageing Research Reviews, 77,* 101619. https://doi.org/10.1016/j.arr.2022.101619

Bhatia, S., Rawal, R., Sharma, P., Singh, T., Singh, M., & Singh, V. (2022). Mitochondrial dysfunction in Alzheimer's disease: Opportunities for drug development. *Current Neuropharmacology, 20*(4), 675–692. https://doi.org/10.2174/1570159X19666210517114016

Buccellato, F. R., D'Anca, M., Fenoglio, C., Scarpini, E., & Galimberti, D. (2021). Role of oxidative damage in Alzheimer's disease and neurodegeneration: From pathogenic mechanisms to biomarker discovery. *Antioxidants (Basel, Switzerland), 10*(9), 1353. https://doi.org/10.3390/antiox10091353

Dan, L., & Zhang, Z. (2023). Alzheimer's disease: An axonal injury disease? *Frontiers in aging neuroscience, 15,* 1264448. https://doi.org/10.3389/fnagi.2023.1264448

Deng, Q., Wu, C., Parker, E., Liu, T. C., Duan, R., & Yang, L. (2024). Microglia and astrocytes in Alzheimer's disease: Significance and summary of recent advances. *Aging and Disease, 15*(4), 1537–1564. https://doi.org/10.14336/AD.2023.0907

Doshi, V., Joshi, G., Sharma, S., & Choudhary, D. (2024). Gene therapy: An alternative to treat Alzheimer's disease. *Naunyn-Schmiedeberg's Archives of Pharmacology, 397*(6), 3675–3693. https://doi.org/10.1007/s00210-023-02873-z

Florescu, D. N., Boldeanu, M. V., Şerban, R. E., Florescu, L. M., Serbanescu, M. S., Ionescu, M., ... & Vere, C. C. (2023). Correlation of the pro-inflammatory cytokines IL-1β, IL-6, and TNF-α, inflammatory markers, and tumor markers with the diagnosis and prognosis of colorectal cancer. *Life, 13*(12), 2261.

Gal, J., Katsumata, Y., Zhu, H., Srinivasan, S., Chen, J., Johnson, L. A., Wang, W. X., Golden, L. R., Wilcock, D. M., Jicha, G. A., Cykowski, M. D., & Nelson, P. T. (2022). Apolipoprotein E proteinopathy is a major dementia-associated pathologic biomarker in individuals with or without the APOE Epsilon 4 allele. *The American Journal of Pathology, 192*(3), 564–578. https://doi.org/10.1016/j.ajpath.2021.11.013

Ganguly, U., Kaur, U., Chakrabarti, S. S., Sharma, P., Agrawal, B. K., Saso, L., & Chakrabarti, S. (2021). Oxidative stress, neuroinflammation, and NADPH oxidase: Implications in the pathogenesis and treatment of Alzheimer's disease. *Oxidative Medicine and Cellular Longevity, 2021,* 7086512. https://doi.org/10.1155/2021/7086512

Heneka, M. T., van der Flier, W. M., Jessen, F., Hoozemanns, J., Thal, D. R., Boche, D., Brosseron, F., Teunissen, C., Zetterberg, H., Jacobs, A. H., Edison, P., Ramirez, A., Cruchaga, C., Lambert, J. C., Laza, A. R., Sanchez-Mut, J. V., Fischer, A., Castro-Gomez, S., Stein, T. D., Kleineidam, L., …, & Riechers, S. P. (2025). Neuroinflammation in Alzheimer disease. *Nature Reviews Immunology, 25*(5), 321–352. https://doi.org/10.1038/s41577-024-01104-7

Ichimata, S., Yoshida, K., Li, J., Rogaeva, E., Lang, A. E., & Kovacs, G. G. (2024). The molecular spectrum of amyloid-beta (Aβ) in neurodegenerative diseases beyond Alzheimer's disease. *Brain pathology (Zurich, Switzerland), 34*(1), e13210. https://doi.org/10.1111/bpa.13210

Johnson, J., Mercado-Ayon, E., Mercado-Ayon, Y., Dong, Y. N., Halawani, S., Ngaba, L., & Lynch, D. R. (2021). Mitochondrial dysfunction in the development and progression of neurodegenerative diseases. *Archives of Biochemistry and Biophysics, 702,* 108698. https://doi.org/10.1016/j.abb.2020.108698

Kabir, M. T., Uddin, M. S., Setu, J. R., Ashraf, G. M., Bin-Jumah, M. N., & Abdel-Daim, M. M. (2020). Exploring the role of PSEN mutations in the pathogenesis of Alzheimer's disease. *Neurotoxicity Research, 38*(4), 833–849. https://doi.org/10.1007/s12640-020-00232-x

Kapasi, A., Leurgans, S. E., Arvanitakis, Z., Barnes, L. L., Bennett, D. A., & Schneider, J. A. (2021). Aβ (Amyloid Beta) and tau tangle pathology modifies the association between small vessel disease and cortical microinfarcts. *Stroke, 52*(3), 1012–1021. https://doi.org/10.1161/STROKEAHA.120.031073

Ke, J., Tian, Q., Xu, Q., Fu, Z., & Fu, Q. (2021). Mitochondrial dysfunction: A potential target for Alzheimer's disease intervention and treatment. *Drug Discovery Today, 26*(8), 1991–2002. https://doi.org/10.1016/j.drudis.2021.04.025

Khan, B., Iqbal, M. K., Khan, M. A., Khan, H., Kiyani, M. M., Bashir, S., & Li, S. (2025). Unraveling the complexity of Alzheimer's disease: Insights into etiology and advancements in treatment strategies. *Journal of Molecular Neuroscience: MN, 75*(2), 57. https://doi.org/10.1007/s12031-025-02337-4

Khan, S. S., Khatik, G. L., & Datusalia, A. K. (2023). Strategies for treatment of disease-associated dementia beyond Alzheimer's disease: An update. *Current Neuropharmacology*, *21*(2), 309–339. https://doi.org/10.2174/1570159X20666220411083922

Kotarba, S., Kozłowska, M., Scios, M., Saramowicz, K., Barczuk, J., Granek, Z., Siwecka, N., Wiese, W., Golberg, M., Galita, G., Sychowski, G., Majsterek, I., & Rozpędek-Kamińska, W. (2024). Potential mechanisms of tunneling nanotube formation and their role in pathology spread in Alzheimer's disease and other proteinopathies. *International Journal of Molecular Sciences*, *25*(19), 10797. https://doi.org/10.3390/ijms251910797

Kumari, S., Dhapola, R., & Reddy, D. H. (2023). Apoptosis in Alzheimer's disease: Insight into the signaling pathways and therapeutic avenues. *Apoptosis: An International Journal on Programmed Cell Death*, *28*(7–8), 943–957. https://doi.org/10.1007/s10495-023-01848-y

Leuzy, A., Ashton, N. J., Mattsson-Carlgren, N., Dodich, A., Boccardi, M., Corre, J., Drzezga, A., Nordberg, A., Ossenkoppele, R., Zetterberg, H., Blennow, K., Frisoni, G. B., Garibotto, V., & Hansson, O. (2021). 2020 update on the clinical validity of cerebrospinal fluid amyloid, tau, and phospho-tau as biomarkers for Alzheimer's disease in the context of a structured 5-phase development framework. *European Journal of Nuclear Medicine and Molecular Imaging*, *48*(7), 2121–2139. https://doi.org/10.1007/s00259-021-05258-7

Marsh, J. A., Huang, G., Bowling, K., Renton, A. E., Ziegemeier, E., Ball, T., Pottier, C., Cruchaga, C., Day, G. S., Bateman, R. J., Llibre-Guerra, J. J., McDade, E., & Karch, C. M. (2025). Evaluating pathogenicity of variants of unknown significance in APP, PSEN1, and PSEN2. *Neurotherapeutics: The Journal of the American Society for Experimental Neurotherapeutics*, *22*(3), e00527. https://doi.org/10.1016/j.neurot.2025.e00527

Matsubara, T., Ishii, K., Saito, Y., Tokumaru, A. M., Arakawa, A., Hara, M., Kurihara, M., Sengoku, R., Kanemaru, K., Iwata, A., Arai, T., Miyashita, A., Ikeuchi, T., Hasegawa, M., Murayama, S., & Saito, Y. (2024). Neurofibrillary tangle-predominant dementia followed by amyloid β pathology: A clinico-radio-pathological case providing insights into current disease-modifying therapeutic strategy. *Acta Neuropathologica Communications*, *12*(1), 98. https://doi.org/10.1186/s40478-024-01815-x

Monteiro, A. R., Barbosa, D. J., Remião, F., & Silva, R. (2023). Alzheimer's disease: Insights and new prospects in disease pathophysiology, biomarkers and disease-modifying drugs. *Biochemical Pharmacology*, *211*, 115522. https://doi.org/10.1016/j.bcp.2023.115522

Moore, K. B. E., Hung, T. J., & Fortin, J. S. (2023). Hyperphosphorylated tau (p-tau) and drug discovery in the context of Alzheimer's disease and related tauopathies. *Drug Discovery Today*, *28*(3), 103487. https://doi.org/10.1016/j.drudis.2023.103487

Musiek, E. S., & Morris, J. C. (2021). Possible consequences of the approval of a disease-modifying therapy for Alzheimer disease. *JAMA Neurology*, *78*(2), 141–142. https://doi.org/10.1001/jamaneurol.2020.4478

Pardo-Moreno, T., González-Acedo, A., Rivas-Domínguez, A., García-Morales, V., García-Cozar, F. J., Ramos-Rodríguez, J. J., & Melguizo-Rodríguez, L. (2022). Therapeutic approach to Alzheimer's disease: Current treatments and new perspectives. *Pharmaceutics*, *14*(6), 1117. https://doi.org/10.3390/pharmaceutics14061117

Pelucchi, S., Gardoni, F., Di Luca, M., & Marcello, E. (2022). Synaptic dysfunction in early phases of Alzheimer's disease. *Handbook of clinical neurology*, *184*, 417–438. https://doi.org/10.1016/B978-0-12-819410-2.00022-9

Plascencia-Villa, G., & Perry, G. (2022). Neuropathologic changes provide insights into key mechanisms of Alzheimer disease and related dementia. *The American Journal of Pathology*, *192*(10), 1340–1346. https://doi.org/10.1016/j.ajpath.2022.07.002

Ravindranath, V., & Sundarakumar, J. S. (2021). Changing demography and the challenge of dementia in India. *Nature Reviews Neurology*, *17*(12), 747–758. https://doi.org/10.1038/s41582-021-00565-x

Rawat, P., Sehar, U., Bisht, J., Selman, A., Culberson, J., & Reddy, P. H. (2022). Phosphorylated tau in Alzheimer's disease and other Tauopathies. *International Journal of Molecular Sciences*, *23*(21), 12841. https://doi.org/10.3390/ijms232112841

Roda, A. R., Serra-Mir, G., Montoliu-Gaya, L., Tiessler, L., & Villegas, S. (2022). Amyloid-beta peptide and tau protein crosstalk in Alzheimer's disease. *Neural Regeneration research*, *17*(8), 1666–1674. https://doi.org/10.4103/1673-5374.332127

Sehar, U., Rawat, P., Reddy, A. P., Kopel, J., & Reddy, P. H. (2022). Amyloid beta in aging and Alzheimer's disease. *International Journal of Molecular Sciences*, *23*(21), 12924. https://doi.org/10.3390/ijms232112924

Silva-Spínola, A., Baldeiras, I., Arrais, J. P., & Santana, I. (2022). The road to personalized medicine in Alzheimer's disease: The use of artificial intelligence. *Biomedicines*, *10*(2), 315. https://doi.org/10.3390/biomedicines10020315

Sun, Y., Islam, S., Michikawa, M., & Zou, K. (2024). Presenilin: A multi-functional molecule in the pathogenesis of Alzheimer's disease and other neurodegenerative diseases. *International Journal of Molecular Sciences*, *25*(3), 1757. https://doi.org/10.3390/ijms25031757

Tahami Monfared, A. A., Byrnes, M. J., White, L. A., & Zhang, Q. (2022). Alzheimer's disease: Epidemiology and clinical progression. *Neurology and Therapy*, *11*(2), 553–569. https://doi.org/10.1007/s40120-022-00338-8

Thal, D. R., & Tomé, S. O. (2022). The central role of tau in Alzheimer's disease: From neurofibrillary tangle maturation to the induction of cell death. *Brain Research Bulletin*, *190*, 204–217. https://doi.org/10.1016/j.brainresbull.2022.10.006

Walia, V., Kaushik, D., Mittal, V., Kumar, K., Verma, R., Parashar, J., Akter, R., Rahman, M. H., Bhatia, S., Al-Harrasi, A., Karthika, C., Bhattacharya, T., Chopra, H., & Ashraf, G. M. (2022). Correction to: Delineation of neuroprotective effects and possible benefits of antioxidants therapy for the treatment of Alzheimer's diseases by targeting mitochondrial-derived reactive oxygen species: Bench to bedside. *Molecular Neurobiology*, *59*(1), 681–682. https://doi.org/10.1007/s12035-021-02665-7

Yang, Y., Bagyinszky, E., & An, S. S. A. (2023). Presenilin-1 (PSEN1) mutations: Clinical phenotypes beyond Alzheimer's disease. *International Journal of Molecular Sciences*, *24*(9), 8417. https://doi.org/10.3390/ijms24098417

Yang, Y., Tapias, V., Acosta, D., Xu, H., Chen, H., Bhawal, R., Anderson, E. T., Ivanova, E., Lin, H., Sagdullaev, B. T., Chen, J., Klein, W. L., Viola, K. L., Gandy, S., Haroutunian, V., Beal, M. F., Eliezer, D., Zhang, S., & Gibson, G. E. (2022). Altered succinylation of mitochondrial proteins, APP and tau in Alzheimer's disease. *Nature Communications*, *13*(1), 159. https://doi.org/10.1038/s41467-021-27572-2

Yeung, A. W. K., Tzvetkov, N. T., Georgieva, M. G., Ognyanov, I. V., Kordos, K., Jóźwik, A., Kühl, T., Perry, G., Petralia, M. C., Mazzon, E., & Atanasov, A. G. (2021). Reactive oxygen species and their impact in neurodegenerative diseases: Literature landscape analysis. *Antioxidants & Redox Signaling*, *34*(5), 402–420. https://doi.org/10.1089/ars.2019.7952

Yu, H., & Wu, J. (2021). Amyloid-β: A double agent in Alzheimer's disease? *Biomedicine & Pharmacotherapy = Biomedecine & Pharmacotherapie*, *139*, 111575. https://doi.org/10.1016/j.biopha.2021.111575

Yu, L., Jin, J., Xu, Y., & Zhu, X. (2022). Aberrant energy metabolism in Alzheimer's disease. *Journal of Translational Internal Medicine*, *10*(3), 197–206. https://doi.org/10.2478/jtim-2022-0024

4 Pharmacological, Non-Pharmacological Management, and Emerging Therapies for the Management of Alzheimer's Disease

Irmak Ferah Okkay and Ufuk Okkay*
Faculty of Pharmacy, Department of Pharmacology, Ataturk University, Erzurum, Turkey

4.1 INTRODUCTION

Alzheimer's disease (AD) is the most common form of dementia, accounting for nearly half of all cases [1]. It is characterized by progressive brain atrophy, enlarged ventricles, and distinctive histological changes in brain tissue. The disease manifests as a gradual decline in intellectual and cognitive functions, initially presenting as memory impairment and loss of insight [2]. Early symptoms often include forgetfulness, confusion, and eventually, the emergence of disease. As the disease advances, further symptoms may arise, including increased excitability, language impairments (aphasia), difficulties with motor planning (apraxia), loss of recognition abilities (agnosia), and disturbances in gait [2–4]. The precise etiology of AD remains unclear, and diagnosis is typically made by excluding other potential causes and confirmed through post-mortem examination [5]. The prognosis is poor, with a decline typically occurring over five to ten years, ultimately leading to death [6, 7].

AD primarily affects the elderly, with the risk increasing significantly with age—up to 20% of individuals over 85 may develop the disease [8, 9]. The socioeconomic impact is considerable, with the annual cost of dementia care in the United States estimated to be between $150 and $215 billion [10]. Both hereditary and sporadic forms of AD exist.

AD is a multifaceted disorder characterized by a complex pathobiological framework (Figure 4.1 and Table 4.1). Early-onset AD is associated with genetic anomalies such as trisomy 21, mutations in the presenilin-1 (PSEN1) gene on chromosome 14, and the presence of the ε4 allele of apolipoprotein E (APOE) on chromosome 19. The ε4 variant, unlike the more common ε2 and ε3 alleles, is strongly linked to the accumulation of amyloid-β (Aβ) deposits, a hallmark of AD pathology [11–13].

Mutations in three specific genes—APP, PSEN1, and PSEN2—are known to cause autosomal dominant early-onset AD. These genes play a crucial role in the generation of Aβ peptides, which are produced through the enzymatic cleavage of amyloid precursor protein (APP) by β-secretase and γ-secretase, with presenilins serving as the catalytic core of γ-secretase [14, 15]. The accumulation of Aβ in the brain, both as soluble oligomers and amyloid plaques, supports the amyloid hypothesis, which posits that Aβ plays a central role in AD pathogenesis [16].

*Corresponding Author. E-mail: irmakferah@atauni.edu.tr

DOI: 10.1201/9781003530510-4

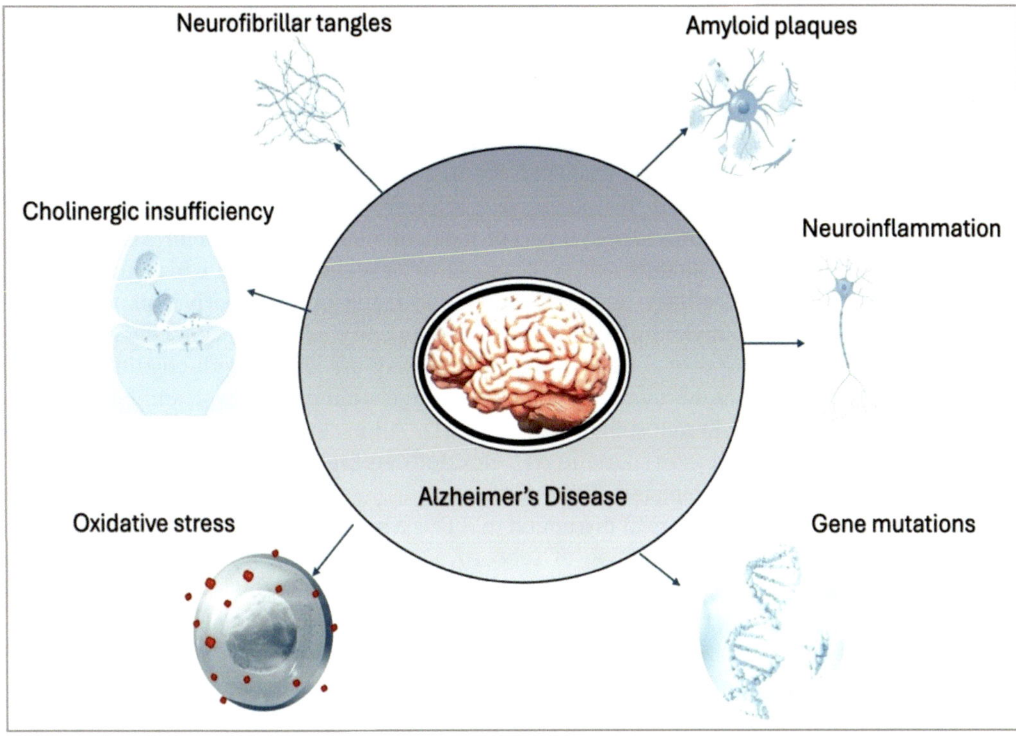

FIGURE 4.1 Major pathological features of Alzheimer's disease. Key mechanisms involved in Alzheimer's disease: neurofibrillary tangles, amyloid plaques, neuroinflammation, gene mutations, oxidative stress, and cholinergic insufficiency. These factors collectively contribute to neurodegeneration and cognitive decline.

TABLE 4.1
Pathophysiology of Alzheimer's Disease

Feature	Description	Significance
Brain atrophy	Progressive loss of brain tissue leading to shrinkage.	Leads to cognitive deficits and is a marker of disease progression.
Ventricular enlargement	Enlargement of brain ventricles due to loss of surrounding brain tissue.	Indicates brain tissue loss and is associated with severe stages of AD.
Amyloid plaques	Extracellular deposits of amyloid-β, a hallmark of AD pathology.	Central to the amyloid hypothesis of AD; thought to trigger tau pathology.
Neurofibrillary tangles	Intracellular aggregates of hyperphosphorylated tau protein.	Correlates with cognitive decline and neurodegeneration in AD.
Neuronal loss	Degeneration and death of neurons contributing to cognitive deficits.	Results in functional impairments and is associated with disease severity.
Genetic mutations	Mutations in APP, PSEN1, and PSEN2 genes, leading to increased Aβ production.	Causes early-onset AD in autosomal dominant cases, rare but significant.
APOE ε4 allele	A major genetic risk factor for late-onset AD, linked to impaired Aβ clearance.	Increases the risk of developing AD and influences treatment response.
Cognitive decline	Gradual deterioration of intellectual and cognitive functions, including memory loss and dementia.	Primary clinical manifestation of AD, progressively worsening over time.

While autosomal dominant cases are rare, the more common sporadic, late-onset form of AD also has a significant genetic component. Numerous genes have been identified with alleles that increase the risk of AD, with APOE ε4 being the most significant [17]. APOE encodes apolipoprotein E (ApoE), a lipid carrier protein, and individuals with the ε4 allele have a more than threefold increased risk of developing AD. Although ε4 carriers make up less than a quarter of the population, they account for over half of all AD cases [17]. Clinical trials indicate that ApoE4 carriers and non-carriers respond differently to therapies, suggesting a pharmacogenetic influence on treatment efficacy. However, routine genetic testing for APOE status is not yet part of standard clinical evaluation for AD.

The defining pathological features of AD consist of amyloid plaques, which form outside cells from Aβ, and neurofibrillary tangles inside cells, made up of hyperphosphorylated tau protein [18, 19]. Although amyloid plaque formation is an early and consistent characteristic of AD, tau tangles accumulate over time and are more closely associated with cognitive decline. Current theories suggest that the accumulation of Aβ triggers tau pathology, leading to neuronal dysfunction and cell death. In autosomal dominant AD, Aβ accumulates due to mutations that increase its production, whereas in sporadic AD, elevated cerebral Aβ levels are thought to result from impaired clearance mechanisms [20].

The most prominent neurochemical disruption in AD is a significant deficiency in acetylcholine (ACh), primarily due to the degeneration of subcortical cholinergic neurons, particularly in the basal forebrain. This "cholinergic hypothesis" suggests that ACh deficiency is critical in the development of AD symptoms [21, 22]. However, AD also involves complex disruptions across multiple neurotransmitter systems, including glutamate, serotonin, and various neuropeptides, affecting both cholinergic neurons and the cortical and hippocampal regions they innervate.

Aβ aggregation is central to AD pathogenesis. While amyloid plaques are composed of highly ordered Aβ fibrils, soluble Aβ oligomers, and possibly even dimers, are now believed to be more pathogenic [23]. Tau also aggregates, forming paired helical filaments that constitute neurofibrillary tangles. Modifications of tau after translation, such as phosphorylation, impair its normal activity and promote its tendency to aggregate [24]. The mechanisms by which Aβ and tau contribute to neuronal dysfunction and death may involve disruptions in synaptic transmission, excitotoxicity, oxidative stress, and neuroinflammation [25]. The selective vulnerability of certain cortical neurons to AD-related pathology remains poorly understood.

Moreover, oxidative stress plays a critical role in the pathophysiology of AD. Neurons in the brain continuously utilize oxygen for energy production, a process that generates reactive oxygen species (ROS), commonly known as free radicals [26]. Under normal conditions, these ROS are neutralized by cellular antioxidant defense mechanisms. However, in AD, this balance is disrupted, leading to the accumulation of ROS and subsequent cellular damage [27]. Oxidative stress particularly affects neuronal lipids, proteins, and DNA, resulting in cellular dysfunction. This damage contributes to the accumulation of Aβ plaques and hyperphosphorylated tau proteins, which, in turn, lead to neuronal death and brain atrophy [27]. Reducing oxidative stress has become a significant therapeutic target in AD, and ongoing research is exploring the potential of antioxidant therapies in mitigating disease progression.

In AD, the medial temporal lobe, including the entorhinal cortex and hippocampus, is particularly vulnerable to neuronal dysfunction and loss. Early symptoms typically include anterograde episodic memory deficits, such as repeated questioning, misplacing items, and missing appointments [28]. These symptoms often lead to a diagnosis of mild cognitive impairment, as they are noticeable but not yet disruptive enough to significantly impair daily functioning. Patients with mild cognitive dysfunction convert to AD at a rate of about 10% annually, though not all mild cognitive dysfunction patients will progress to AD [29]. In AD, cognitive and functional abilities decline progressively, eventually affecting visuospatial and executive functions. The later stages of the disease are marked by increasing dependence and progression to an akinetic-mute state, with death typically occurring within 6–12 years of disease onset, often due to complications such as pneumonia or pulmonary embolism [30].

4.2 THERAPEUTIC APPROACHES

4.2.1 Acetylcholinesterase Inhibitors

Current therapeutic strategies for AD focus on enhancing cholinergic transmission within the central nervous system (CNS). The primary approach involves the use of acetylcholinesterase (AChE) inhibitors, which aim to increase ACh levels by preventing its breakdown in the synaptic cleft [31]. Donepezil, rivastigmine, and galantamine are commonly used for this purpose, while tacrine, though the first AChE inhibitor approved for AD, is now rarely used due to its adverse effect profile. These medications offer a slight enhancement in cognitive abilities for individuals with mild to moderate AD. Rivastigmine is unique among these drugs as it is metabolized by AChE into a carbamylate derivative, avoiding interactions with cytochrome P450 enzymes, unlike donepezil and galantamine, which are metabolized by these enzymes and may interact with other medications. Typical side effects of AChE inhibitors involve nausea, diarrhea, vomiting, loss of appetite, tremors, slowed heart rate, and muscle cramps, which are all associated with increased cholinergic activity. Tacrine is associated with hepatotoxicity, which has limited its clinical use [32] (Figure 4.2 and Table 4.2).

4.2.2 NMDA Antagonists

Another therapeutic approach involves the use of NMDA receptor antagonists, such as memantine. Excessive stimulation of NMDA receptors by glutamate can lead to excitotoxicity and neuronal death, processes implicated in AD. Memantine acts as a non-competitive antagonist of NMDA receptors, partially blocking the associated ion channels and reducing pathological Ca^{2+} influx, thereby protecting neurons without completely inhibiting necessary cellular functions [33] (Table 4.2).

Short-term studies suggest that memantine can slow the progression of memory loss in patients with moderate to severe AD, although it does not prevent or slow neurodegeneration and is not more effective than AChE inhibitors. Memantine is usually well-accepted, though it can cause side effects like confusion, agitation, and restlessness, which can sometimes be hard to distinguish from symptoms of AD. Given its distinct mode of action and possible neuroprotective benefits, memantine is

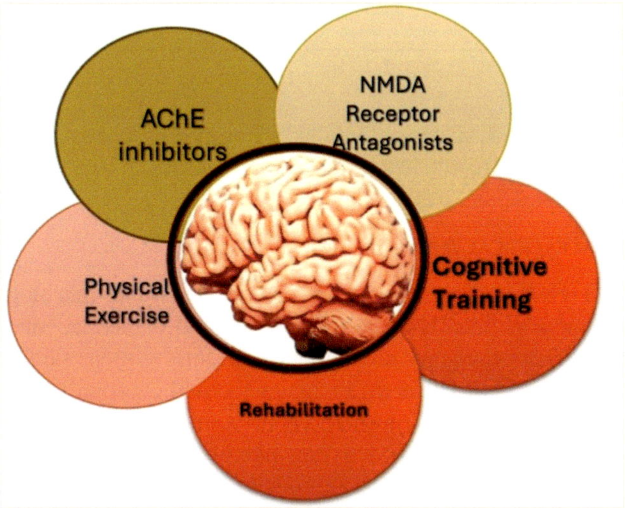

FIGURE 4.2 Therapeutic strategies in Alzheimer's disease. Current treatment approaches for Alzheimer's disease, including AChE inhibitors, NMDA receptor antagonists, cognitive training, rehabilitation, and physical exercise. These strategies aim to alleviate symptoms and maintain cognitive function.

TABLE 4.2
Treatment Options for Alzheimer's Disease

Therapeutic Approach	Mechanism of Action	Common Therapies	Benefits	Challenges/Side Effects
Acetylcholinesterase Inhibitors	Inhibit acetylcholinesterase, increasing acetylcholine levels in the synaptic cleft.	Donepezil, rivastigmine, galantamine, tacrine	Modest improvement in cognitive function in mild to moderate AD.	Nausea, diarrhea, vomiting, bradycardia, muscle cramps, hepatotoxicity (for tacrine).
NMDA receptor antagonists	Block NMDA receptors to reduce excitotoxicity and protect neurons.	Memantine	Slows progression of memory loss in moderate to severe AD; often used in combination with AChE inhibitors.	Side effects include confusion, agitation, restlessness, often indistinguishable from AD symptoms.
Cognitive training and rehabilitation	Enhance cognitive function through targeted exercises focusing on memory, reasoning, and information processing speed.	Brain exercises, social activities like board and card games, use of non-dominant hand	May help preserve cognitive function, improve social interaction, and stimulate brain activity.	Limited evidence for long-term benefits; effectiveness varies among individuals.
Physical exercise	Improves overall brain health by increasing blood flow and enhancing vascular function.	Aerobic exercises like brisk walking, swimming	Reduces risk of cognitive decline, improves mood, and helps maintain independence.	Small or inconclusive studies; specific exercise regimens and intensity may vary in effectiveness.
Emerging treatments	Target specific pathological processes like amyloid plaque formation and gut–brain axis interactions.	Monoclonal antibodies (e.g., donanemab), microbiota–gut–brain axis modulation (e.g., SCFA, FMT)	Potential to slow disease progression by targeting underlying mechanisms.	Long-term efficacy and safety are still under investigation; gender-specific responses to treatments need further research.

often administered alongside an AChE inhibitor [33]. However, there is still a lack of long-term data to fully support this combined treatment approach (Figure 4.2).

4.2.3 Non-pharmacological Approaches

Cognitive training and rehabilitation are gaining traction as non-pharmacological interventions for improving cognitive functions in early-stage AD and vascular dementia, where memory issues are particularly prominent. These therapies are distinct from interventions targeting behavioral, emotional, or physical functions [34]. Different therapies focused on cognitive enhancement, such as brain exercises focusing on reasoning, information processing speed, and memory, are employed to enhance cognitive outcomes [35]. Simple activities like using the non-dominant hand or engaging in new and challenging tasks are recommended to stimulate the brain. Social activities, including board and card games, are also beneficial for cognitive health and social interaction. Although evidence supporting cognitive training in AD is limited, studies suggest that intensive cognitive training over six months may help preserve cognitive function in mild-to-moderate AD [36]. To confirm these findings, future research should involve larger, multicenter trials with extended observation periods. Cognitive training is preferred when drug therapies are contraindicated due to side effects or drug interactions [37].

Regular physical exercise is well-known for promoting overall health, and its advantages extend to those diagnosed with AD, offering specific cognitive benefits. Studies indicate that engaging in moderate physical activity can slow down the cognitive decline associated with AD. For instance, a study conducted by the University of Wisconsin revealed that adults over the age of 60, classified as high-risk for AD, exhibited a significant decrease in memory lapses and other cognitive deficits after engaging in 30 minutes of moderate physical activity five times per week. Furthermore, these individuals exhibited a lowered risk of developing AD compared to their less active counterparts [4, 21].

The cognitive benefits of exercise are further supported by research from the University of Kansas, which revealed that regular physical activity has the potential to enhance memory test scores. More impressively, regular physical activity has been shown to promote enlargement of the hippocampus, a critical brain region responsible for learning and memory processes. The hippocampus is one of the first areas to experience damage in the early stages of AD and preserving or even increasing its size through exercise offers a promising strategy to combat memory loss. Together, these studies highlight the critical role that physical activity plays not only in maintaining general AD disease.

Although many exercise studies have been small or yielded mixed results regarding memory and brain function, low-intensity aerobic activities, like brisk walking or swimming, are believed to be most beneficial for the brain. The exact mechanism is unclear, but it is hypothesized that enhanced vascular health and increased cerebral perfusion of oxygenated blood play a role in improving brain function. Ergotherapy, or occupational therapy, aims to help dementia patients improve self-care, productivity, and leisure, thereby enhancing their functional abilities, social participation, quality of life, and overall happiness [38]. Although symptomatic treatments like cholinesterase inhibitors and memantine can merely slow the progression of symptoms [39], combining behavioral methods with pharmacological treatment may optimize benefits for both patients and caregivers, highlighting the need for non-pharmacological intervention programs [40]. Physical therapy also plays a crucial role in managing mental health issues like anxiety and depression in dementia patients [41]. Routine physical activity enhances mood, decreases reliance on medication, and aids in managing emotional symptoms such as agitation, anger, and hostility. Aside from its cognitive and emotional benefits, physical therapy provides notable social benefits as well. It helps lower social anxiety, strengthens interpersonal relationships, and supports individuals with dementia in preserving their independence for as long as possible [42] (Figure 4.2 and Table 4.2).

4.2.4 Emerging Treatments

Recent studies have underscored the crucial impact of the microbiota–gut–brain axis on neurodegenerative disorders, especially AD. Astrocytes, the predominant glial cells within the CNS, play a vital role in preserving CNS homeostasis, and impairments in their activity are associated with various neurological diseases. The two-way interaction between astrocytes and microglia has been recognized as a major factor contributing to neuroinflammation and neurodegeneration [43, 44]. Emerging studies have revealed that alterations in gut microbiota can influence astrocyte behavior in AD. Disruptions in gut microbiota have been associated with a decrease in reactive astrogliosis, the maintenance of astrocyte homeostasis, and enhanced protection against conditions such as amyloidosis and tau-induced neurodegeneration. Research by Chandra et al. [45] and Seo et al. [46] reveals that these benefits are notably more significant in male mice. The response of astrocytes to gut microbiota imbalances exhibits a form of sexual dimorphism, similar to the variations seen in microglia [45, 46]. This suggests a critical need to consider gender differences in future investigations. Furthermore, the neuroprotective advantages provided by gut microbiota depletion were found to diminish following microbiota restoration and supplementation with short-chain fatty acids, indicating that these therapeutic interventions can alter the effects of microbiota imbalances. Specifically, in antibiotic-treated

APP/PS1-21 mice, fecal microbiota transplantation from age-matched control mice reversed astrogliosis, while in germ-free TE4 mice, supplementation with short-chain fatty acids reinstated gliosis and tau pathology [45, 46]. While these animal studies are still in their early stages, they indicate that the gut microbiota may influence the development and progression of AD, particularly by modulating astrocytic activity.

One of the most notable developments is the use of monoclonal antibodies like donanemab, which have shown promise in slowing cognitive decline by targeting amyloid plaques in the brain. In 2023, these treatments gained significant attention as they demonstrated the ability to reduce amyloid levels and slow disease progression in clinical trials [47].

These developments highlight the importance of a multifaceted approach in treating AD, combining both traditional and novel therapeutic strategies to manage and potentially alter the course of the disease. As research progresses, these treatments may offer new hope for patients and enhance the well-being of individuals living with AD (Table 4.2).

Ongoing clinical trials for AD management are increasingly focused on disease-modifying therapies, particularly those targeting amyloid and tau pathologies. Notably, monoclonal antibodies such as lecanemab and donanemab have shown promise in slowing cognitive decline by reducing Aβ plaques and are currently in advanced Phase III trials [48, 49]. In parallel, anti-tau therapies, neuroinflammation modulators, and gene-based interventions are under investigation [50]. Trials exploring novel delivery systems, such as intranasal insulin or mRNA-based therapeutics, are also gaining traction. Additionally, multi-domain lifestyle interventions combining diet, exercise, and cognitive training are being evaluated for their preventive potential. Despite progress, challenges remain in identifying optimal patient subgroups, improving early diagnosis, and balancing efficacy with safety in long-term use.

4.3 CONCLUSION AND PROSPECTS

AD remains a formidable neurodegenerative disorder, characterized by complex pathological features, including Aβ aggregation, tau protein hyperphosphorylation, and neurotransmitter disruptions. Despite significant advances in understanding its molecular basis, AD's precise etiology continues to elude researchers. Current therapeutic approaches, such as acetylcholinesterase inhibitors and NMDA receptor antagonists, offer symptomatic relief but are ineffective in stopping the progression of the disease. Emerging treatments, such as monoclonal antibodies targeting amyloid plaques and interventions involving the microbiota–gut–brain axis, hold promise in slowing cognitive decline and addressing underlying pathology. Looking ahead, a multidimensional strategy will be crucial in tackling AD. This approach must combine pharmacological interventions with non-pharmacological therapies, such as cognitive training, physical exercise, and social engagement, to enhance quality of life and cognitive function. Advances in genetic and microbiome research, coupled with innovative pharmacological agents, offer potential new avenues for personalized treatments, particularly as we gain deeper insights into how genetic predispositions, like the APOE ε4 allele, influence disease progression. Future research should focus on long-term, multicenter trials to confirm the efficacy of emerging treatments, particularly in diverse patient populations. Furthermore, addressing gender differences in neurodegeneration and therapeutic responses could optimize treatment strategies. With continued innovation, there is hope for not only improving symptom management but also fundamentally altering the trajectory of AD, offering a brighter future for those affected by this devastating disease.

REFERENCES

1. Tozlu ÖÖ, Türkez H, Okkay U, Ceylan O, Bayram C, Hacimüftüoglu A, Mardinoglu A (2022) Assessment of the neuroprotective potential of d-cycloserine and l-serine in aluminum chloride-induced experimental models of Alzheimer's disease: And studies. Front Nutr, 9.
2. Knopman DS, Amieva H, Petersen RC, Chételat G, Holtzman DM, Hyman BT, Nixon RA, Jones DT (2021) Alzheimer disease. Nat Rev Dis Primers, 7.

3. Kramer JH, Miller BL (2000) Alzheimer's disease and its focal variants. Semin Neurol, 20:447–454.
4. Breijyeh Z, Karaman R (2020) Comprehensive review on Alzheimer's disease: Causes and treatment. Molecules, 25.
5. Manap ASA, Almadodi R, Sultana S, Sebastian MG, Kavani KS, Lyenouq VE, Shankar A (2024) Alzheimer's disease: A review on the current trends of the effective diagnosis and therapeutics. Front Aging Neurosci, 16.
6. Chen WT, Li JH, Guo JX, Li L, Wu H (2024) Diagnosis and therapy of Alzheimer's disease: Light-driven heterogeneous redox processes. Adv Colloid Interfac, 332.
7. Khan FA, Khan A, Imran M, Ahmad A, Jeon G (2024) Advancements in deep learning for Alzheimer's disease diagnosis: A comprehensive exploration and critical analysis of neuroimaging approaches. Expert Syst, *42*.
8. Cespedes MI, Fripp J, McGree JM, Drovandi CC, Mengersen K, Doecke JD (2017) Comparisons of neurodegeneration over time between healthy ageing and Alzheimer's disease cohorts via Bayesian inference. BMJ Open, 7.
9. Xia X, Jiang QL, McDermott J, Han JDJ (2018) Aging and Alzheimer's disease: Comparison and associations from molecular to system level. Aging Cell, 17.
10. Hurd MD, Martorell P, Langa K (2015) Future monetary costs of dementia in the United States under alternative dementia prevalence scenarios. J Popul Ageing, 8:101–112.
11. Ali J, Choe K, Park JS, Park HY, Kang H, Park TJ, Kim MO (2024) The interplay of protein aggregation, genetics, and oxidative stress in Alzheimer's disease: Role for natural antioxidants and immunotherapeutics. Antioxidants-Basel, 13.
12. Chhetri A, Goel K, Ludhaidch A, Singh P, Munshi A (2024) Role of imaging genetics in Alzheimer's disease: A systematic review and current update. CNS Neurol Disord-Drug Targets, 23:1143–1156.
13. Öztan G, Issever H (2023) Molecular mechanisms and genetics of Alzheimer's disease. Turk J Biochem, 48:218–229.
14. Gao Y, Ren RJ, Zhong ZL, Dammer E, Zhao QH, Shan S, Zhou Z, Li X, Zhang YQ, Cui HL, Hu YB, Chen SD, Chen JJ, Guo QH, Wang G (2019) Mutation profile of and in Chinese familial Alzheimer's disease. Neurobiol Aging, 77:154–157.
15. Xiao XW, Liu H, Zhou L, Liu XX, Xu TY, Zhu Y, Yang QJ, Hao XL, Liu YZ, Zhang WW, Zhou YF, Wang JL, Li JC, Jiao B, Shen L, Liao XX (2023) The associations of, and genes with Alzheimer's disease: A large case-control study in Chinese population. CNS Neurosci Ther, 29:122–128.
16. Gulisano W, Maugeri D, Baltrons MA, Fà M, Amato A, Palmeri A, D'Adamio L, Grassi C, Devanand DP, Honig LS, Puzzo D, Arancio O (2019) Role of amyloid-β and tau proteins in Alzheimer's disease: Confuting the amyloid Cascade (vol 64, pg S611, 2018). J Alzheimers Dis, 68:415–415.
17. El-Tallawy HN, Saadeldin HM, Ezzeldin AM, Tohamy AM, Eltellawy S, Bathalath AM, Shehab MM (2022) Genetic, clinical, and biochemical aspects of patients with Alzheimer disease. Egypt J Neurol Psych, 58.
18. Dávila-Bouziguet E, Targa-Fabra G, Avila J, Soriano E, Pascual M (2019) Differential accumulation of Tau phosphorylated at residues Thr231, Ser262 and Thr205 in hippocampal interneurons and its modulation by Tau mutations (VLW) and amyloid-β peptide. Neurobiol Dis, 125:232–244.
19. Rajmohan R, Reddy PH (2017) Amyloid-Beta and phosphorylated tau accumulations cause abnormalities at synapses of Alzheimer's disease neurons. J Alzheimers Dis, 57:975–999.
20. Nasb M, Tao WC, Chen N (2024) Alzheimer's disease puzzle: Delving into pathogenesis hypotheses. Aging Dis, 15:43–73.
21. Chen ZR, Huang JB, Yang SL, Hong FF (2022) Role of cholinergic signaling in Alzheimer's disease. Molecules, 27.
22. Wu CK, Stopa E, Geula C, Thal L (2004) Neurofibrillary degeneration is linked with an apoptotic signal-fas-associated death domain in the basal forebrain cholinergic neurons of Alzheimer's disease. Neurobiol Aging, 25:S425–S426.
23. Bharadwaj PR, Dubey AK, Masters CL, Martins RN, Macreadie IG (2009) Aβ aggregation and possible implications in Alzheimer's disease pathogenesis. J Cell Mol Med, 13:412–421.
24. Chiasseu M, Destroismaisons L, Velde CV, Leclerc N, Di Polo A (2015) Tau accumulation in the somato-dendritic compartment of retinal ganglion cells plays a critical role in glaucomatous neurodegeneration. Invest Ophth Vis Sci, 56.
25. Sato N, Morishita F (2015) The roles of lipid and glucose metabolism in modulation of β-amyloid, tau, and neurodegeneration in the pathogenesis of Alzheimer disease. Front Aging Neurosci, 7.
26. Orfali R, Alwatban AZ, Orfali RS, Lau L, Chea N, Alotaibi AM, Nam YW, Zhang M (2024) Oxidative stress and ion channels in neurodegenerative diseases. Front Physiol, 15.

27. Domanskyi A, Parlato R (2022) Oxidative stress in neurodegenerative diseases. Antioxidants-Basel, 11.
28. Igarashi KM (2023) Entorhinal cortex dysfunction in Alzheimer's disease. Trends Neurosci, 46:124–136.
29. Khan UA, Liu L, Provenzano FA, Berman DE, Profaci CP, Sloan R, Mayeux R, Duff KE, Small SA (2014) Molecular drivers and cortical spread of lateral entorhinal cortex dysfunction in preclinical Alzheimer's disease. Nat Neurosci, 17:304–311.
30. Wilson RS, Segawa E, Boyle PA, Anagnos SE, Hizel LP, Bennett DA (2012) The natural history of cognitive decline in Alzheimer's disease. Psychol Aging, 27:1008–1017.
31. Ahmad M, Uddin I, Murad M, Maryam L, Subhan Z, Khan MS, Nabi K, Nabi M, Rahim F, Ullah H, Rehman ZU, Bekhit MM, Alnbaheen MS, Algarni AS, Aghayeva S (2024) Cholinesterase inhibitors for the treatment of Alzheimer's disorder: Synthesis, biological analysis and molecular modelling studies of iso-quinoline derivatives. Results Chem, 9.
32. Hussain R, Ullah H, Rahim F, Sarfraz M, Taha M, Iqbal R, Rehman W, Khan S, Shah SAA, Hyder S, Alhomrani M, Alamri AS, Abdulaziz O, Abdelaziz MA (2022) Multipotent cholinesterase inhibitors for the treatment of Alzheimer's disease: Synthesis, biological analysis and molecular docking study of benzimidazole-based thiazole derivatives. Molecules, 27.
33. Danysz W, Parsons CG (2012) Alzheimer's disease, β-amyloid, glutamate, NMDA receptors and memantine – Searching for the connections. Brit J Pharmacol, 167:324–352.
34. Mandolesi L, Polverino A, Montuori S, Foti F, Ferraioli G, Sorrentino P, Sorrentino G (2018) Effects of physical exercise on cognitive functioning and wellbeing: Biological and psychological benefits. Front Psychol, 9.
35. Harvey PD (2019) Domains of cognition and their assessment. Dialogues Clin Neuro, 21:227–237.
36. Trebbastoni A, Imbriano L, Podda L, Rendace L, Sacchetti ML, Campanelli A, D'Antonio F, de Lena C (2018) Cognitive training in patients with Alzheimer's disease: Findings of a 12-month randomized controlled trial. Curr Alzheimer Res, 15:452–461.
37. Giuli C, Papa R, Lattanzio F, Postacchini D (2016) The effects of cognitive training for elderly: Results from. Rejuv Res, 19:485–+.
38. Green M (2024) Occupational therapy interventions for older adults with Alzheimer's disease and related dementias. Am J Occup Ther, 78.
39. Hogan DB (2014) Long-term efficacy and toxicity of cholinesterase inhibitors in the treatment of Alzheimer disease. Can J Psychiat, 59:618–623.
40. Magierski R, Sobow T, Schwertner E, Religa D (2020) Pharmacotherapy of behavioral and psychological symptoms of dementia: State of the art and future progress. Front Pharmacol, 11.
41. Orgeta V, Leung P, del-Pino-Casado R, Qazi A, Orrell M, Spector AE, Methley AM (2022) Psychological treatments for depression and anxiety in dementia and mild cognitive impairment. Cochrane Db Syst Rev, 4: CD009125.
42. Jia RX, Liang JH, Xu Y, Wang YQ (2019) Effects of physical activity and exercise on the cognitive function of patients with Alzheimer disease: A meta-analysis. Bmc Geriatr, 19.
43. Lee HG, Wheeler MA, Quintana FJ (2022) Function and therapeutic value of astrocytes in neurological diseases. Nat Rev Drug Discov, 21:339–358.
44. Patani R, Hardingham GE, Liddelow SA (2023) Functional roles of reactive astrocytes in neuroinflammation and neurodegeneration. Nat Rev Neurol, 19:395–409.
45. Chandra S, Di Meco A, Dodiya HB, Popovic J, Cuddy LK, Weigle IQ, Zhang XQ, Sadleir K, Sisodia SS, Vassar R (2023) The gut microbiome regulates astrocyte reaction to Aβ amyloidosis through microglial dependent and independent mechanisms. Mol Neurodegener, 18.
46. Seo DO, Donnell D, Jain N, Ulrich JD, Herz J, Li YH, Lemieux M, Cheng JY, Hu H, Serrano JR, Bao X, Franke E, Karlsson M, Meier M, Deng S, Desai C, Dodiya H, Lelwala-Guruge J, Handley SA, Kipnis J, Sisodia SS, Gordon JI, Holtzman DM (2023) ApoE isoform- and microbiota-dependent progression of neurodegeneration in a mouse model of tauopathy. Science, 379:155–+.
47. Cummings J (2023) Anti-amyloid monoclonal antibodies are transformative treatments that redefine Alzheimer's disease therapeutics. Drugs, 83:569–576.
48. van Dyck CH, Sabbagh M, Cohen S (2023) Lecanemab in early Alzheimer's disease. Reply. N Engl J Med, 388:1631–1632.
49. Mintun MA, Wessels AM, Sims JR (2021) Donanemab in early Alzheimer's disease. Reply. N Engl J Med, 385:667.
50. Yiannopoulou KG, Papageorgiou SG (2013) Current and future treatments for Alzheimer's disease. Ther Adv Neurol Disord, 6:19–33.

5 Mechanism of Neurodegeneration in Parkinson's Disease

*Gajendra T A, Suchismita Patra, and Sairam Krishnamurthy**
Department of Pharmaceutical Engineering and Technology,
IITBHU, Varanasi, India

5.1 INTRODUCTION

Parkinson's disease (PD) is a long-term, irreversible, and advancing neurodegenerative disorder (ND). The defining pathological feature of this disease is the degeneration of dopaminergic neurons in the substantia nigra (SN) pars compacta (SNpc). Beyond this, the disorder often extends its impact to various brain areas, including the brainstem, hippocampus, and cerebral cortex [1]. While some regressions of these neurons are a normal part of aging, most individuals retain more than 80% of them. In PD, however, this loss is much more pronounced. Without treatment, PD worsens over the years, eventually leading to a rigid, akinetic state [2]. Patients die because of complications from immobilization, aspiration pneumonia, or pulmonary arrest [3].

Clinical aspects of PD encompass four cardinal motor manifestations: movement slowness (bradykinesia), muscular stiffness, static tremor, and positional imbalance. These kinetic symptoms profoundly affect the quality of the patient's life, and non-motor symptoms such as insomnia, cognitive impairment, and mood disorders are often accompanied by them. Neurodegenerative mechanisms are essential for creating focused therapeutics and personalized medicine [4]. Understanding molecular and cellular processes helps researchers to develop personalized treatments for diseases like Parkinson's, Alzheimer's, and other NDs [5]. Biomarker discoveries can aid in early diagnosis and treatment. Improved disease models, such as animal and cellular models and human brain organoids, can be created based on molecular mechanisms [6].

Knowing the progression and phases of disease can help to determine the time and strategy of treatment interventions. Gene therapy and neuroprotective methods are examples of innovative treatment approaches that can be developed. Neurodegeneration research broadens scientific understanding by adding to basic neuroscience knowledge and supporting multidisciplinary approaches [7].

5.2 PATHOPHYSIOLOGY OF PD ATTRIBUTED VIA NEURODEGENERATION

Being a catecholamine neurotransmitter, dopamine holds a pivotal role in the brain's communication system, including one crucial for reward-motivated behavior. This "feel-good" neurotransmitter is involved in reward, pleasure, and motivation. However, its functions are far more diverse and complex. Dopamine is vital for diverse cognitive processes, such as attention, acquisition, memory retention, and mood regulation. Likewise, it holds significant importance in motor control, influencing movement initiation, coordination, and fine-tuning [8].

Through a cascade of enzymatic reactions, dopamine is derived from the tyrosine amino acid in dopaminergic neurons. Those are primarily located in the SNpc and the ventral tegmental area

*Corresponding Author. E-mail: ksairam.phe@itbhu.ac.in

DOI: 10.1201/9781003530510-5

(VTA) of the midbrain. The dopaminergic neurons of SNpc innervate all parts of the striatum. Caudate and putamen circuits modulate various aspects of behavior and cognition [9].

The outflow from the striatal neurons propagates via two key pathways: the striato-nigral (direct) pathway and the striato-subthalamic (indirect) pathway [10]. The direct pathway relates to the excitatory D1 receptor, and the indirect pathway features the dopaminergic D2 inhibitory receptor [11]. Dopamine released in the striatum enhances the activity of the direct pathway while diminishing the activity of the indirect pathway. But in PD, the opposite effect occurs due to the gradual degeneration of dopaminergic neurons in the SN, resulting in a marked decrease in dopamine levels in the brain [12]. This depletion in dopamine level disrupts the intricate equilibrium of neurotransmission within the basal ganglia, a group of interconnected structures critical for regulating voluntary movement [13].

Misfolded and aggregated proteins are the most common feature shared by every neurodegenerative disease. For PD it is alpha-synuclein (α-syn). It is a tiny, highly charged protein of the CNS (Central nervous system). Encoded by the SNCA gene, it participates in synaptic plasticity and neurotransmitter release [14]. Recent research shows that neural activity controls the release of endogenous α-syn. Under physiological conditions, endogenous α-Syn maintains a balanced state among unfolded monomers and α-helically structured tetramers with very little aggregation [15]. An imbalance between tetramer and monomer ratios induces an upsurge in α-syn unfolded monomers, which promotes aggregation. α-Syn aggregates to form oligomers, protofibrils, and insoluble fibrils, which eventually build up into Lewy bodies (LBs) because of a conformational change into a β-sheet shape [16]. Recent investigations show that oligomers and protofibrils, generated during initial stages of aggregation, are the most hazardous types of α-Syns, contributing to cell death in PD (Figure 5.2) [15].

5.3 FACTORS CONTRIBUTING TO NEURODEGENERATION

The depletion of dopaminergic neurons in the nigrostriatal region is the primary cause of motor malfunction in PD patients. Among several factors of neurodegeneration, age stands out as the foremost contributor. Additionally, genetics and environmental factors play crucial roles in the development of PD through neuronal degeneration. These factors collectively influence the progression of the disease by impacting the health and function of neurons in the brain.

5.3.1 Aging Process

Aging is an inevitable natural process. Therefore, it is the primary risk factor for neurodegeneration, involving impairment of multiple pathways, like mitochondrial dysfunction, free radical damage, autophagy, and neural inflammation [17]. Tyrosine hydroxylase (TH) enzyme catalyzes the first and rate-limiting phase in the biosynthetic pathway of catecholamines, such as dopamine, and is therefore an important marker for DA neurons. Evidence from nonhuman primates indicates an age-related decrease in staining for TH. Disrupted intracellular clearance processes are most notably observed in the ventral SN in aged nonhuman primate brains [18].

With aging, there's a progressive reduction of brain mass and dopaminergic neurons in the ventral region of the SN, which plays a crucial role in motor function [19]. This decline is also associated with a higher level of α-synuclein inside cells, autooxidation of dopamine outside the synaptic vesicle, and iron accumulation in the ventral SN that eventually triggers microglia activation, causing nigral neurons to be more vulnerable to damage and neuroinflammation in PD [20, 21]. Although age is a foremost factor, it alone cannot cause PD. Dopaminergic neuronal degeneration occurs with aging, but the rate is much lower than seen in PD patients. Only 2% of people above 80 years suffer from PD [22]. So, along with age, other environmental and genetic factors are also required to shift from normal age-related neuronal erosion to the accelerated dopaminergic neuronal degeneration seen in PD.

5.3.2 Genetic Factors

Gene influences both the onset and progression of disease. For example, the SNCA mutation was discovered to be the major cause of autosomal dominant PD, as it encodes for the alpha-synuclein protein [23]. Many other genes were eventually revealed to be mutated and cause PD. Among these, the LRRK2 gene mutation is particularly significant, as it is a major contributor to hereditary forms of the disease [10]. The LRRK2 gene produces a protein kinase that regulates immune cell signaling inside the brain and further suggests the microglial involvement in the late onset of PD [24]. Other genes like PARK2, PINK1, and PARK7 mutations are also involved in the early stage of PD, but they follow an autosomal recessive inheritance pattern where both copies of the gene should be altered for the onset of the disease. Primarily, these genes are responsible for controlling and regulating mitochondrial homeostasis. Consequently, their mutation generates mitochondrial stress, which leads to neuronal cell death [25].

5.3.3 Environmental Factors

PD is mostly sporadic [26]. Emerging evidence suggests that various external factors can contribute to its development. Traumatic brain injuries, infectious diseases, and environmental toxins, such as pesticides and heavy metals, have all been implicated in causing PD. For example, toxic metals like copper, iron, manganese, mercury, and lead can disrupt neurotransmission and induce oxidative stress, especially affecting mitochondria [27]. Oxidative radicals generated by these toxins interact with cellular macromolecules like DNA, proteins, and lipids, which leads to cellular dysfunction and neuronal degeneration [28].

For instance, iron is a heavy metal that produces free hydroxyl radicals through the Fenton reaction and is also implicated in alpha-synuclein toxicity. Similarly, copper can also generate free radicals either through the Fenton–Haber–Weiss reaction or by participating in the 6-hydroxydopamine (6-OHDA) redox cycle [27, 29]. Even copper accumulation in the brain can reduce dopamine levels [30]. Exposure to manganese (Mn) can happen through diet, work-related activities, and environmental sources. Its toxicity can increase PD risk, causing motor impairment via damage to the SN and basal ganglia nuclei [31].

Chronic occupational exposure to lead can result in numerous neurological deficits, including significant swelling and neuron loss in both the central and peripheral nervous systems. These impairments often manifest as motor dysfunction, notably characterized by the loss of voluntary muscle movement [28]. Mercury, another neurotoxic heavy metal, is present in inorganic and organic forms and can be microbially transformed into methylmercury, resulting in possible transfer through the food chain. Mercury can reduce brain neuron numbers and cause movement disorders [28].

In the 1980s, scientists discovered that MPTP (N-methyl-4-phenyl-1,2,3,6-tetrahydropyridine) is metabolized by astrocytes to MPP+ and targets SN, causing advanced PD similar to paraquat, which is a pesticide [32]. This finding has prompted hypotheses regarding the possible involvement of pesticides in the development of PD. Insecticides and herbicides are globally used to combat disease-carrying vectors and agricultural pests. However, their usage has prompted investigations into their neurological effects, particularly their potential to induce detrimental changes in the brain. Rotenone, a major organophosphate, is known to elevate the risk of PD. These findings emphasize the critical need for a delicate balance between pesticide use in agricultural fields and human health [33].

5.3.4 Other contributors

PD can also be attributed to the use of certain illicit drugs. Among these, three notable psychostimulants linked to PD are methamphetamine, amphetamine, and cocaine. Higher doses of these excitatory drugs, such as amphetamine, elevate reactive oxygen levels that can lead to harm to dopaminergic neurons and axon terminals in the human brain, resulting in defective morphology of the SN. This anatomical change is significant as the SN regulates motor function, thereby contributing to PD symptoms [34].

5.4 CELLULAR AND MOLECULAR MECHANISMS OF NEURODEGENERATION

5.4.1 Mitochondrial Dysfunction and Oxidative Stress

Following the discovery that Parkinson's-like symptoms could be induced by MPTP, an opioid, researchers uncovered a crucial link between mitochondrial dysfunction and the disease. MPTP is acquired by dopamine neurons through the dopamine transporter and is oxidized to form MPP+ (1-methyl-4-phenylpyridinium) [35]. Inside dopamine neurons, MPP+ accumulates specifically within mitochondria and disrupts the function of complex I (NADH ubiquinone oxidoreductase), which plays a crucial role in the electron transport chain responsible for generating cellular energy in the form of ATP [36].

Mitochondrial malfunction happens in both autosomal recessive and dominant PD. PARKIN and PINK1 genes are associated with autosomal recessive PD, whereas autosomal dominant PD is associated with SNCA and LRRK2 mutations [37]. PARKIN is an adaptable neuroprotective E3 ubiquitin ligase recognized as a gene responsible for autosomal recessive Parkinsonism. It has a broad protective role, defending against cell death under distinct types of stress, including mitochondrial stress, endoplasmic reticulum (ER) stress, excitotoxicity, and proteotoxic stress.

Therefore, the PARKIN-deprived cells are more prone to oxidative damage. PINK1 is another mitochondrial gene linked to PD. It is found at the external and internal mitochondrial membranes and within the cytosol. Mitochondria can function as both a source and target of reactive oxygen species (ROS). Inhibiting complex I reduces mitochondrial ATP synthesis and elevates the production of ROS, damaging mitochondrial DNA (mtDNA) and respiratory chain components [38]. This model is well-known for explaining the susceptibility of the dopamine neurons in SNpc. These neurons have a significant oxidative burden due to the autoxidation of cytoplasmic dopamine coupled with low antioxidant capacity. Iron accumulation in sporadic PD patients' brains leads to elevated ROS generation, transcriptional enhancement of SNCA, and α-syn aggregation. Mitochondria can exchange iron with the cytoplasm, and inhibition of complex I through rotenone, MPTP, and paraquat toxicity results in iron accumulation that shows heightened apoptotic cell death without elevation in ROS formation or oxidative damage [39].

Mitochondrial maintenance processes, such as dynamic fusion and fission, regulated mitophagy, and biogenesis, are crucial for maintaining a healthy mitochondrial population and energy production. Disruption of these pathways leads to mitochondrial fragmentation. Impaired mitophagy associated with mutations in PINK1 and PARKIN, dysfunctional lipid pathways, and reduced expression of PGC1α (a critical regulator of mitochondrial biogenesis and neuroprotection) in sporadic PD brains highlight the intricate mechanisms contributing to PD pathophysiology [40, 41]. Alterations in mitochondrial DNA maintenance genes are associated with Parkinsonism, but they can contribute only when provoked by other insults. Studies show complexes I and II deficiencies; mtDNA deletions reduced copy numbers in SN neurons. This suggests that mtDNA mutations and respiratory chain malfunctions contribute to PD progression [42].

5.4.2 Protein Aggregation and Misfolding

Protein aggregation and misfolding are common among all the neurodegenerative diseases in PD. Under normal physiological conditions, α-synuclein exists as a natively unfolded monomer or adopts a helical conformation when associated with lipid membranes. However, in PD, α-synuclein undergoes pathological misfolding, adopting β-sheet-rich structures that are highly prone to aggregation [43]. This misfolding is influenced by a variety of factors, including genetic mutations, oxidative stress, environmental toxins, and impairments in cellular protein management systems. Aggregated forms of α-synuclein progress through a continuum, forming oligomers, protofibrils, and eventually insoluble fibrils that accumulate in neuronal cells, particularly in the form of LBs and Lewy neurites [44]. Mutations or multiplications in the SNCA gene, which encodes α-synuclein, increase the protein's expression and aggregation propensity, strongly linking it to familial PD. Environmental factors such as pesticide exposure contribute to oxidative stress, further promoting α-synuclein misfolding [45].

The aggregation of α-synuclein is highly neurotoxic, with soluble oligomers being particularly harmful. These oligomers disrupt cellular membrane integrity, leading to calcium dysregulation, and impair mitochondrial function, resulting in reduced energy production and increased oxidative stress through ROS [46]. Furthermore, aggregated α-synuclein interferes with synaptic function and axonal transport, critical processes for neuronal survival and communication. Adding to its toxicity is the prion-like behavior of α-synuclein, where misfolded proteins act as seeds, inducing misfolding and aggregation of native α-synuclein in neighboring cells. This self-propagating mechanism spreads pathological aggregates throughout the brain, contributing to the progressive neurodegeneration seen in PD [47].

The accumulation of aggregated α-synuclein is compounded by dysfunction in cellular protein degradation systems, such as the ubiquitin-proteasome system (UPS) and autophagy-lysosomal pathways. Normally, these systems clear misfolded or aggregated proteins, but in PD, their efficiency is significantly impaired. Aggregated α-synuclein not only resists degradation but also further inhibits these systems, creating a feedback loop that accelerates protein accumulation [48]. Genetic mutations in autophagy-related genes, such as LRRK2 and GBA, exacerbate this dysfunction, heightening the risk of PD. This interplay between α-synuclein aggregation, impaired degradation pathways, and cellular toxicity underlies the progressive neuronal damage characteristic of PD [49].

5.4.3 Autophagy-lysosomal Pathway

Autophagy is an innate cellular process that removes dysfunctional or unnecessary components through a lysosome-mediated pathway. This process enables the orderly degradation and recycling of cellular materials. There are several types of autophagy, such as macroautophagy, chaperone-mediated autophagy, and endosomal microautophagy, that are crucial for sustaining cellular balance by removing defective or toxic materials [50].

Neurons rely on autophagy for eliminating damaged components and organelles, which is critical for neurotransmission and proteomic integrity. Autophagy declines with age, making neurons more susceptible to stress and death. Autophagy dysfunction has been associated with several neurodegenerative illnesses, including Alzheimer's, Parkinson's, Huntington's, and amyotrophic lateral sclerosis (ALS). Studies, notably those on Atg5 or ubiquitin-deficient mice, highlight autophagy's role in motor impairments and cytoplasmic inclusions. Mutations in autophagy-related genes such as PARKIN and PINK1 are connected to these disorders, as well as irregularities in protein expression, such as gamma-aminobutyric acid receptor-associated protein-like 1/glandular epithelial cell 1 (GABARAPL1/GEC1), which are involved in the proliferation of cells, invasion, and autophagic influx [51].

PD is characterized by the presence of LBs containing aberrant assembly of α-synuclein (α-syn) protein. Mutations in the SNCA gene are rare but strongly contribute to the onset and advancement of PD. Any malfunction in the degradative processes, whether directly or indirectly impacting components, can disrupt autophagy pathways. The UPS destroys monoubiquitinated α-syn, while macroautophagy processes deubiquitinated α-syn. In PD, both lysosomes and mitochondria perform important roles in breakdown pathways [52].

All types of autophagy contribute to PD. However, lysosomes play a vital role in α-synuclein (α-syn) breakdown. Changes in lysosomal enzymes, notably hydrolases, influence protein breakdown, resulting in aggregation buildup and neuronal injury. Lysosomal hydrolase levels in cerebrospinal fluid (CSF) are lower in PD patients [53]. GBA is another lysosomal enzyme crucial for degrading α-synuclein (α-syn). Reduced GBA activity or levels in the brain lead to α-syn accumulation [54]. GBA1 and ATP10B genes relate to PD and play a direct role in lysosomal activity and α-syn processing. TFEB, a critical regulator of lysosomal biogenesis and autophagy, may be a therapeutic target for increasing α-syn clearance. Strategies that improve lysosomal function may thus have promise for slowing the progression of PD [55].

5.4.4 Neuroinflammation

5.4.4.1 Role of Microglia and Astrocytes

Microglia, the brain's basic innate immune cells, play a significant role in homeostasis and host defense systems. They sense environmental changes, maintain physiological homeostasis, and defend against harmful stimuli. Microglia release pro-inflammatory cytokines to recruit cells and eliminate pathogenic substances, but chronic neuroinflammation can cause neurotoxicity and neurodegeneration. Imaging and fluid indicators can help determine microglial activity. Microglia within the central nervous system may act in a pro-inflammatory or neuroprotective manner, depending on their activation status. Pro-inflammatory cytokines induce resting microglia to express molecules associated with inflammation, which are deleterious in NDs [56]. Conversely, neuroprotective microglia activate microglia, causing the production of neuroprotective and tissue repair factors. IL-4 is recognized for suppressing the release of pro-inflammatory cytokines [57]. Astrocytes, the brain's most common glial cells, play crucial, vital roles in brain stability, managing blood circulation, sustaining the blood–brain barrier, and controlling neurotrophin secretion. They can be assessed using biomarkers like GFAP and S100B. Reactive astrogliosis, a key feature of CNS pathology, may be detected by changes in astrocyte molecular expression and morphology. Astrocytes can have pro-inflammatory and immunoregulatory subpopulations, with anti-inflammatory cytokines potentially triggering neuroprotective activation. STAT3, a transcription factor, is involved in reactive astrogliosis [58]. Microglia and astrocytes' interaction is crucial for neuroinflammation and CNS homeostasis, causing pathological processes in neurodegenerative diseases like Alzheimer's, ALS, and Parkinson's [59].

5.4.4.2 Cytokine Involvement

Cytokines are small polypeptides that regulate local cellular processes, such as endurance, development, and differentiation. They are amplified in response to disease, injury, and infection and play a crucial role in tissue rebuilding. Chronic microglial production of pro-inflammatory cytokines, such as IL-1, IL-6, tumor necrosis factor (TNF)-α, and IFN-γ, has been linked to NDs. These cytokines may play a role in neurodegeneration and have potential as therapeutic targets in various central nervous system diseases. However, their role in neuroprotection and neurodegeneration remains complex. Recent research suggests that microglial activation and pro-inflammatory cytokines are promising targets for ND treatment [60].

5.4.4.3 Glutamate Toxicity

CNS maintains proper function by maintaining an equilibrium among excitatory and inhibitory neuronal connections. Glutamate, the predominant neurotransmitter, mediates most excitatory signals and is vital for cognitive, motor, sensory, and autonomic functions. Disrupted glutamate balance can lead to significant neuropathological effects and has been linked to neurological conditions such as epilepsy, multiple sclerosis, Alzheimer's disease, Huntington's disease, ALS, and PD. Glutamate is released and reabsorbed through the glutamate–glutamine cycle, with two primary excitatory amino acid transporters (EAATs) being most prevalent in the CNS. EAAT1 is found on astrocytes, while EAAT2 is responsible for 90% of glutamate reuptake in the brain. The distribution of EAATs on glial cells is critical for maintaining glutamate balance in both healthy and diseased states [61].

Glutamate accumulation at synapses triggers apoptotic cell death due to Ca^{2+} overload, a hallmark of NDs, including PD. This excitotoxicity is linked to alterations in glutamate transporter and receptor expression, potentially contributing to neuroinflammation. Chemokines impact glutamate homeostasis, downregulating EAATs expression and upregulating receptors, thereby potentiating glutamatergic neurotransmission and excitotoxicity [62].

5.4.4.4 Apoptosis and Cell Death Pathway

Neurodegenerative diseases involve two major death pathways: apoptosis and necrosis. Apoptosis is a type of programmed cell death (PCD) characterized by cell shrinkage, chromatin condensation, and DNA fragmentation. It can be identified using terminal deoxynucleotidyl transferase dUTP nick end labeling (TUNEL) assays, performed either in vivo or in situ [63].

Necrosis, an alternate form of cell death, is marked by cell swelling, leading to the breakdown of the cell membrane and leakage of intracellular contents. Although DNA fragmentation occurs during the degradation process, necrosis does not involve chromatin condensation. The TNF can induce necrosis, leading to recruitment of specific death receptors or Toll-like receptors. Uncontrolled and controlled necrosis are still ambiguous, with some neurons experiencing uncontrolled necrosis immediately after ischemia. Severe ischemia deprives neurons of oxygen and glucose supply, leading to ATP depletion and Na^+/K^+-activated ATPase inhibition. Excitotoxicity and Ca^+ influxes have also been found in different neurodegenerative diseases, where it was believed that neurons died in a regulated manner. The differences between uncontrolled and controlled neuronal deaths lie only in the differences between cell death initiators, which depend on whether the initiators can be controlled [64].

Apoptosis is suggested as the primary death pathway in PD brains, with every LB-positive neuron positive for pro-apoptotic Bax staining. Pro-apoptotic mediators like p53 and MAP kinase distribution are altered in SN neurons. Necroptosis, a potent inhibitor of RIP1, may also be involved in PD death pathways, as demonstrated in a recent study [65]. α-Synuclein is a major pathogenic protein in PD, as its aggregates form the core of Lewy bodies. The correlation between an α-synuclein burden and neuronal loss and clinical symptom progression is still under debate [64].

5.5 EMERGING MECHANISMS AND THEORIES

5.5.1 Gut–Brain Axis

The gut–brain axis in PD is a bidirectional communication between the gastrointestinal tract and the central nervous system, playing a crucial role in the disease's development and progression. Early gastrointestinal symptoms, microbial dysbiosis, and α-synuclein aggregation in the gut are linked to neurodegeneration in PD. The gut microbiota in PD is altered, with increased pro-inflammatory bacteria and decreased beneficial ones. This can induce systemic and CNS inflammation, contributing to neurodegeneration. Chronic GI inflammation can promote α-synuclein aggregation and its spread to the CNS. Epidemiological studies, animal models, and gut microbiome analysis support the gut–brain axis in PD. Therapeutic implications include microbiome-based therapies, targeting the vagus nerve, anti-inflammatory strategies, and dietary interventions. However, questions remain about causality, mechanisms, and intervention timing (Figure 5.1) [66].

5.5.2 Prion-like Propagation of Alpha-Synuclein

The prion-like propagation of α-synuclein in the gut–brain axis is a hypothesis that suggests misfolded α-synuclein aggregates, a hallmark of PD, may originate in the gut and spread to the brain

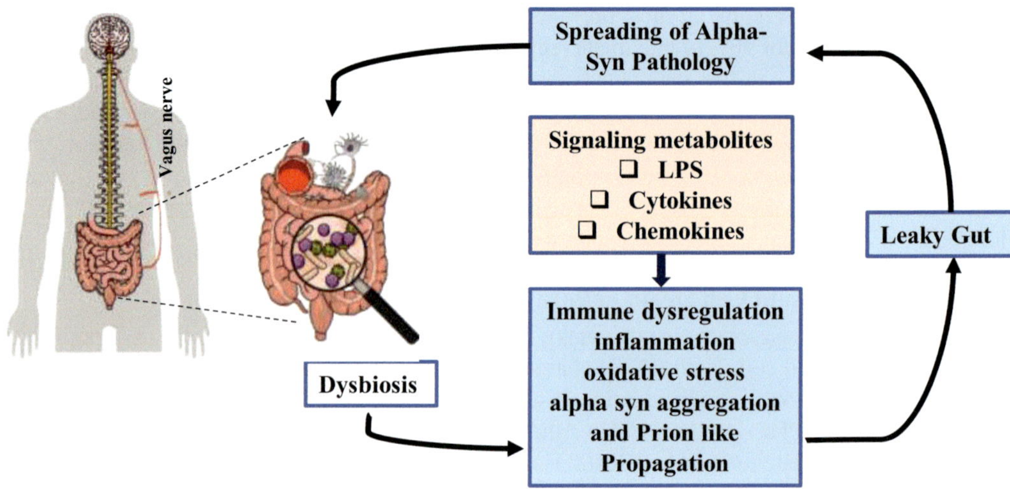

FIGURE 5.1 Schematic of gut–brain axis in PD, showing α-syn propagation via the vagus nerve. Leaky gut permits LPS, cytokines, and chemokines, causing immune dysregulation, inflammation, oxidative stress, and α-syn aggregation with prion-like spread.

through neural pathways, particularly the vagus nerve. This hypothesis is supported by early symptoms of PD, evidence from experimental models, and human studies. Prion-like mechanisms of α-synuclein aggregates include seeding and templating, transcellular propagation, and aggregate toxicity. Factors influencing propagation include age, gut dysbiosis, and environmental exposures. Therapeutic implications include targeting aggregation, modulating the microbiome, enhancing autophagy, and blocking transmission via vagus-nerve-targeted therapies. However, unresolved questions remain, such as whether α-synuclein pathology always originates in the gut or starts

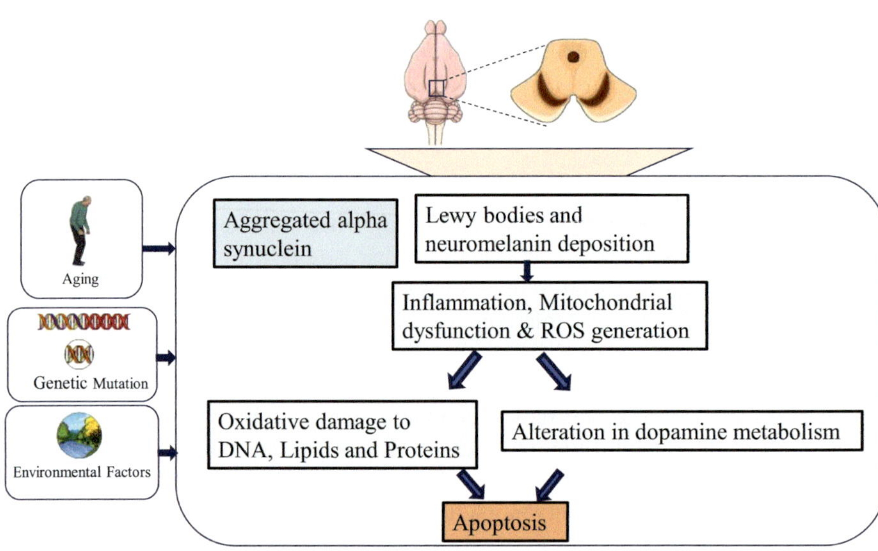

FIGURE 5.2 Schematic representation of PD pathology. Aging and environmental or genetic factors lead to aggregated alpha-synuclein, forming Lewy bodies and neuromelanin deposits. These trigger inflammation, mitochondrial dysfunction, and ROS generation, resulting in oxidative damage to DNA, lipids, and proteins, along with altered dopamine metabolism. These processes culminate in neuronal apoptosis.

independently in the brain, and what environmental or genetic factors predispose individuals to gut-derived α-synuclein aggregation [67].

5.6 CONCLUSION

5.6.1 Summary of Key Mechanisms

Dopamine, a neurotransmitter, is crucial for brain communication, cognitive processes, and motor control. It is synthesized in dopaminergic neurons in the SNpc and VTA. PD is a chronic, irreversible neurodegenerative condition that causes severe dopaminergic neuron loss in the SNpc. Therefore, it causes motor symptoms such as bradykinesia, muscle weakness, resting tremor, and postural instability, along with non-motor symptoms such as sleeplessness and cognitive dysfunction. The study of molecular and cellular pathways is critical for creating targeted treatments, personalized medicine, and gene therapy.

Factors contributing to neurodegeneration include aging, genetics, and environmental influences. Aging is the leading risk factor, involving mitochondrial dysfunction, oxidative stress, impaired autophagy, and neuroinflammation. Genetic mutations, such as those in the SNCA gene, disrupt mitochondrial homeostasis, leading to oxidative stress and neuronal death. Environmental factors like traumatic brain injuries, infections, and exposure to toxins also contribute to neurodegeneration. Illicit drug use also damages dopaminergic neurons. Mitochondrial dysfunction disrupts complex I of the electron transport chain, leading to reduced ATP production and elevated ROS. This oxidative stress damages mitochondrial DNA and respiratory components, exacerbating neuronal vulnerability, especially in dopamine-rich regions. Dysregulation in mitochondrial quality control mechanisms, including mitophagy and biogenesis, is crucial for the pathophysiology of PD. Impaired lysosomal function through targets like TFEB offers therapeutic promise. Neuroinflammation in PD is mediated by microglia and astrocytes, with cytokines like IL-1, IL-6, TNF-α, and IFN-γ, contributing to chronic inflammation and neurodegeneration. Addressing excitotoxicity through modulation of glutamate homeostasis could mitigate neurodegeneration. Targeting apoptosis and cell death pathways may reduce neuronal loss. According to recent scientific research, early symptoms like microbial dysbiosis and misfolded α-synuclein aggregation in the gut are linked to neurodegeneration. The gut microbiota in PD is altered, leading to systemic and CNS inflammation. α-Synuclein aggregates propagate through prion-like mechanisms, affecting age, gut dysbiosis, and environmental exposures. Therapeutic implications include targeting misfolded protein aggregation, microbiome modulation, autophagy enhancement, and blocking transmission via vagus-nerve-targeted therapies.

REFERENCES

1. Flores-Ponce, X. and I.J.M. Velasco, *Dopaminergic neuron metabolism: relevance for understanding Parkinson's disease*. 2024. **20**(6): Metabolomics p. 116.
2. Zhou, Z.D., et al., *Role of dopamine in the pathophysiology of Parkinson's disease*. 2023. **12**(1): Translational Neurodegeneration p. 44.
3. Gerlach, O.H., A. Winogrodzka, and W.E.J.M.D. Weber, *Clinical problems in the hospitalized Parkinson's disease patient: systematic review*. 2011. **26**(2): Movement Disorders p. 197–208.
4. Mejía, C.A.A., *Therapeutic Exercise for Parkinson's Disease and Related Disorders: A Comprehensive Clinical Strategy for Preventing and Reversing Neurodegeneration*. 2024: Springer Nature.
5. Strianese, O., et al., *Precision and personalized medicine: how genomic approach improves the management of cardiovascular and neurodegenerative disease*. 2020. **11**(7): Genes p. 747.
6. Faravelli, I., et al., *Back to the origins: human brain organoids to investigate neurodegeneration*. 2020. **1727**: Brain Research p. 146561.

7. Uyyala, S.J.I.J.o.M. and H. Sciences, *The development of new treatments for neurological disorders: insights, innovations, and ethical foundations*. 2023. Indus Journal of Medical and Health Sciences. **1**(01): p. 45–55.
8. Shnider, D.H., *Bottom Up: An Integrated Neurological and Cognitive Behavioural Book Which Addresses the Key Principles of Neuropsychotherapy. Five Important Emotional Parts of the Brain with Skills Training*. 2018: AuthorHouse.
9. Rao, J.J.H.o.C.N., *Neurochemistry of Parkinson's disease*. 2007. **83**: Handbook of Clinical Neurology p. 153–204.
10. Ramakrishna, K., et al., *Relevance of Genetic Model and Linked Therapeutics in Parkinson's Disease*, in *Neurodegenerative Diseases*. 2025: CRC Press. p. 42–52.
11. Melleu, F.F. and N.S.J.C.N. Canteras, *Pathways from the superior colliculus to the basal ganglia*. 2024. **22**(9): Cerebral Cortex p. 1431–1453.
12. Ayon-Olivas, M., et al., *Dopaminergic input regulates the sensitivity of indirect pathway striatal spiny neurons to brain-derived neurotrophic factor*. 2023. **12**(10): Biology p. 1360.
13. Ramesh, S. and A.S.P.M.J.A.n. Arachchige, *Depletion of dopamine in Parkinson's disease and relevant therapeutic options: a review of the literature*. 2023. **10**(3): AIMS Neuroscience p. 200.
14. Gómez-Benito, M., et al., *Modeling Parkinson's disease with the alpha-synuclein protein*. 2020. **11**: Frontiers in Pharmacology p. 356.
15. Nuber, S., et al., *Abrogating native α-synuclein tetramers in mice causes a L-DOPA-responsive motor syndrome closely resembling Parkinson's disease*. 2018. **100**(1): Neuron p. 75–90. e5.
16. Ramakrishna, K., et al., *Importance of Newer 3D Organoids Model-Based Therapeutic Approach in Parkinson's Disease*, in *Neurodegenerative Diseases*. 2025, CRC Press. p. 53–59.
17. Bennett, D.A., et al., *Prevalence of parkinsonian signs and associated mortality in a community population of older people*. 1996. **334**(2): The New England Journal of Medicine p. 71–76.
18. Kanaan, N.M., J.H. Kordower, and T.J.J.J.o.C.N. Collier, *Age-related accumulation of Marinesco bodies and lipofuscin in rhesus monkey midbrain dopamine neurons: relevance to selective neuronal vulnerability*. 2007. **502**(5): The Journal Of Comparative Neurolog p. 683–700.
19. Stark, A., B.J.C. Pakkenberg, and t. research, *Histological changes of the dopaminergic nigrostriatal system in aging*. 2004. **318**(1): Cell and Tissue Research p. 81–92.
20. Kanaan, N., J. Kordower, and T.J.E.J.o.N. Collier, *Age-related changes in dopamine transporters and accumulation of 3-nitrotyrosine in rhesus monkey midbrain dopamine neurons: Relevance in selective neuronal vulnerability to degeneration*. 2008. **27**(12): The Journal of Comparative Neurology p. 3205–3215.
21. Zecca, L., et al., *Iron, neuromelanin and ferritin content in the substantia nigra of normal subjects at different ages: consequences for iron storage and neurodegenerative processes*. 2001. **76**(6): The Journal of Neurochemistry p. 1766–1773.
22. Pringsheim, T., et al., *The prevalence of Parkinson's disease: a systematic review and meta-analysis*. 2014. **29**(13): Movement Disorders p. 1583–1590.
23. Spillantini, M., et al., *Nerve growth factor mRNA and protein increase in hypothalamus in a mouse model of aggression*. 1989. **86**(21): Proc Natl Acad Sci USA p. 8555–8559.
24. Moehle, M.S., et al., *LRRK2 inhibition attenuates microglial inflammatory responses*. 2012. **32**(5): Neurobiology of Disease p. 1602–1611.
25. Mullin, S. and A.H.J.N.c. Schapira, *Pathogenic mechanisms of neurodegeneration in Parkinson disease*. 2015. **33**(1): Neurologic Clinics p. 1–17.
26. Modi, P., et al., *Understanding Pathophysiology of Sporadic Parkinson's Disease in Drosophila Model: Potential Opportunities and Notable Limitations*. 2016: IntechOpen.
27. Paris, I., et al., *Copper-dopamine complex induces mitochondrial autophagy preceding caspase-independent apoptotic cell death*. 2009. **284**(20): Journal of Biological Chemistry p. 13306–13315.
28. Caudle, W.M., et al., *Industrial toxicants and Parkinson's disease*. 2012. **33**(2): NeuroToxicology p. 178–188.
29. Cruces-Sande, A., E. Méndez-Álvarez, and R.J.J.o.n. Soto-Otero, *Copper increases the ability of 6-hydroxydopamine to generate oxidative stress and the ability of ascorbate and glutathione to potentiate this effect: potential implications in Parkinson's disease*. 2017. **141**(5): Journal of Neurochemistry p. 738–749.
30. Yu, W.-R., et al., *Copper (Cu2+) induces degeneration of dopaminergic neurons in the nigrostriatal system of rats*. 2008. **24**(2): Neurotoxicology p. 73.
31. Olanow, C.W.J.A.o.t.N.Y.A.o.S., *Manganese-induced parkinsonism and Parkinson's disease*. 2004. **1012**(1): Annals of the New York Academy of Sciences p. 209–223.

32. Langston, J.W.J.J.o.P.s.d., *The MPTP story*. 2017. **7**(s1): Journal of Parkinson's Disease p. S11–S19.
33. Dawson, T.M. and V.L.J.S. Dawson, *Molecular pathways of neurodegeneration in Parkinson's disease*. 2003. **302**(5646): Science p. 819–822.
34. Ball, N., et al., *Parkinson's disease and the environment*. 2019. **10**: F1000Research p. 421551.
35. Srikanth, Y., et al., *Unveiling the multifaceted pharmacological actions of indole-3-carbinol and diindolylmethane: a comprehensive review*. 2025. **14**(5): Plants p. 827.
36. Langston, J., et al., *Evidence of active nerve cell degeneration in the substantia nigra of humans years after 1-methyl-4-phenyl-1, 2, 3, 6-tetrahydropyridine exposure*. 1999. **46**(4): Annals of Neurology p. 598–605.
37. Lill, C.M.J.M. and c. probes, *Genetics of Parkinson's disease*. 2016. **30**(6): Molecular and Cellular Probes p. 386–396.
38. Abou-Sleiman, P.M., M.M. Muqit, and N.W.J.N.R.N. Wood, *Expanding insights of mitochondrial dysfunction in Parkinson's disease*. 2006. **7**(3): Nature Reviews Neuroscience p. 207–219.
39. Muñoz, Y., et al., *Parkinson's Disease: The Mitochondria-Iron Link*. 2016. **2016**(1): BioMed Research International p. 7049108.
40. Pickrell, A.M. and R.J.J.N. Youle, *The roles of PINK1, parkin, and mitochondrial fidelity in Parkinson's disease*. 2015. **85**(2): Neuron p. 257–273.
41. Eschbach, J., et al., *Mutual exacerbation of PGC-1α deregulation and α-synuclein oligomerization*. 2014. **77**(1): Neurobiology of Aging p. 15.
42. Grünewald, A., et al., *Mitochondrial DNA depletion in respiratory chain–deficient Parkinson disease neurons*. 2016. **79**(3):Annals of Neurology p. 366–378.
43. Tan, J.M., et al., *Protein misfolding and aggregation in Parkinson's disease*. 2009. **11**(9): Cellular Signalling p. 2119–2134.
44. Zhang, G., et al., *New perspectives on roles of alpha-synuclein in Parkinson's disease*. 2018. **10**: Neural Regeneration Research p. 370.
45. Stefanis, L.J.C.S.H.p.i.m., *α-Synuclein in Parkinson's disease*. 2012. **2**(2): Cold Spring Harbor Perspectives in Medicine p. a009399.
46. Du, X.-y., X.-x. Xie, and R.-t.J.I.j.o.m.s. Liu, *The role of α-synuclein oligomers in Parkinson's disease*. 2020. **21**(22): International Journal of Molecular Sciences p. 8645.
47. Jan, A., et al., *The prion-like spreading of alpha-synuclein in Parkinson's disease: update on models and hypotheses*. 2021. **22**(15): International Journal of Molecular Sciences p. 8338.
48. Lee, J., et al., *Targeted degradation of α-synuclein aggregates in Parkinson's disease using the AUTOTAC technology*. 2023. **18**(1): Nature Biomedical Engineering p. 41.
49. Zhang, H., C. Duan, and H.J.M.N. Yang, *Defective autophagy in Parkinson's disease: lessons from genetics*. 2015. **51**: Molecular Neurobiology p. 89–104.
50. Parzych, Katherine R., and Daniel J. Klionsky. "*An overview of autophagy: morphology, mechanism, and regulation*." Antioxidants & redox signaling 20.3 (2014): 460–473.
51. El Haddad, S., et al., *Disturbed expression of autophagy genes in blood of Parkinson's disease patients*. 2020. **738**: Neuroscience Letters p. 144454.
52. Rott, R., et al., *α-Synuclein ubiquitination and novel therapeutic targets for Parkinson's disease*. 2014. **13**(4): Nature Reviews Neurology p. 630–637.
53. Tokuda, T., et al., *Decreased α-synuclein in cerebrospinal fluid of aged individuals and subjects with Parkinson's disease*. 2006. **349**(1): Neuroscience Letters p. 162–166.
54. Mazzulli, J.R., et al., *Gaucher disease glucocerebrosidase and α-synuclein form a bidirectional pathogenic loop in synucleinopathies*. 2011. **146**(1): Cell p. 37–52.
55. Smolders, S. and C.J.A.n.c. Van Broeckhoven, *Genetic perspective on the synergistic connection between vesicular transport, lysosomal and mitochondrial pathways associated with Parkinson's disease pathogenesis*. 2020. **8**(1): Acta Neuropathologica Communications p. 63.
56. Zhan, Y., et al., *Deficient neuron-microglia signaling results in impaired functional brain connectivity and social behavior*. 2014. **17**(3): Nature Neuroscience p. 400–406.
57. Park, K., et al., *Neuroprotective role of microglia expressing interleukin-4*. 2005. **81**(3): Journal of Neuroscience Research p. 397–402.
58. Kwon, H.S. and S.-H.J.T.n. Koh, *Neuroinflammation in neurodegenerative disorders: the roles of microglia and astrocytes*. 2020. **9**(1): The Neuroscientist p. 42.
59. Liu, W., Y. Tang, and J.J.L.s. Feng, *Cross talk between activation of microglia and astrocytes in pathological conditions in the central nervous system*. 2011. **89**(5–6): Life Sciences p. 141–146.
60. Smith, J.A., et al., *Role of pro-inflammatory cytokines released from microglia in neurodegenerative diseases*. 2012. **87**(1): Neurobiology of Disease p. 10–20.

61. O'Donovan, S.M., C.R. Sullivan, and R.E.J.n.S. McCullumsmith, *The role of glutamate transporters in the pathophysiology of neuropsychiatric disorders.* 2017. **3**(1): Neuropsychopharmacology p. 32.
62. Haroon, E., A.H. Miller, and G.J.N. Sanacora, *Inflammation, glutamate, and glia: a trio of trouble in mood disorders.* 2017. **42**(1): Nature Reviews Neuroscience p. 193–215.
63. Elmore, S.J.T.p., *Apoptosis: a review of programmed cell death.* 2007. **35**(4):Toxicologic Pathology p. 495–516.
64. Fricker, M., et al., *Neuronal cell death.* 2018. **98**(2): The Journal of Cell Biology p. 813–880.
65. Christofferson, D.E. and J.J.C.o.i.c.b. Yuan, *Necroptosis as an alternative form of programmed cell death.* 2010. **22**(2): Current Opinion in Cell Biology p. 263–268.
66. Tan, A.H., S.Y. Lim, and A.E.J.N.R.N. Lang, *The microbiome–gut–brain axis in Parkinson disease—from basic research to the clinic.* 2022. **18**(8): Nature Reviews Neurology p. 476–495.
67. Chen, Y., et al., *Prion-like propagation of* α*-synuclein in the gut-brain axis.* 2018. **140**: Acta Neurologica Scandinavica p. 341–346.

6 Pharmacological, Non-Pharmacological Management, and Emerging Therapies for the Management of Parkinson's Disease

Anuj Babber[#], Kokkonda Jackson Sugunakara Chary[#], Malhar Desai[#], Gurpreet Kaur Grewal, and Anuradha Sharma[*]

Department of Molecular Biology and Genetic Engineering, School of Bioengineering and Biosciences, Lovely Professional University, Phagwara, Punjab, India

6.1 INTRODUCTION

Parkinson's disease (PD) is the fastest growing neurodegenerative condition, with an increase of 118% in patients suffering from PD from 1990 to 2015 [1]. It is marked by the early demise of dopaminergic neurons in the brain's substantia nigra (SN). While manifesting with both motor and non-motor symptoms, the clinical syndrome of "Parkinsonism" has been mainly characterized by four motor symptoms: tremor, rigidity, akinesia (bradykinesia), and instability in posture (TRAP) [2, 3]. PD is widely accepted to be an age-dependent disease, with occurrences increasing in people with advanced ages. Roughly 1% of the population older than 60 is known to suffer from PD, which increases to around 4% of the population over the age of 80 [4]. While there is no current consensus regarding the cause behind PD, it is accepted that PD has a multifaceted etiology, with both environmental and genetic factors playing a role, with multiple studies having linked prolonged exposure to pesticides to a higher risk of PD [5], and smoking and coffee consumption being associated with a reduced risk of PD [6]. The main pathophysiology of PD includes the death and loss of dopaminergic neurons in the brain, mainly the substantia nigra pars compacta (SNpc) region. While the reason behind the death of the neurons is not known for certain, various mechanisms have been proposed for the same, with the main one being the oligomerization and accumulation of proteins called α-synuclein (α-Syn), leading to the formation of inclusion bodies called Lewy bodies inside the nerve cells [7, 8]. Figure 6.1 illustrates the Braak hypothesis, which postulates that the Lewy body-like pathology caused by α-Syn can spread to the ventral midbrain region from the gastrointestinal (GI) tract through the vagus nerve [9].

Other than Lewy bodies, many other processes like impairment of mitochondrial function, neuroinflammation, microglial activation (microgliosis), and oxidative stress have also been associated with the pathogenesis of PD [10]. One of the earliest genes associated with PD was the alpha-synuclein (SNCA) gene, which is responsible for coding for the α-Synuclein protein, which, as

[#]Contributed equally.
[*]Corresponding Author. annusha254@gmail.com

DOI: 10.1201/9781003530510-6

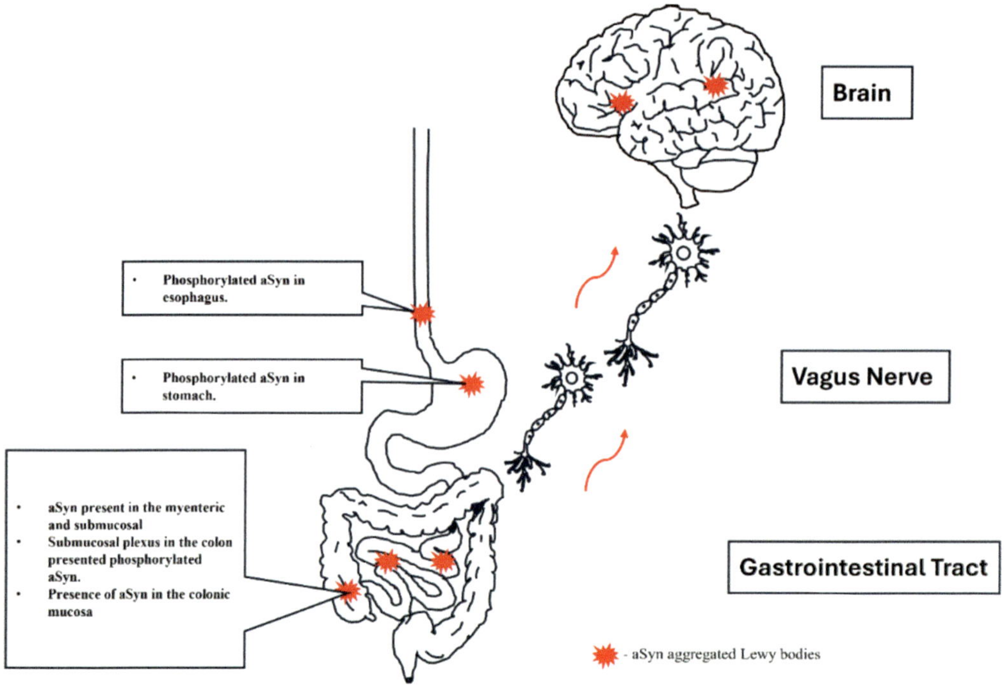

FIGURE 6.1 The Braak hypothesis [9, 10] and the distribution of α-Syn in the GI tract.

mentioned previously, is linked with the development of Lewy bodies – one of the major proposed causes of PD. Not only is the cause behind PD difficult to identify, but the preemptive diagnosis of PD is also quite challenging due to the fact that by the time the cardinal symptoms of PD are apparent, the damage in the brain is already quite substantial. Such brain damage is often mediated through the immune system, with studies showing evidence of contribution of T-cells in PD pathology [11]. Figure 6.2 provides a concise overview of the α-Syn-mediated T-cell activation [12].

This chapter explores the pharmacological and non-pharmacological management strategies of PD, while also focusing on some of the emerging therapies for Parkinson's.

6.2 PHARMACOLOGICAL MANAGEMENT OF PARKINSON'S DISEASE

PD is a common chronic disorder that affects the nervous system. Current treatments for PD aim to restore dopamine activity in the dopamine-depleted striatum, which helps improve motor symptoms [13]. Unfortunately, there are few pharmacological options for managing the non-motor symptoms, which can often be the most disabling aspect of the disease. No established treatments can slow, stop, or modify the progression of PD. Various pharmacological drugs can help manage motor symptoms, improving movement as the disease progresses. Commonly used medications for PD include those that increase dopamine levels, such as levodopa (L-DOPA) and dopamine agonists, and those that inhibit the breakdown of dopamine, such as catechol-O-methyltransferase (COMT) and monoamine oxidase (MAO)-B inhibitors (Figure 6.3) [14]. Although L-DOPA can cause significant side effects, most patients eventually require it. While a few major advancements have occurred since the introduction of L-DOPA, several experimental therapies are being researched. These drugs hold promise for disease modification in the near future.

Additionally, regenerative approaches, such as stem cell and gene therapies, are being explored. These new therapies could lead to significant advances in PD treatment, providing clinicians with novel and effective options in the coming years.

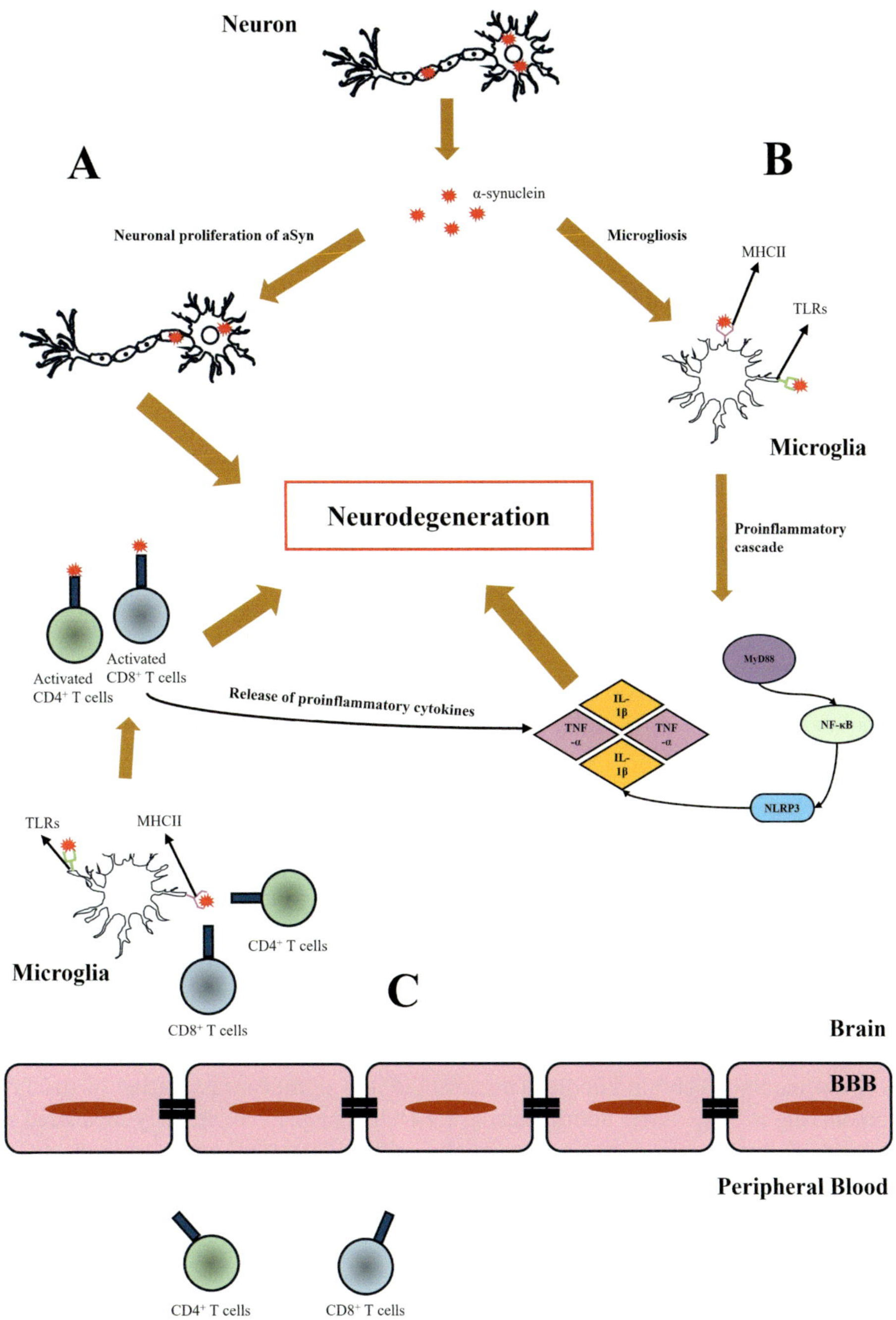

FIGURE 6.2 Aggregation of misfolded α-Syn proteins in neurons can lead to the development of PD pathology in multiple ways. (A) α-Syn proteins are released from the neurons and can propagate to other neurons leading to neurodegeneration. (B) α-Syn gets recognized by microglial cells, leading to activation of the microglia (microgliosis). α-Syn is recognized by TLRs, leading to the activation of a pro-inflammatory cascade via the adapter protein myeloid differentiation primary response 88 (MyD88), activating NF-κB, which recruits the inflammasome component NLRP3, causing the release of neuroinflammatory cytokines like interleukin-1β (IL-1β) and tumor necrosis factor α (TNF-α), leading to neurodegeneration. (C) Activated microglia act as antigen presenting cells and present α-Syn via MHCII. $CD4^+$ and $CD8^+$ T-cells infiltrate through the damaged blood–brain barrier (BBB), get activated after recognizing α-Syn, and cause the release of pro-inflammatory cytokines, leading to neurodegeneration.

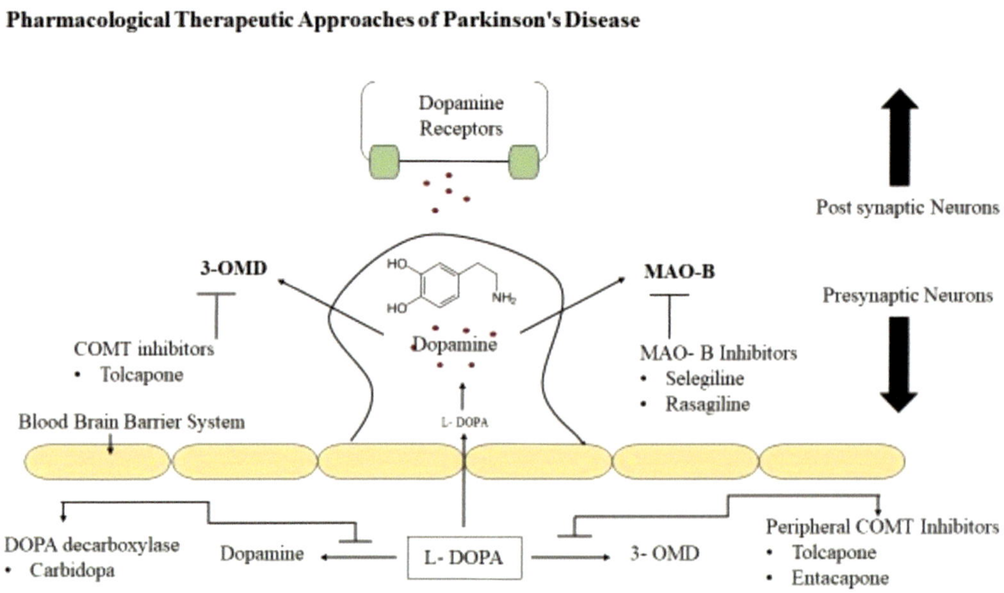

FIGURE 6.3 The figure shows the important pharmacological drugs to treat Parkinson's disease: amantadine, dopamine agonists (bromocriptine, pramipexole, ropinirole), COMT inhibitors (tolcapone, entacapone), DDC inhibitors (carbidopa), and MAO-B inhibitors (selegiline, rasagiline). These drugs act on several routes and receptors and target both peripheral and central mechanisms across the blood–brain barrier, increasing dopamine availability and activity.

6.2.1 Levodopa (L-DOPA)

L-DOPA acts as a precursor to dopamine, which aims to replace dopamine in the dopamine-deficient striatum. In addition to being effective for treating symptomatic parkinsonism brought on by carbon monoxide intoxication and post-encephalitic parkinsonism, L-DOPA can cross the blood–brain barrier, or BBB, and is then transformed into dopamine by the enzyme known as DOPA decarboxylase in both the central nervous system (CNS) and the peripheral nervous system [15]. The study conducted by Chao et al. investigated the acute behavioral and neurochemical consequences of administering L-DOPA intravenously (IN) to rats that had unilateral lesions of the medial forebrain bundle. This resulted in a significant reduction in striatal dopamine, amounting to 97%. After administering IN-DOPA or the vehicle, they measured turning behavior in a wide field, foot slips on a horizontal grid, and postural motor asymmetry in a cylinder, either with or without benserazide pre-treatment. Ten to twenty minutes after treatment, ipsilateral turnings were slightly reduced by IN L-DOPA without benserazide pre-treatment, while contralateral turnings were slightly increased. Opposite forelimb slips on the grid were decreased by IN L-DOPA with saline pre-treatment, but no effects were seen in the cylinder test. These findings lend credence to the theory that L-DOPA can reduce behavioral deficits in hemiparkinsonian animals by avoiding the BBB through the IN route [16].

L-DOPA's effectiveness can be seen quickly, in the beginning stages of PD. However, as the disease progresses, the duration of its effects shortens, necessitating more frequent dosing. Significant side effects include dyskinesias (involuntary movements) and motor fluctuations, particularly in advanced stages. Dyskinesias typically occur at peak drug levels and may require dosage adjustments. Advanced PD patients may experience rapid motor function changes, known as the "on-off" phenomenon, where the "on" state provides good symptom control, but the effect of L-DOPA quickly wears off, leaving patients in the "off" state with severe motor symptoms. These fluctuations result from variable drug absorption and dopamine level changes in the brain [17].

Chronic neuroinflammation plays a prominent role in PD progression. A preclinical research study explored the effect of L-DOPA on NLRP3 (nucleotide-binding domain, leucine-rich-containing family, pyrin domain-containing-3) inflammasome activation and tyrosine hydroxylase (TH) expression in 1-methyl-4-phenyl-1,2,3,6-tetrahydropyridine (MPTP)-induced parkinsonism and suggested the amelioration of NLRP3 inflammasome activation by L-DOPA treatment. Levels of TH, an important enzyme for dopamine synthesis, were reported to decline in striatal lesions, which were found upregulated by L-DOPA treatment in the PD mice. Modulated NLRP3 and TH levels were further suggested to be responsible for improved behavioral outcomes in this study as indicated by improved performance in pole, balanced beam, and rotarod test [18].

Despite the availability of L-DOPA, PD side effects, such as nausea and vomiting, orthostatic hypertension, and neuropsychiatric symptoms, make managing the condition extremely difficult, such as anxiety and hallucinations due to dopamine's effects outside the targeted brain regions. The current therapy is often limited in their ability to adequately control the disease's motor and non-motor manifestations, particularly in advanced stages. To minimize L-DOPA's side effects, strategies include using the lowest effective dose, dividing doses throughout the day, and considering alternative dopaminergic treatments. L-DOPA is taken with DOPA decarboxylase inhibitors (DDCIs), such as benserazide and carbidopa, to lessen peripheral adverse effects. These drugs do not cross the BBB but instead stop L-DOPA from being converted to dopamine outside of the brain [19]. Levodopa/carbidopa/entacapone (LCE), a combination of L-DOPA, carbidopa, and entacapone, has shown promise in addressing the drawbacks of conventional PD treatment.

6.2.2 Levodopa/Carbidopa/Entacapone Combinational Therapy

LCE is a combined therapy that deals with the wearing off symptoms in patients on L-DOPA/carbidopa treatment. Approved by the Food and Drug Administration (FDA) in 2003, LCE comprises three major components, including L-DOPA (a dopamine precursor), carbidopa (an aromatic L-amino acid decarboxylase inhibitor), and entacapone (a catechol-O-methyltransferase inhibitor). While L-DOPA and carbidopa have long been fundamental in treating PD, adding entacapone slows L-DOPA degradation in peripheral tissues, increasing its plasma concentration and enhancing its transport to CNS [20]. This therapy has been suggested to improve the management of PD symptoms and prolonged periods without any symptoms as compared to other therapies for patients.

This triple combination simplifies treatment by allowing patients to take just one medication as opposed to separate doses of LCE, improving convenience and compliance. LCE is prescribed for patients experiencing the fluctuations of end-dose motor and the "wearing-off phenomenon" on the levodopa/carbidopa dual agent. It is indicated for treating fluctuations in motors of adult patients with PD either as a single agent with the equivalent strength of its three components for those using these products separately or as a replacement for levodopa/carbidopa in patients experiencing the "wearing-off phenomenon" without dyskinesia on a daily L-DOPA dosage of 600 mg or less [20]. Its mechanism of action is based on L-DOPA acting as a dopamine precursor, carbidopa inhibiting L-DOPA metabolism in the periphery, and entacapone further preventing L-DOPA breakdown which enhances the drug availability in CNS [20].

Sleep impairment is considered to be associated with advanced stages of the disease as well as the severity of symptoms in PD. LCE was found to improve different parameters of sleep quality such as sleep onset, sleep consolidation, and rapid eye movement along with the management of different motor and non-motor symptoms in PD patients taking LCE at bedtime [21]. This combination has further been exploited to make LCE intestinal gel aiming to increase its bioavailability and is available in the market in European countries. An old double-blind, randomized study also suggested improvement in PD symptoms based on PDQ-8 scores (PD questionnaire). A total of 184 PD patients participated in this study. They were on levodopa/benserazide or levodopa/carbidopa treatment and were randomly divided into two groups and given either LCE or levodopa/carbidopa treatment for 12 weeks. This comparative study documented the significant improvement in quality of life with improved non-motor symptoms, including personal relationships, stigma, communication, and depression as compared to other treatment group and exhibited minimal or no motor symptom fluctuations [22].

While there aren't many side effects linked to LCE, they include nausea, vomiting, diarrhea, hypotension in the legs, and symptoms of the CNS such as hallucinations and impulse control disorders are some of the observed side effects which are attributed to its components. However, careful consideration is required while applying in patients who also had other disorders such as stomach ulcers, cardiovascular conditions, and cancers. LCE may elevate the risk of GI hemorrhage, cardiovascular complications, and severe conditions like rhabdomyolysis or neuroleptic malignant syndrome in such circumstances. Additional precautions should involve vigilant monitoring for dermatological lesions, hepatic dysfunction, and psychiatric disorders [20].

6.2.3 Dopamine Agonists

Pramipexole: Pramipexole is a non-ergot type dopamine receptor agonist with specific binding affinity to D3 and Dopamine 2 (D2) receptors used for the treatment of restless leg syndrome and PD. It was first approved for use in PD in 1997, followed by approval for usage in restless leg treatment in 2006 [23]. This agonist has been reported to have strong neuroprotection effects in various *in vitro* and *in vivo* studies [24]. In PD, it is used either alone or in conjunction with L-DOPA or carbamazepine to increase its efficacy and reduce its dosage along with the management of adverse symptoms.

Following oral treatment, pramipexole has a 90% bioavailability and is effectively absorbed. Its pharmacokinetic profile includes rapid absorption, extensive distribution, minimal metabolism, and primary renal excretion. Most side effects can be prevented with appropriate dosage optimization, and monotherapy has a good percentage of survival up to 34 months with no sign of efficacy decrease. The range of effective daily dosages is 1.5–6.0 mg/day. However, the increased frequency of adverse effects at 6.0 mg/day without corresponding improvement in symptom control indicates that this dosage is not recommended [24]. Lemke et al. conducted a prospective, observational open study to evaluate the safety and effectiveness of pramipexole as an add-on therapy to L-DOPA for treating anhedonia and mental depression in PD patients. The study found a significant decrease in the frequency of anhedonia, and amelioration of mild depression which were significantly correlated. Drop-outs from this drug due to adverse effects occurred in only 3.5% of subjects. However, 86% of patients were also receiving classic antidepressant medications, such as Selective Serotonin Reuptake Inhibitors (SSRIs) and Tricyclic Antidepressants (TCAs), which may have influenced the study's outcomes [25].

Beyond its primary indications, pramipexole has shown potential in treating bipolar depression, treatment-resistant depression, and essential tremors, although further studies are required to confirm its efficacy in these areas. Recent research also suggests pramipexole's effectiveness in reducing morphine tolerance and withdrawal symptoms, indicating its potential to restore morphine's analgesic effects. Mechanistically, pramipexole's efficacy is attributed to its high affinity for D3 and D2 dopaminergic receptors, with minimal activity at D1 receptors. Pramipexole's dosing varies based on the condition being treated and the patient's renal function, with specific considerations for pregnancy, breastfeeding, and adverse effects such as somnolence, orthostatic hypotension, and impulse control disorders [23].

Ropinirole: According to Rewane et al., it is a new non-ergoline dopamine D2 agonist that is recommended for the treatment of both early and severe PD [26]. It is mostly processed by the liver, and it takes around 5.8 hours for it to be eliminated. In the early stages of PD, ropinirole works well as a monotherapy to reduce symptoms. Early PD treatment involves L-DOPA and ropinirole. Compared to L-DOPA alone, administration to individuals suffering motor fluctuations has been demonstrated to result in a decreased incidence of dyskinesias [26]. Positron emission tomography research using 18F-dihydroxyphenylalanine revealed that ropinirole may slow the development of dopamine loss in neurons in comparison to L-DOPA therapy; however, this is still up for debate.

Among ropinirole's side effects are edema, nausea, somnolence, orthostatic hypotension, paranoia, and dyskinesia. A once-daily version of ropinirole is presently being developed, with the goal of improving its convenience, tolerability, and effectiveness [26].

Korczyn et al. conducted a study to establish the ideal ropinirole (ReQuip®) dosage for the treatment of PD in its early stages. Three months' worth of data from PD patients' monotherapy studies revealed that 75% of patients showed a therapeutic response to ropinirole at very low doses of ≤7.5 mg/day. However, an increase in dose up to 10.4 mg/day at the end of a three-year trial study and 14.9 mg/day at the end of a five-year clinical trial showed better improvements as compared to the patients with lower dose, suggesting continued dose titration may benefit most early PD patients [26].

Rotigotine: All dopamine (DA) receptors exhibit agonistic action when rotigotine (Neupro®) is administered. Rotigotine is a member of the non-ergolinic dopamine agonist class. The non-ergolinic dopamine agonist rotigotine is also recommended for use as a transdermal patch formulation either as a monotherapy for PD in its early stages or as an adjuvant to L-DOPA in all phases of the disease [27]. When it comes to managing PD, transdermal rotigotine is an efficient and typically well-tolerated supplement. The once-daily transdermal patch approach has several advantages, including the potential to prevent pulsatile dopaminergic activation [28]. Clinical trials have shown that transdermal rotigotine is more effective than a placebo in individuals with both early-stage and severe PD; however, there has not always been consistent evidence of its noninferiority when compared to oral dopamine agonists like ropinirole or pramipexole.

Standford and Scott conducted a study on using rotigotine transdermal patch on a daily basis, which demonstrated consistent release and uptake of the drug, reaching steady-state concentrations in one and two days [29]. A meta-analysis that reviewed different clinical trial reports of the transdermal application of rotigotine confirmed its efficacy in treating PD and suggested improved motor and non-motor symptoms as well as sleep quality [28]. Rotigotine is applicable as continuous drug delivery for PD treatment as its levels are stable in plasma up to 24 hours. Further, innovative delivery methods like polymer micelles, chitosan nanoparticles, and subcutaneous injections are being explored for rotigotine, with some showing potential neuroprotective effects. Although rotigotine is generally well tolerated, side effects such as headaches, GI problems, application-site reactions, and somnolence are prevalent [28, 29].

Apomorphine: Apomorphine is an aporphine alkaloid that is produced when morphine is acidified. It has a broad-spectrum affinity for all D1- and D2-like receptors and is a strong dopamine receptor agonist [30]. It is an acidified derivative of morphine and also possesses antagonist activity to serotonergic receptors. Due to fast hepatic metabolism, apomorphine has limited bioavailability (< 4%) upon oral administration; thus, it is primarily administered via either continuous subcutaneous infusion or intermittent subcutaneous injections. For patients suffering unpredictable "off" episodes, intermittent subcutaneous injections, when combined with oral therapy, offer significant comfort. Additionally, the coadministration of an antiemetic medication has enhanced tolerability greatly. When apomorphine is continuously infused, the amount of time spent in "off" periods is significantly reduced without increasing dyskinesia [30].

In terms of efficacy, apomorphine shows almost overlapping effects like dopamine or L-DOPA for PD motor symptoms. A double-blind study conducted by Kempster et al. [31] reported no significant difference between the effect of apomorphine and L-DOPA on different motor outcomes such as tremor severity, walking, dyskinesia, Webster disability scale, and scores of tapping hands; however, the apomorphine effects were observed for shorter duration as compared to L-DOPA, and these findings were later on confirmed by another study by Merello et al. Different studies also checked the effect of apomorphine injections given intermittently to PD patients along with L-DOPA and suggested the remarkable reduction in routine "off" periods and associated symptoms like pain, anhedonia, and urinary system-related problems [30].

Researchers are attempting to improve the length of effective durations for apomorphine. In a recent study, they attempted to develop a eutectic-based system that allows self-emulsification and entrapment of apomorphine post-subcutaneous injection and enhances its bioavailability up to 48 hours in serum [32]. The system was tested on male Wistar albino rats and pigs, where enhanced pharmacokinetics was suggested by the presence of apomorphine in plasma after 24 hours in rats and 48 hours in pigs [32].

Amantadine: Amantadine is an aliphatic primary amine that is licensed to treat PD and drug-induced movement disorders (DIMDs). It has complicated effects on neurotransmitter systems in the basal ganglia. In 1966, the US Food and Drug Administration approved amantadine as a preventive measure against Asian influenza. Later, it was licensed for the treatment of adult patients' respiratory illnesses brought on by influenza A virus strains [33]. This significantly impacts clinical outcomes and patient quality of life, particularly in those receiving antipsychotic treatments. Revealed evidence from multiple small controlled trials support amantadine's efficacy in drug-induced parkinsonism. Amantadine demonstrates a more favorable profile of tolerability in these individuals [33]. Observational studies and case reports suggest potential benefits as a second-line treatment for tardive dyskinesia, neuroleptic malignant syndrome, and catatonia, as well as to prevent dystonic reactions, although evidence specific to akathisia is lacking. Based on robust phase III trial data, demonstrating significant amantadine-DR/ER has been licensed by the FDA to be used as an adjuvant drug with L-DOPA or carbidopa in patients with PD who are suffering episodes of reduced off-time and dyskinesia. These cuts to the off-peak hours, along with improvements in dyskinesia, establish amantadine-DR/ER as the only medication approved for treating both types of motor complications. The analyses further validate these findings, revealing clinically relevant reductions in OFF time, sustained over 100 weeks of open-label treatment [34]. These reductions were observed independent of changes in L-DOPA dosing, underscoring amantadine-DR/ER's direct impact on motor fluctuations.

Significant benefits were also noted in patients experiencing $\geq$ 2.5 hours of OFF time at baseline, with amantadine-DR/ER outperforming placebo and demonstrating efficacy comparable to or exceeding other adjunct therapies. Moreover, even patients with less baseline OFF time showed significant treatment effects. The sustained therapeutic response observed over two years highlights the durability of amantadine-DR/ER in managing both dyskinesia and OFF time, thereby offering a valuable treatment option for PD patients with complex motor fluctuations [34].

The other study conducted by Caroff et al. is alternative for treating dyskinesia in PD patients receiving L-DOPA-based therapy, either in conjunction with or apart from concurrent dopaminergic medicines. These two clinical trials showed that amantadine medication significantly reduced dyskinesia, improved "on" time without bothersome dyskinesia, and decreased "off" time in individuals with PD who had dyskinesia [33]. The mechanism of action of this drug in parkinsonism is suggested to be a combination of dopaminergic and glutamatergic properties, which increase the release of more dopamine and the management of PD-related motor and non-motor symptoms [35].

6.2.4 Monoamine Oxidase-B Inhibitors

The oxidative degradation of amine-based neurotransmitters is catalyzed by MAO in brain, including dopamine. There are two MAOs, MAO-A and MAO-B, with respective 20% and 80% contribution to total MAO activity in the brain. MAO inhibitors, particularly MAO-B inhibitors, block or lower the dopamine catabolism in the synaptic cleft and promote dopamine signaling [36].

A significant role of the MAO-B antagonists in the treatment of PD, particularly in the early stages of the illness, is much talked about for its neuroprotective role. Recent studies on endogenous MAO-B regulation, including mRNA, protein, and activity levels, and the influence of DNA methylation, provide insights into its expression in healthy and degenerative conditions [36]. The most commonly used MAO-B inhibitors are rasagiline, selegiline, and safinamide. These inhibitors exhibit beneficial effects by regulating MAO-B activity, slowing down the clinical progression of the disease, modulating oxidative stress, and slowing down the α-Syn aggregation [36].

Rasagiline (N-propargyl-1(R)-aminoindan): This is a highly specific inhibitor of MAO-B which is used for PD management in early and late-stage PD. It slows down dopamine metabolism and provides symptomatic relief as well as exhibits neuroprotective properties by dealing with neurotoxic molecules and increasing the expression of neuroprotective proteins [36]. Speiser et al. conducted a study on a strong inhibitor of MAO-B in development, TVP-1012, for neurological disorders which indicated that memory and learning improved with long-term rasagiline treatment in both adult and senescent rats under normal or hypoxic conditions. To explore further, postnatal anoxia-lesioned rats were used as a model of cholinergic dysfunction. Rasagiline mesylate was administered to these rats from day 1 to 60, initially through nursing and later in drinking water (0.5 mg/kg/day). CNS access to the drug was confirmed via MAO activity analysis at 21 days. Rasagiline therapy reduced adolescent hyperactivity by day 28, normalized passive avoidance test performance by day 40, and restored spatial memory in the Morris water maze by day 50. Untreated anoxia rats failed these tasks, performing worse than rasagiline-treated groups and the regular control group. These results indicate that rasagiline improves memory and learning tasks, correlating with increased choline acetyltransferase activity in cholinergic brain regions [37].

Selegiline: Selegiline is also a selective, irreversible MAO-B inhibitor, among the first ever used as antiparkinsonian medication, and provides symptomatic relief. In a recent study, selegiline-associated clinical studies were reviewed and suggested this drug exhibits better UPDRS (unified PD rating scale) scores as compared to placebo with an increase in treatment duration; however, its high doses are associated with increased risk of adverse effects, prepared by Wang et al., and Sridhar et al. prepared a selegiline formulation as chitosan nanoparticles and assessed the drug's pharmacokinetics and pharmacodynamics when given intraperitoneally to rats as opposed to orally [38]. Over 90% of the medication was loaded into spherical nanoparticles with consistent *in vitro* and *ex vivo* drug release, thanks to the improved formulation. Compared to oral dosing, intranasal delivery resulted in significantly higher amounts of selegiline in brain tissue and plasma – by a factor of 20 and 12, respectively. Along with noticeably higher levels of dopamine, catalase activity, and the amount of glutathione in the brain, intranasal nanoparticles also improved performance in tests of locomotor activity, catalepsy, and stride length. As a result, compared to oral administration, intranasally delivered selegiline nanoparticles show better therapeutic potential and could be a promising treatment for PD.

Safinamide: Safinamide, also known as Xadago®, is an oral, selective, reversible MAO-B inhibitor with both dopaminergic and non-dopaminergic (glutamatergic) effects. It is authorized for use in the European Union to treat mid- to late-stage fluctuating PD as an adjuvant therapy to a stable dose of L-DOPA, either by itself or in conjunction with other PD drugs [38]. In a placebo-controlled research study, safinamide showed a significant increase in daily "on" time without dyskinesia when a fixed or flexible dose of 50–100 mg/day was administered for a 24-week period in individuals suffering from mid- to late-stage PD with motor fluctuations. With safinamide, other outcomes such as motor function, overall medical condition, and health-related quality of life were generally improved treatment [39].

Safinamide reduces concurrent pain management and improves pain-related scores in the Parkinson's Disease Quality of Life Questionnaire (PDQ-39). After six months of medication, path analysis was done to look at both direct and indirect relationships among safinamide and PDQ-39 pain-related variables. The percentage of patients in the safinamide group who did not get pain medication was found to be much lower than that of the placebo group. At 100 mg/day, safinamide dramatically decreased the average individual use of painkillers by about 24% and enhanced two of the three pain-related PDQ-39 items in the "Bodily discomfort" domain. The route analysis showed that approximately 80% of the overall benefit was due to safinamide's direct effect on pain [40].

6.2.5 COMT Inhibitors

In individuals with severe PD, catechol-O-methyltransferase inhibitors (COMT-Is) are advised as first-line L-DOPA add-on therapy to reduce end-of-dose motor fluctuations. At the moment, two periphery COMT-Is are regarded as first-line options: entacapone (ENT) and opicapone [41].

Entacapone: It is a specific and potent COMT inhibitor that is clinically indicated for use in PD patients as a supplement to L-DOPA or carbamazepine and was first approved in 1998. In a randomized control study involving 205 patients on L-DOPA therapy randomly divided into placebo and entacapone treatment groups, the treatment group was given 200 mg entacapone for 24 weeks along with each L-DOPA dose. Entacapone-supplemented L-DOPA treatment was suggested to increase the response time as suggested by reduced "off" timings as compared to the patients who received L-DOPA alone (placebo group) and helped in motor symptom management [41]. Entacapone was also suggested to be safe in an open-label extension study for three years when given as an additional therapy along with L-DOPA and dopamine decarboxylase inhibitor treatment. Initially, entacapone administration caused minor adverse effects such as nausea, insomnia, dizziness, and hallucinations, which were managed by L-DOPA dose adjustment; however, this additional therapy significantly increased the effect of a single dose of L-DOPA up to 2.8 hours which was stable throughout the study [42]. A systemic meta-analysis conducted by Li et al. also suggested that entacapone, when used as an adjuvant therapy, increased L-DOPA "on" time, reduced "off" periods, and significantly reduced effective L-DOPA dose [43]. Based upon its improved outcome, entacapone is now widely used in different combination therapies for PD management.

Opicapone: This is a potent third-generation, COMT inhibitor that got approval for clinical use in 2016 by a European medical agency. It is recommended as an adjunctive treatment to PD patients on L-DOPA/DDCI therapy whose motor symptoms are not improving with ongoing treatment. As it is impermeable through BBB, it is referred to as a peripheral COMT inhibitor. A clinical study focusing on the pharmacokinetics and pharmacodynamics of opicapone showed COMT inhibition up to 72 hours after dosing. However, the opicapone activity was lower after calorie-rich food intake, suggesting that it should be taken before meals for better effects. Preclinical studies in rodents suggested a significant increase of 1.3-fold in L-DOPA concentration upon administration with oral opicapone and suggested offering better COMT inhibition as compared to entacapone and tolcapone in homogenates of the liver and brain [44]. Based on clinical trials, a daily dose of 50 mg opicapone was found to be noninferior to entacapone and, when compared to a placebo, lowered the mean-off time by almost 50 minutes. The most frequent adverse events resulting from treatment were raised blood creatine phosphokinase levels, falls, dyskinesia, and sleeplessness. Because it is well tolerated, administered once per day, and not linked to the risk of liver failure, opicapone addresses the drawbacks of other COMT inhibitors, making it relevant to medical treatment and clinical practice. Additionally, there was a 39.3-minute decrease in "off" time when opicapone was substituted for entacapone. In conclusion, patients with PD who have end-of-dose motor fluctuations may benefit from opicapone, a third-generation COMT inhibitor [45].

Clinical studies assessing opicapone for PD patients with motor fluctuations, phase II results showed that opicapone boosted plasma L-DOPA levels and decreased peak 3-OMD (3-O-methyldopa, a product of L-DOPA breakdown by COMT) levels, leading to better motor responses and longer "on" times, with mostly mild to moderate side effects. Follow-up research by Ferreira et al. [46] confirmed that opicapone enhanced L-DOPA bioavailability and provided steady COMT inhibition, significantly cutting down the "off" time. The phase III BIPARK I trial further established that 50 mg Opicapone (OPC) effectively reduced OFF time and extended ON time compared to placebo, with a manageable safety profile, including dyskinesia, insomnia, and constipation as common side effects, but no deaths. The BIPARK II study supported these outcomes, highlighting substantial reductions in "off" time and with the increased "on" time, strengthening OPC's effectiveness and safety [47].

6.3 NON-PHARMACOLOGICAL APPROACHES TO MANAGE PARKINSON'S DISEASE

To address the challenges associated with the pharmacological approaches, researchers and clinicians have emphasized the importance of non-pharmacological approaches and alternative therapies alongside medication, especially in the early stages of PD [48]. Exercise and psychological

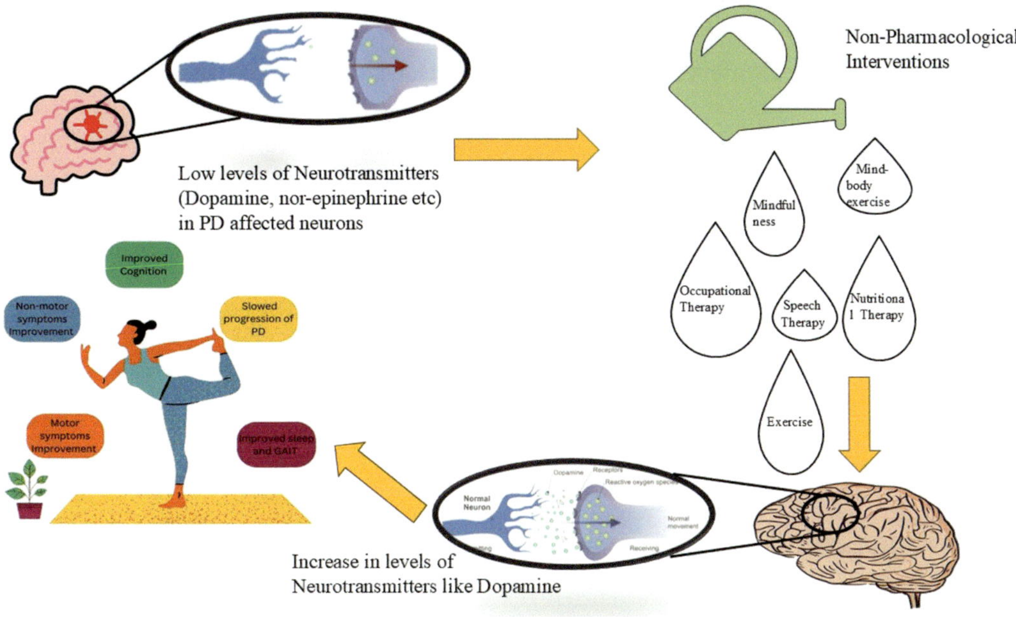

FIGURE 6.4 Non-pharmacological methods of Parkinson's disease management are shown in this graphic. The main aim of these interventions is to raise dopamine and norepinephrine levels in PD-affected neurons, which are reported to be reduced in PD patients.

support, as well as physical, occupational, and speech therapy, among other non-pharmacological methods, have an impactful role in the management of PD (Figure 6.4). These therapies improve mobility, enhance daily functioning, and maintain overall well-being by addressing both non-motor symptoms like tremors and stiffness and motor ones like depression and cognitive decline. By focusing on the holistic needs of patients, non-pharmacological strategies not only improve the standard of living for those who have PD but also empower them to maintain independence and engage more fully in daily activities, ensuring comprehensive and personalized approaches in managing the complexities of the PD [48]. Various non-pharmacological approaches are discussed in this section in the context of the management of PD and are summarized in Table 6.1.

6.3.1 Exercise

In individuals suffering from PD, exercise is emphasized to have a positive impact on cognitive and motor functions. It also improves executive skills, including decision-making, organizing, and problem-solving. The association between physical activity (PA) and PD was first seen to be advantageous in 1992 by Sasco and colleagues, who discovered that males who participated in sports during their college and adult years were less likely to develop PD. This result has been repeatedly confirmed in other epidemiological investigations. For instance, in the Swedish National March cohort study involving 43,368 participants, men who engaged in more than 6 hours weekly of household and commuting activities had a 43% decreased risk of PD than those who spent less than 2 hours/week on such activities. Similarly, the Finnish survey on health examinations at mobile clinics also reported a reduced risk of PD among individuals who participated in heavy leisure-time PA [49].

This format of exercise is reported to have an impact on neuroprotection and neuro-restoration by triggering changes in the synaptic plasticity of the brain. According to a study, the UPDRS maximum oxygen consumption, pace of walking, axial functions, and rigidity score were all improved by aerobic activities performed in a common environment. Better motor skills, happier moods,

TABLE 6.1
Non-Pharmacological Interventions Used in Parkinson's Disease and Their Reported Outcomes

Non-Pharmacological Intervention	Outcome	Molecular Mechanism Involved	References
	Exercise		
Gait training	Improvement in motor functions	Increased levels of Dopamine 2 (D2) receptors	[51]
Treadmill	Slows down the progression of PD	An increase in SIRT1 levels, lowers caspase-3 and BAX protein levels	
Skilled exercise	Enhancement of motor functions	Increased levels of BDNF in the striatum and an overall increase in AMPA receptors	
	Nutritional Therapy		
Inclusion of Mediterranean–DASH diet	Reduces the risk of PD or slows down the progression of PD	Unknown	[56]
Inclusion of whey protein	Improved motor function	Increased glutathione synthesis	[56, 57]
Inclusion of essential vitamins	Improved sleep and PD symptoms	Reduction in inflammatory markers, increased antioxidants, and glutathione production	[56, 59]
	Speech Therapy		
Expiratory muscle strength training (EMST)	Improved dysphagia	Decreased damage of basal ganglia	[70]
LSVT LOUD	Improved articulation, pitch, loudness, and pronunciation	Increased neuroplasticity by the release of neurotransmitters	[62, 63]
LSVT big	Improved balance, and improvement in doing job-related things	Increase of neurotransmitters generating new neuron connection	
	Mind–Body Exercise		
Tai-Chi	Improved motor and non-motor functions	Makes structural and functional changes to prefrontal, parietal, and temporal lobes	[66]
Qigong	Improvement in thinking, decision-making, and memory capacity	Qigong needs a high amount of concentration, and this leads to the involvement of active parts of the brain	[64]
Yoga	Improved motor function, working memory as well as non-motor functions	Increased network connectivity in anterior, posterior, and medial prefrontal regions. Increase in BDNF, antioxidant levels, and telomerase activity.	[60]
Occupational therapy	Improvement in hand strength and daily activities	Unknown	[71]

reduced fatigue, and an all-around higher quality of life were demonstrated in these improvements in PD patients. Cycling demonstrated overall improvement in the UPDRS, but stretching had little to no effect on the same scale, according to another study that compared cycling to stretching [46].

As the stage of PD progresses, the gait of the patient becomes more impaired, resulting in noticeable variations in step length, stride length, and speed. Walking, especially Nordic walking, has been found to improve these various gait characteristics [50].

Treadmill walking and gait training have been shown to increase the caudal basal ganglia's D2 receptor levels, as observed in PET scans using F-fallypride. Additionally, treadmill workout exerts a positive neuroprotective effect in PD by enhancing SIRT1 activity, which enhances mitochondrial function and lowers neuroinflammation. This exercise reduces the presence of microglia and astrocytes, potentially slowing the progression of the disease. Furthermore, treadmill exercise decreases the levels of apoptotic proteins caspase-3 and BAX while simultaneously upregulating autophagy expression-related proteins like P62, IC3-II, and Beclin-1, contributing to its overall neuroprotective impact [51].

The study conducted by Bastioli et al. (2022) on the levels of BDNF in striatum and SN is also suggested to be enhanced by routine skilled exercise, which enhances motor function. Additionally, it increases AMPA receptor expression, which boosts neuroplasticity. Because PD overexpresses indicators of inflammation like TNF-α, IL-β, and NF-κB, it causes muscular atrophy [52]. Frequent exercise increases the antioxidant capacity of cells, which lowers the pro-inflammatory cytokine expression and the rate at which reactive oxygen species (ROS) are generated. Neuronal mitochondrial function is also impacted by PD. Exercise is enhanced by upregulating dynamin-like protein (DLP) and mitofusin 1/2 (MFN1/2); mitochondrial fusion and fission are regulated. Mitochondria produce ATP, which is utilized for cellular processes. By raising the number of copies of mitochondrial DNA and lowering the frequency of point mutations, exercise also lessens the depletion of mitochondrial DNA [50].

6.3.2 Nutritional Therapy

PD patients are considered to have a high risk of malnutrition due to impaired peristaltic movements affecting food movement and digestion, impaired digestion due to altered gut microbiota, altered gut–brain bidirectional communication, and altered sensory stimuli (smell, taste) resulting in decreased appetite [53]. The intensity of these symptoms can vary due to ongoing mainstream treatments such as L-DOPA combination therapies [53]. Nutrition therapy thus comes as a supportive approach, not to cure PD but to improve other aspects of patient health.

The enteric nervous system (ENS) is suggested to be responsible for dopamine production in the GI tract, which plays an important role in the protection of intestinal mucosa and GI mobility regulation [53]. The ENS and CNS bidirectional communication is regulated by the gut–brain axis, thus exerting emotional and cognitive control over gut function which further regulates enteroendocrine signaling, immune activation, intestinal permeability, and enteric reflex [54]. Reduced dopamine production in PD could result in altered permeability to undigested food particles, toxins, and microbes due to compromised gut protection, thus eliciting inflammatory reactions [53]. Although the onset of PD is considered majorly in CNS, Braak's classic hypothesis suggests the peripheral extension of PD onset. Pathogens in GI upon altered mucosal protection penetrate the intestinal mucosa and induce the formation of α-Syn aggregates. Two different categories of PD individuals have been documented with one category in which aggregates were detected in CNS first followed by spread to ENS (and another peripheral nervous system) and the other category in which α-Syn aggregates were detected in ENS first followed by spread in CNS [54].

The research studies examining the relationship between dietary patterns and PD suggest a significant relationship between the severity of PD symptoms with high intake of processed, packaged, high-fat, and sugar-rich diets; however, reduced severity is related to the uptake of foods rich in antioxidants, flavonoids, and medium-chain triglycerides [55]. As nutritional therapy is used as a supportive therapy with pharmacological treatment, it is mainly focused on improving overall patient health, reducing difficulties related to digestion, appetite, sleep, and cognition as well as slowing down the disease progression. Foods with neuroprotective and anti-inflammatory properties are recommended for maintaining overall brain health which include nutrient-rich nuts (walnuts, cashews, almonds, pistachios, and Brazil nuts), spices (turmeric and cinnamon), anti-inflammatory fat-rich foods (olive oil, coconut oil, rosemary, oily fish, salmon, and tuna), and leafy green vegetables (spinach, kale, collard greens, etc.) [56]. Antioxidant-rich foods such as green tea, raspberries, blueberries, and other fruits containing pigments like anthocyanins, epigallocatechin, vitamins (C and E), and carotenoids also protect neuronal health by neutralizing oxidative stress [56]. These dietary antioxidants are reported to exhibit myriad effects rather than a single process such as vitamins and polyphenols directly counteracting free radicals; minerals act as cofactors of antioxidative enzymes, thus increasing their activity. Some of these antioxidants regulate downstream signaling and alleviate the effect of oxidative stress-induced damage. Polyphenols like resveratrol, curcumin, and catechin inhibit the Lewy body formation, thus promoting neuronal health [56].

The foods rich in water and fiber content, such as dates, apples, figs, kiwi, prunes, radishes, nuts, berries, and beans are suggested to deal with PD symptoms related to digestion as these foods increase the food mobility in the gut. Probiotic and prebiotic intake is also suggested to deal with dysbiosis as it adds healthy bacteria to the gut and also helps to reduce inflammation [53]. Further, literature reports indicate the deficiency of CoQ10, glutathione, and vitamins C and D in the individuals suffering from PD, often lacking, and suggest that intake of CoQ10 (300–1,200 mg daily) can improve motor neuron symptoms and inflammation. While glutathione supplements are not effective, their production can be boosted by a Mediterranean diet rich in specific vegetables and possibly N-acetyl cysteine (NAC). A small study showed that high doses of NAC (1,200–8,000 mg/day) increased serum glutathione in PD patients without adverse effects at 1,200 mg daily [57, 58].

Deficiencies of vitamin B and vitamin D are common in PD. Optimal vitamin B (B1 and B2) levels are associated with improved motor function, cognition, mood, sleep, and motivation. Vitamins B16, B12, and folate are beneficial in reducing homocysteine levels in patients taking L-DOPA [59]. Although vitamin D and vitamin C doses are not specifically studied for PD, maintenance of optimal levels after specific examination is suggested through dietary inclusion. However, vitamins along with fats interfere with drug absorption, so dose, time of taking the drug, etc. are critically reviewed after thorough nutrition-focused examination by clinicians. Further, the National Institute of Health (NIH) suggests the uptake of omega-3 fatty acids to lessen inflammation and improve motor skills in PD patients, with purified supplements recommended due to contamination risks. Increasing omega-3 intake to 1–4 g daily may be beneficial [53, 59].

6.3.3 Speech Therapy

Research indicates that 89% of people with PD experience speech-related problems known as hypokinetic dysarthria. This motor speech disorder arises from damage to the basal ganglia, a brain region responsible for muscle movement. In PD, hypokinetic dysarthria affects articulation, pitch, loudness, and speech stress. Additionally, the slower muscle movements associated with this condition can impair swallowing, a condition known as dysphagia [60].

Conventional speech therapy for PD management focuses on multiple attributes, like articulation, respiration, loudness, and rate, and is delivered once or twice a week without specifically focusing on sensory and cuing attributes of speech. A groundbreaking speech therapy, known as "Lee Silverman Voice Treatment (LSVT LOUD®)," is reported to be delivered with higher efficacy and help to manage the complex etiology of PD. LSVT LOUD majorly focuses on the improvement of voice by increasing vocal output in an intensive treatment regimen which also focuses on the sensory component of speech, including self-monitoring and cueing [61].

A clinical trial on PD compared two treatments aimed at increasing vocal loudness: LSVT LOUD voice treatment and a respiratory treatment. The results showed that LSVT LOUD significantly boosted vocal loudness, with improvements lasting up to 24 months, outperforming the respiratory treatment [62]. Another study also confirmed that LSVT LOUD's beneficial impacts were noticeably greater than those in untreated PD patients and persisted for six months. The range of effect sizes was 0.65–2.03, emphasizing that focusing on voice rather than only breathing is crucial for better outcomes [63]. In a recent study, this group evaluated the voice- and articulation-focused LSVT LOUD and LSVT-ARCTIC with an untreated condition. The results showed that LSVT LOUD increased the sound pressure level more than LSVT-ARCTIC and the untreated condition. Both LSVT LOUD and LSVT-ARCTIC enhanced the communication effectiveness index, but only LSVT LOUD demonstrated long-term advantages [61].

6.3.4 Complementary and Alternative Therapies

The evidence of people opting for complementary and alternative medicine (CAM) therapies for PD management along with a pharmacological approach or alone is increasing. As per "National Centre for Complementary Alternative Medicine," CAM includes a diverse group of practices, including mind–body exercises (yoga, meditation, Tai Chi, acupuncture, massage, etc.), use of natural products (probiotics, herbs, minerals, and vitamins), and use of alternative medicine systems like homeopathy, Chinese medicinal system, and Ayurveda [64].

Mind and body exercises, integrating cognitive focus, breath regulation, and physical movement, are beneficial for neurocognitive health. A mini review by Zhang et al. (2022) quoted the meta-analyses demonstrating interventions like Pilates, dancing, yoga, and tai chi enhance older individuals' cognitive function, regardless of levels of cognitive impairment [65]. The popularity of these practices is increasing in the United States, with over 10% of adults having adopted mind–body exercises. These exercises are evidenced to ameliorate motor symptom impairments in PD; however, limited evidence is available regarding its impact on non-motor symptoms, particularly those related to cognitive deficits.

Tai Chi, a moderate-intensity aerobic exercise, is reported to improve different cognitive parameters such as memory, attention, execution as well as social participation [66]. A clinical study involving 60 PD patients with impaired cognition randomly divided patients into control and Tai Chi group. The control group was receiving the conventional drug; on the other hand, Tai Chi group received 12 weeks long Tai Chi training, and after 12 weeks, Tai Chi group showed significant improvement in attention, memory, cognition, and computing activities as compared to control subjects [66]. The underlying mechanism behind improved memory and cognition is suggested to be structural changes in the brain induced by long-term Tai Chi practice, which includes the thickening of different brain regions and increased volume of the whole brain [66].

Qigong is another Chinese traditional therapy combining meditation with breathing, movement, and postural activities suggested to be effective in the improvement of motor symptoms like stride and impairment of gait in addition to non-motor symptoms, including constipation, depression, and disturbed sleep [64]. A recent follow-up clinical study showed that qigong significantly improved balance and functional coordination in moderate-stage PD patients who were practicing this therapy. These results were suggested by improved scores in the one-legged blind balancing test with the time up and go test (TUG), which are important indicators for motor coordination and balance [67]. A meta-analysis conducted by Chen et al. (2020) considering seven studies involving 325 participants (145 females, 180 males) also showed that in PD patients, qigong markedly improved walking ability, balance, and motor symptoms. The impact on motor symptoms and walking was prominent ($P < 0.01$) when the undergoing group was compared with the passive PD individuals' group; however, with walking abilities as opposed to a passive approach ($P < 0.01$), there was no discernible change observed w.r.t active control group. Furthermore, qigong exercise significantly improved balance in comparison to both active and passive PD controls ($P < 0.05$). This study also revealed age had an impact on how qigong affected motor symptoms [68].

Yoga is another mind–body therapy that originated in India and emphasizes the integration of physical skills to achieve a philosophical idea of oneness. In other words, yoga aims to achieve harmony by integrating physical and brain activities. Yoga is being used as adjunctive or rehabilitation therapy for different diseases like heart disease, arthritis, hypertension, cancer, neurodegenerative diseases, and epilepsy [60]. Different types of yoga are reported to improve rigidity, dyskinesia, balance-related symptoms, mood disorders as well as other issues like sleep, fatigue, and pain. Iyengar Yoga, comprising postural techniques, relaxation exercises, and meditation, was shown to improve muscle strength, motion, and flexibility in PD patients, whereas Power Yoga improved the stiffness and rigidity as well as muscle strength in lower limbs. Various stretching exercises practiced in yoga enhance muscle lengthening and improve joint flexibility by activating the stretch receptors in joints, muscles, and ligaments [60].

Breathing and meditation techniques of yoga are beneficial in non-motor symptoms like anxiety, depression, and stress which greatly impact the mental health of PD patients, worsening their movement disorders. A randomized control trial compared mindfulness yoga and standard stretching and resistance techniques with 79 idiopathic PD patients in each group. In this study both the groups received intervention for eight weeks. Yoga with mindfulness reduced motor symptoms as effectively as stretching and resistance techniques; however, the additional benefits of yoga in terms of reduced anxiety and depression were also observed [60].

Herbal medicine is being explored as the adjunct therapy in managing PD, potentially complementing conventional treatments. Some herbs, such as *Ginkgo biloba* and *Curcuma longa* (turmeric), are noted for their neuroprotective properties, possibly slowing PD progression through antioxidant and anti-inflammatory effects. *Mucuna pruriens*, containing natural L-DOPA, may help alleviate motor symptoms, while herbs like *Bacopa monnieri* (brahmi) and *Panax ginseng* are thought to improve cognitive function [69]. However, challenges include inconsistent potency, potential interactions with standard PD medications, and limited scientific evidence. Patients considering herbal medicine should consult healthcare professionals to ensure safe and integrated use within their overall treatment plan.

6.3.5 Occupational Therapy

This therapy involves the practices of everyday activities, such as work, play, and self-care. Occupational therapists understand the intricate interplay that affects patients with PD due to personal, environmental, occupational, and task performance aspects. Through the restoration of lost abilities or bodily functions and the provision of compensating techniques like the use of assistive technology, this therapy enhances occupational performance. A study conducted assessed effectiveness and satisfaction through post-program semi-structured interviews. In 2018, 23 of the 77 participants in the program had been diagnosed with early PD and were seen within three years of diagnosis, placing them in stages 1–2 of the disease. Deficits in the strength of the hands (60.0 ± 23.4 pounds) and coordination (right hand 30.0 ± 8.0 seconds) were observed in the participants, and 65.2% of them indicated difficulties with daily tasks. A total of 16 participants were interviewed, and 75% expressed great satisfaction and 60% of those who remembered doing at-home workouts said they continued to follow the program. The program was delivered with fidelity in 96% of cases. Just two occupational therapy practitioners at a single office had embraced the approach after a year, which remains in the organization. The study suggests that tailored OT programs effectively reach people with early PD and highlight the need for initiatives for sustainability and scaling in this demographic [69].

6.4 EMERGING THERAPIES

Emerging therapies for PD management are surfacing as a new hope for PD patients, caregivers as well as families. The limitations associated with existing therapies like the palliative nature of traditional pharmacological approaches and limited efficacy as well as the variability of alternative therapies strengthen the demand for emerging therapies which include immunotherapy, gene therapy, stem cell therapy, fecal microbiota transplantation (FMT), and deep brain stimulation (DBS) therapy.

6.4.1 Immunotherapy

PD is thought to be characterized by α-Syn aggregation. This protein is present in the peripheral nervous system, which includes the blood, as well as in both neuronal and non-neuronal cells. Various theories, supported by *in vivo* and *in vitro* experiments, suggest the movement of α-Syn from intracellular to extracellular spaces, affecting surrounding cells, particularly synapses [72].

α-Syn also impacts other brain cells, including glial cells, which constitute a significant portion of the brain, thus eliciting various immunogenic responses. The transfer of α-Syn aggregates between different brain cells disrupts lysosomal function and damages mitochondria, triggering pro-inflammatory responses and cellular death on a large scale [72].

These findings helped the researchers to design a hypothesis that intracellular or extracellular reduction of α-Syn aggregates could prevent further damage, resulting in the development of immunotherapies aimed at eliminating extracellular α-Syn in order to stop the spread of cell damage. There are primary adopted techniques, including active immunization (vaccination) and passive immunization. While passive immunization includes the delivery of antibodies precisely targeting α-Syn, active immunization induces an individual's immune system to create antibodies against α-Syn [73].

In active immunization studies, Masliah et al. (2005) used recombinant human α-Syn to immunize transgenic mice overexpressing α-Syn, resulting in the production of antibodies against α-Syn. Reduced α-Syn aggregation and decreased neurodegeneration were observed in the immunized mice as compared to controls [74]. Clinical trials for active immunization against α-Syn aggregates also suggest promising immunological responses [75]. A tiny peptide mimicking the C-terminal section of α-Syn (AFFITOPE® PD01A) was administered subcutaneously to early-stage PD patients, generating an immunological response in 86% of the vaccinated individuals, and among them 63% produced serum antibodies against PD01A, which showed specific detection of responses and reduction of oligomeric and fibrillar α-Syn levels. Further, this immunization resulted in the reduction of oligomeric α-Syn levels in CSF and plasma of early-stage PD patients in the 26th week after treatment [76].

For passive immunization, antibodies directed against α-Syn have been tested, and various antibodies are in clinical trials. In one such work, transgenic mice overexpressing human wild-type α-Syn were injected with a monoclonal antibody, 9E4, which targets the C-terminal of α-Syn. This treatment reduced the accumulation of cleaved α-Syn in hippocampal and cortical neurons. Further research on C-terminal and N-terminal antibodies has shown reduced neurodegeneration and improved behavior in treated mice. Another antibody, which has entered human trials, which includes PRX002, a humanized variant of the C-terminal guided murine monoclonal antibody 9E4 [77], is an IgG1 monoclonal. These findings highlight the potential of immunotherapies in managing PD by targeting and reducing harmful α-Syn aggregates.

6.4.2 Stem Cell Therapy

Stem cell therapy is based on replacing lost dopaminergic neurons with dopamine producing cells, which can include aborted human fetal tissue or embryonic stem cells (ESCs). Clinical studies evidence the reduced neurological symptoms and improved motor symptoms upon transplantation of fetal midbrain tissue in PD patients. In a study by Tronci et al. (2015), ventral mesencephalon tissue from fetus was transplanted to volunteer early PD patients, and regular follow-up was taken up to four years, including PET and MRI examinations. This study reported the persistent relief from motor symptoms; however, this technique is associated with technical and ethical challenges as this treatment needed six to eight aborted human fetuses per patient [78].

Pluripotent stem cells, such as induced pluripotent stem cells and ESCs, cells (iPSCs) further, have the great promise for treating PD through neural grafting, especially to replace degenerated dopaminergic neurons in midbrain. Early on, there were challenges in consistently producing enough of these neurons and ensuring they were the right type. However, advancements in techniques have improved their production, though results can still vary. Effective protocols exist for differentiating hPSCs into specific neural types in both 2D and 3D cultures. Using dual-SMAD inhibition and WNT activation, hPSCs can be directed to become DA neurons by expressing markers like FOXA2 and LMX1A. Over 80 days in a neuronal maturation medium, these cells develop into functional DA neurons [77]. Now, with clinical trials for ESC-based therapies approaching, there's hope for a significant step forward in PD treatment [79].

iPSCs, discovered in 2007, offer another exciting option by reprogramming adult cells into stem cells. This method allows for creating personalized grafts from a patient's own cells, reducing the need for immune system suppression which addresses the limitation of allogenic grafting of ESCs. iPSC-derived neural grafts have shown positive results in primate studies, improving motor function without forming tumors [80]. Major drawback of using hiPSCs is the cost. Approx. $800,000 are needed to treat one patient with an autologous iPSC-derived cell product. To reduce the cost, once isolated, stem cells can target or treat multiple individuals by HLA typing. In this HLA, donor can be matched with the recipient, and the grafting/transplantation can be done in the most matched HLA-type patient, and these cells, once isolated, can be given to multiple patients whose HLA matches the most [80].

6.4.3 Gene Therapy

As discussed previously, PD onset has been associated to both environmental and genetic risks. Accumulation of misfolded aggregates of α-Syn protein, which is encoded by gene SNCA, is major hallmark of PD. In early studies of PD, patients' missense mutation was detected in this gene, and later studies identify point mutations in this gene which are reportedly A53T, A30P, E46K, G51D, H50Q, A53V, A53E, and A30G) [81]. In addition to SNCA, gene mutations in LRRK2, PARK7, PARK2, PINK1, and GBA which are involved in dopaminergic signaling, neurogenesis, remodeling of cytoskeletal architecture, mitochondrial and lysosomal function, calcium homeostasis, etc. are also responsible for PD onset and progression [81].

Gene therapy involves transfer of correct gene to replace the mutant or malfunctioning gene by various means of transfer. *Ex vivo* gene therapy involves genomic editing of cultivated cells and their subsequent delivery to the patient. *In vivo* gene therapy involves delivery of genetic material (RNA or DNA) with the help of vectors or by genome modifying technology such as CRISPR–Cas to correct the mutation or regulate the gene. The vectors used for delivering genetic material can be viral or non-viral. Most common viral vectors used for gene therapy or gene delivery are adenoviruses or lentiviruses. The gene of interest is encoded into these viral vectors by removing their own pathogenic genes, and then these viruses are allowed to infect the host cell, and our gene of interest is integrated into the host cell [80].

Various literature studies suggest the potential of gene therapy in PD treatment. SNCA gene mutation which is responsible for production of misfolded α-Syn aggregates can be knocked out by shRNA or anti-sense oligonucleotides. Delivery of these constructs via adenoviruses has been reported to prevent neurodegeneration in PD models [82, 83]. Use of CRISPR-interference by Heman-Ackah et al. (2016) also knocked down SNCA gene successfully, having triplication mutation in iPSCs [83].

Anti-sense oligonucleotides when injected directly against LRRK2 mutation into cerebellum showed decrease of fibril-induced α-Syn inclusions within the LRRK2 protein levels in brain cells of mice, and no additional abnormal characteristic in any other tissues was found. First-phase clinical trials of this technique are going on patients with PD [84]. Mutation in GBA gene is associated with lysosomal dysfunction, and there are studies that have targeted this gene by direct injection of adeno-associated viruses having GBA1 gene, and reduced α-Syn levels were reported in rodent models of PD [85]. Intracisternal injections of adeno-associated virus 9 having GBA1 gene are currently under Phase 1/2 clinical trials [86].

Another approach to gene therapy is doing neural regeneration. This therapy is based on the idea of converting astrocytes into induced dopamine releasing neurons. Two teams of scientists have shown that by using shRNA and CRISPR–Cas PTB, an RNA binding protein is inhibited, which suppress neuronal differentiation, and using these approaches mice were rescued from motor and neurochemical deficits [86].

6.4.4 Fecal Microbiota Transplantation (Gut–Brain Axis)

It is observed that PD develops majorly in old age, and studies have found that aging influences the change in the composition of the gut microbiome of the bacteria that produce butyrate and spread α-Syn [87]. A study by Holmqvist et al. (2014) shown that α-Syn can travel from the gut to the brain by injecting human α-Syn into a mouse, which was sufficient to cause α-Syn aggregation in the brainstem and vagus nerve [88]. There are some gut-related symptoms which can be detected years before the PD diagnosis in the patients which include over 80% of PD patients experiencing constipation, bloating, nausea, dysphagia, sialorrhea, vomiting, and gastroparesis [89]. Changes in the gut microbiota can contribute to the onset or worsening of PD symptoms through increased intestinal permeability, inflammation, aberrant fibril α-aggregation, oxidative stress, and reduced neurotransmitter synthesis [89]. Intestinal permeability is measured by fecal zonulin and α-1-antitrypsin, both of which are elevated in PD patients [90]. Research has revealed that PD patients have higher levels of calprotectin, a fecal indicator of intestinal inflammation [90].

With the increase of intestinal permeability, microbial production of lipopolysaccharides (LPS) increases which induces intestinal inflammation, oxidative stress, neuroinflammation, and microglial activation, thus leading to accumulation of neuropathological α-Syn aggregates [91]. It has been found that numerous neurotransmitters present in the human brain, including dopamine, serotonin, GABA, and norepinephrine, are produced by the gut bacteria [92]. L-DOPA, a drug that can cross BBB, is used in treatment of PD; however, its bioavailability is questionable as bacterial tyrosine decarboxylase converts L-DOPA into dopamine [93]. This enzyme is produced by various bacterial species, including *Lactobacillus*, *Enterococcus*, *Clostridium*, *Bifidobacterium*, *Prevotella*, and *Ruminococcus* found in gut [92].

Improved knowledge of gut–brain axis and microbial role in this context enabled researchers to introduce new interventions for PD patients, including administration of FMT, probiotics, psychobiotics, prebiotics, synbiotics, and postbiotics, dietary modification as well as Chinese medicine to modulate gut microbiota [89]. Psychobiotics are category of probiotics that assist people's mental well-being who consume it [94]. FMT is an ancient therapy involving reconstruction of intestinal microbiota through the transplantation of healthy donor's microbiota into patient's intestine [95]. A case study has showed that FMT via colonoscopy improved the PD motor, non-motor symptoms, and constipation at sixth month of the treatment [95]. Use of combination of four different bacterial strains, namely probiotics *Bifidum*, *Lactobacillus*, *Acidophilus*, and *Fermentum* improved scores on the Movement Disorders Society's UPDRS (MDS-UPDRS) [96]. Another study by Srivastav et al. (2019) showed the use of probiotic combination containing *Lactobacillus rhamnosus* GG (LGG), *Bifidobacterium animalis* lactis, and *Lactobacillus acidophilus* raised butyrate levels, which can protect dopaminergic neurons from MPTP and rotenone-induced neurotoxicity [92].

To improve gut health, exercise has also shown the decrease in pro-apoptotic substances and inflammatory cytokines as well as the increased activity of the antioxidant enzymes (catalase and glutathione peroxidase). Exercise can therefore result in a general reduction in intestinal inflammation [57].

6.4.5 Deep Brain Stimulation (DBS), Non-Invasive Brain Stimulation (NIBS), and Bright Light Therapy (BLT)

DBS has emerged as highly effective therapy since last 30 years. DBS involves implantation of battery-operated electronic device known as neurostimulator in specific areas of patients' brain which are majorly involved in motor functions. It has been observed that DBS of the subthalamic nucleus has positive benefits on both motor and depressive symptoms. DBS administration is generally indicated to the patients with L-DOPA wearing off symptoms (bradykinesia), tremors, and tolerance/resistance to dopaminergic agonist treatments [97].

Non-invasive brain stimulation (NIBS) works on the principle of altering neurophysiological activity of brain by using mild magnetic pulses or currents applied to the brain. Transcranial direct current stimulation (tDCS) and reflex testing are used for giving NIBS. In rTMS a coil producing electromagnetic field is placed on the patient's head, and it produces a powerful, focused pulsed electromagnetic field resulting in depolarization of neurons underneath. In tDCS by applying a little current through scalp electrodes, neurons depolarize, neuronal excitability is modulated, and cerebral blood flow is changed [95]. Depression is common symptom of PD patients, and it is believed that it occurs through reduced regional cerebral blood flow in prefrontal cortex of patients with PD. In a study cerebral blood flow was found to be higher after giving rTMS [96].

PD is often linked with many symptoms or complications, one of which is sleep disorder. Sleep disorder can be associated with cardiac rhythm disturbances which can be due to imbalance in neurotransmitter levels. In bright light therapy (BLT), retina or individual with sleep disorder is exposed to bright light which suppresses the release of melatonin from pineal gland and inhibits sleep. PD patients have lack of alertness; exposure to bright light suppresses the release of melatonin, hence making the individual more alert. BLT also stimulates the arousal-related monoaminergic system and also inhibits the ventrolateral preoptic nucleus's activity, which causes sleep [98].

6.4.6 Acupuncture and Electroconvulsive Therapy

Giving stimulations on particular body areas using pressure, electricity, lasers, or needles is known as acupuncture. Acupuncture has also historically been used to treat a range of neurological and mental conditions. It has been reported that acupuncture gave relief from PD patients' motor and non-motor symptoms. It has been revealed in a rat model study that acupuncture treats depressive behavior in rats treated with this therapy, and possible reason for this can be upregulation of mTOR pathway, inhibiting autophagy, raising dopamine, decreasing α-Syn aggregates, and improving synaptic protein expression [99].

Electroconvulsive therapy (ECT) uses small electric currents under effect of anesthesia, which intentionally causes a quick convulsion. ECT seems to change the chemistry of the brain, and these changes can quickly relieve the symptoms of many mental health conditions. Several case studies, observations, and one randomized controlled trial have shown improvement in depressive symptoms [100].

6.5 CONCLUSION

This chapter summarizes the available pharmacological, non-pharmacological, and emerging interventions for management of PD. Initially, it has been concluded that PD is not a monogenic disease as multiple genes as well as various non-genetic factors are involved in progression of this disease. PD treatment is advancing with continuously evolving pharmacological and non-pharmacological approaches. Pharmacological approaches using drugs like L-DOPA, dopamine agonists, or enzymatic inhibitors that enhance dopamine availability greatly provide the symptomatic relief; however, non-pharmacological approaches such as exercise, nutritional therapy, and occupational therapy help to manage the functional outcomes and improve quality of life of the patients. Further, emerging therapies like stem cell therapy, gene therapy, deep DBS, and fecal microbiota transformation hold promise for addressing the underlying pathophysiological mechanisms of PD, potentially altering the course of the disease and offering new hope for patients. The synergy between established treatments and innovative research underscores the importance of a multifaceted approach to PD management. Further, PD management in future could be seen as more promising upon continued collaboration between researchers and clinicians. Amalgamation of both current and emerging therapies can lead us toward more effective and individualized care strategies that not only manage symptoms but also improve overall patient well-being.

REFERENCES

1. Feigin, V. L., Krishnamurthi, R. V., Theadom, A. M., Abajobir, A. A., Mishra, S. R., Ahmed, M. B., et al. Global, regional, and national burden of neurological disorders during 1990–2015: A systematic analysis for the Global Burden of Disease Study 2015. Lancet Neurol 2017;6:877. https://doi.org/10.1016/S1474-4422(17)30299-5
2. Ogawa, T., Fujii, S., Kuya, K., Kitao, S. I., Shinohara, Y., Ishibashi, M., et al. Role of neuroimaging on differentiation of Parkinson's disease and its related diseases. Yonago Acta Med 2018; 61:145. https://doi.org/10.33160/YAM.2018.09.001

3. Greenberg, D. A., Aminoff, M. J., & Simon, R. P. Movement disorders: Clinical Neurology, 11th ed., New York, NY: McGraw-Hill Education; 2021.
4. De Lau, L. M. L., & Breteler, M. M. B. Epidemiology of Parkinson's disease. Lancet Neurol 2006. https://doi.org/10.1016%2FS1474-4422%2806%2970471-9
5. Tanner, C. M., Goldman, S. M., Ross, G. W., & Grate, S. J. The disease intersection of susceptibility and exposure: Chemical exposures and neurodegenerative disease risk. Alzheimer's Dement 2014;10. https://doi.org/10.1016/J.JALZ.2014.04.014
6. Hernán, M. A., Takkouche, B., Caamaño-Isorna, F., & Gestal-Otero, J. J. A meta-analysis of coffee drinking, cigarette smoking, and the risk of Parkinson's disease. Ann Neurol 2002; 52:276–284. https://doi.org/10.1002/ANA.10277
7. Goedert, M., Spillantini, M. G., Del Tredici, K., & Braak, H. 100 years of Lewy pathology. Nat Rev Neurol 2013; 9:13–24. https://doi.org/10.1038/NRNEUROL.2012.242
8. Dickson, D. W., Braak, H., Duda, J. E., Duyckaerts, C., Gasser, T., Halliday, G. M., et al. Neuropathological assessment of Parkinson's disease: Refining the diagnostic criteria. Lancet Neurol 2009; 8:1150–1157. https://doi.org/10.1016/S1474-4422(09)70238-8
9. Braak, H., Ghebremedhin, E., Rüb, U., Bratzke, H., & Del Tredici, K. Stages in the development of Parkinson's disease-related pathology. Cell Tissue Res 2004; 318:121–134. https://doi.org/10.1007/S00441-004-0956-9
10. Tansey, M. G., & Goldberg, M. S. Neuroinflammation in Parkinson's disease: Its role in neuronal death and implications for therapeutic intervention. Neurobiol Dis 2010; 37:510. https://doi.org/10.1016/J.NBD.2009.11.004
11. Karikari, A. A., McFleder, R. L., Ribechini, E., Blum, R., Bruttel, V., & Knorr, S., et al. Neurodegeneration by α-synuclein-specific t cells in AAV-A53T-α-synuclein Parkinson's disease mice. Brain Behav Immun 2022; 101:194–210. https://doi.org/10.1016/j.bbi.2022.01.007
12. Su, R., & Zhou, T. Alpha-synuclein induced immune cells activation and associated therapy in Parkinson's disease. Front Aging Neurosci 2021; 13:769506. https://doi.org/10.3389/fnagi.2021.769506
13. Zahoor, I., Shafi, A., & Haq, E. Pharmacological treatment of Parkinson's disease. Exon Publ 2018; 129–144. DOI:10.15586/codonpublications.parkinsonsdisease.2018.ch7
14. Smulders, K., Dale, M. L., Carlson-Kuhta, P., Nutt, J. G., & Horak, F. B. Pharmacological treatment in Parkinson's disease: Effects on gait. Parkinsonism Relat Disord 2016; 31:3–13. https://doi.org/10.1016/j.parkreldis.2016.07.014
15. National Collaborating Centre for Chronic Conditions (UK). Parkinson's disease: National clinical guideline for diagnosis and management in primary and secondary care. London: Royal College of Physicians; 2006. https://www.nice.org.uk/guidance/cg35
16. Chao, O. Y., Mattern, C., Silva, A. M. D. S., Weßler, J., Ruocco, L. A., Nikolaus, S., …, & Pum, M. E. Intranasally applied L-DOPA alleviates Parkinsonian symptoms in rats with unilateral nigro-striatal 6-OHDA lesions. Brain Res Bull 2012;87(2-3):340–345. https://doi.org/10.1016/j.brainresbull.2011.11.020
17. Nord, M. Levodopa pharmacokinetics-from stomach to brain: A study on patients with Parkinson's disease. LiU E-Press 2019;1567
18. Chen, X., Wang, Z., Yang, W., & Fu, Y. Levodopa improves behavioral deficits of mice with Parkinson's disease symptoms via curbing NLRP3 inflammasome activation and enhancing tyrosine hydroxylase levels in the striatum and substantia nigra. J Integr Neurosci 2024;23(1):2. https://doi.org/10.31083/j.jin2301002
19. Gandhi, K. R., & Saadabadi, A. Levodopa (L-dopa). In StatPearls [Internet]. Treasure Island (FL): StatPearls Publishing; 2023. https://www.ncbi.nlm.nih.gov/books/NBK534787/
20. Hauser, R. A. Levodopa/carbidopa/entacapone (Stalevo). Neurology 2004; 62. https://doi.org/10.1212/wnl.62.1_suppl_1.s64
21. Park, K. W., Jo, S., Lee, S. H., Hwang, Y. S., Lee, D., Ryu, H.-S., & Chung, S. J. Therapeutic effect of levodopa/carbidopa/entacapone on sleep disturbance in patients with Parkinson's disease. J Mo Disord 2020;13(3):205. DOI: https://doi.org/10.14802/jmd.20055
22. Fung, V. S. C., & Herawati, L., Wan, Y., Movement Disorder Society of Australia Clinical Research and Trials Group, & Quality of Life Study Group. Quality of life in early Parkinson's disease treated with levodopa/carbidopa/entacapone. Mov Disord 2009; 24(1):25–31. https://doi.org/10.1002/mds.22269
23. Singh, R., & Parmar, M. Pramipexole. In StatPearls [Internet]. Treasure Island (FL): StatPearls Publishing; 2023. https://www.ncbi.nlm.nih.gov/books/NBK534792/
24. Bennett, J. P., Jr, & Piercey, M. F. Pramipexole—A new dopamine agonist for the treatment of Parkinson's disease. J Neurol Sci 1999;163(1):25–31. DOI: 10.1016/s0022-510x(98)00307-4

25. Lemke, M. R., Brecht, H. M., Koester, J., Kraus, P. H., & Reichmann, H. Anhedonia, depression, and motor functioning in Parkinson's disease during treatment with pramipexole. J Neuropsychiatry Clin Neurosci 2005; 17(2):214–220. https://doi.org/10.1176/jnp.17.2.214
26. Korczyn, A. D., Thalamas, C., & Adler, C. H. Dosing with ropinirole in a clinical setting. Acta Neurol Scand 2002; 106(4):200–204. DOI: 10.1034/j.1600-0404.2002.01343.x
27. Frampton, J. E. Rotigotine transdermal patch: A review in Parkinson's disease. CNS Drugs 2019; 33:707–718. DOI: 10.1007/s40263-019-00646-y
28. Raeder, V., Boura, I., Leta, V., Jenner, P., Reichmann, H., Trenkwalder, C., Klingelhoefer, L., & Chaudhuri, K. R. Rotigotine transdermal patch for motor and non-motor Parkinson's disease: A review of 12 years' clinical experience. CNS Drugs 2021; 35(2):215–231. DOI: 10.1007/s40263-020-00788-4
29. Sanford, M., & Scott, L. J. Rotigotine transdermal patch: A review of its use in the treatment of Parkinson's disease. CNS Drugs 2011; 25:699–719. https://doi.org/10.2165/11593370-000000000-00000
30. Agbo, F., Isaacson, S. H., Gil, R., Chiu, Y.-Y., Brantley, S. J., Bhargava, P., et al. Pharmacokinetics and comparative bioavailability of apomorphine sublingual film and subcutaneous apomorphine formulations in patients with Parkinson's disease and "OFF" episodes: Results of a randomized, three-way crossover, open-label study. Neurol Ther 2021;10(2):693–709. doi: 10.1007/s40120-021-00251-6
31. Kempster, P. A., Frankel, J. P., Stern, G. M., & Lees, A. J. Comparison of motor response to apomorphine and levodopa in Parkinson's disease. J Neurol Neurosurg Psychiatry 1990;53(11):1004–1007. https://doi.org/10.1136/jnnp.53.11.1004
32. Kim, J., Gao, Y., Zhao, Z., Rodrigues, D., Tanner, E. E. L., Ibsen, K., et al. A deep eutectic-based, self-emulsifying subcutaneous depot system for apomorphine therapy in Parkinson's disease. Proc Natl Acad Sci USA 2022;119(9). https://doi.org/10.1073/pnas.2110450119
33. Caroff, S. N., Jain, R., & Morley, J. F. Revisiting amantadine as a treatment for drug-induced movement disorders. Ann Clin Psychiatry 2020;32(3):198–208. https://doi.org/10.1016/j.psc.2019.08.003
34. Hauser, R. A., Lytle, J., Formella, A. E., & Tanner, C. M. Amantadine delayed release/extended-release capsules significantly reduce OFF time in Parkinson's disease. NPJ Parkinson's Dis 2022;8(1):29. https://doi.org/10.1038/s41531-022-00291-1
35. Rascol, O., Fabbri, M., & Poewe, W. Amantadine in the treatment of Parkinson's disease and other movement disorders. Lancet Neurol 2021; 20(12):1048–1056. DOI: 10.1016/S1474-4422(21)00249-0
36. Tong, J., Rathitharan, G., Meyer, J. H., Furukawa, Y., Ang, L.-C., Boileau, I., et al. Brain monoamine oxidase B and A in human parkinsonian dopamine deficiency disorders. Brain 2017; 140(9):2460–2474. https://doi.org/10.1093/brain/awx183
37. Speiser, Z., Katzir, O., Rehavi, M., Zabarski, T., & Cohen, S. Sparing by rasagiline (TVP-1012) of cholinergic functions and behavior in the postnatal anoxia rat. Pharmacol Biochem Behav 1998; 60(2):387–393. DOI: 10.1016/s0091-3057(97)00603-5
38. Wang, K., Liu, Z. H., Li, X. Y., Li, Y. F., Li, J. R., Hui, J. J., et al. Efficacy and safety of selegiline for the treatment of Parkinson's disease: A systematic review and meta-analysis. Front Aging Neurosci 2023; 15:1134472. https://doi.org/10.3389/fnagi.2023.1134472
39. Blair, H. A., & Dhillon, S. Safinamide: A review in Parkinson's disease. CNS Drugs 2017; 31:169–176. DOI: 10.1007/s40263-017-0408-1
40. Cattaneo, C., Barone, P., Bonizzoni, E., & Sardina, M. Effects of safinamide on pain in fluctuating Parkinson's disease patients. J Parkinsons Dis 2017; 7(1):95–101. DOI: 10.3233/JPD-160911
41. Parkinson Study Group. Entacapone improves motor fluctuations in levodopa-treated Parkinson's disease patients. Ann Neurol 1997; 42:747–755. https://doi.org/10.1002/ana.410420515
42. Larsen, J. P., Worm-Petersen, J., Sidén, Å, Gordin, A., Reinikainen, K., Leinonen, M., et al. The tolerability and efficacy of entacapone over 3 years in patients with Parkinson's disease. Eur J Neurol 2003; 10(2):137–146. DOI: 10.1046/j.1468-1331.2003.00559.x
43. Li, J., Lou, Z., Liu, X., Sun, Y., & Chen, J. Efficacy and safety of adjuvant treatment with entacapone in advanced Parkinson's disease with motor fluctuation: A systematic meta-analysis. Eur Neurol 2017; 78(3-4):143–153. https://doi.org/10.1159/000481685
44. Bonifacio, M. J., Torrao, L., Loureiro, A. I., Palma, P. N., Wright, L. C., & Soares-da-Silva, P. Pharmacological profile of opicapone, a third-generation nitrocatechol catechol-O-methyl transferase inhibitor, in the rat. Br J Pharmacol 2015; 172(7):1739–1752. https://doi.org/10.1111/bph.13029
45. Scott, L. J. Opicapone: A review in Parkinson's disease. Drugs 2016; 76:1293–1300. DOI: 10.1007/s40265-016-0623-y
46. Shu, H. F., Yang, T., Yu, S. X., et al. Aerobic exercise for Parkinson's disease: A systematic review and meta-analysis of randomized controlled trials. PLoS One 2014; 9(7):100503. https://doi.org/10.1371/journal.pone.0100503

47. Ferreira, J. J., Lees, A., Rocha, J. F., Poewe, W., Rascol, O., Soares-da-Silva, P., & Bi-Park 1 investigators. Opicapone as an adjunct to levodopa in patients with Parkinson's disease and end-of-dose motor fluctuations: A randomised, double-blind, controlled trial. Lancet Neurol 2016; 15(2):154–165. https://doi.org/10.1016/S1474-4422(15)00336-1
48. Zhen, K., Zhang, S., Tao, X., Li, G., Lv, Y., & Yu, L. A systematic review and meta-analysis on effects of aerobic exercise in people with Parkinson's disease. npj Parkinson's Dis 2022; 8(1):146. https://doi.org/10.1038/s41531-022-00418-4
49. Wang, Y. L., Wang, Y. T., Li, J. F., Zhang, Y. Z., Yin, H. L., & Han, B. Body mass index and risk of Parkinson's disease: A dose-response meta-analysis of prospective studies. PLoS One 2015; 10(6):0131778. https://doi.org/10.1371/journal.pone.0131778
50. Fayyaz, M., Jaffery, S. S., Anwar, F., Zil-E-Ali, A., & Anjum, I. The effect of physical activity in Parkinson's disease: A mini-review. Cureus 2018; 10(7). https://doi.org/10.7759/cureus.2995
51. Fisher, B. E., Li, Q., Nacca, A., et al. Treadmill exercise elevates striatal dopamine D2 receptor binding potential in patients with early Parkinson's disease. NeuroReport 2013; 24(10):509–514. https://doi.org/10.1097/WNR.0b013e328361dc13
52. Bastioli, G., Arnold, J. C., Mancini, M., et al. Voluntary exercise boosts striatal dopamine release: Evidence for the necessary and sufficient role of BDNF. J Neurosci 2022; 42(23):4725–4736. https://doi.org/10.1523/JNEUROSCI.2273-21.2022
53. Lister, T. Nutrition and lifestyle interventions for managing Parkinson's disease: A narrative review. J Mov Disord 2020; 13(2):97–104. https://doi.org/10.14802/jmd.20006
54. Montanari, M., Imbriani, P., Bonsi, P., Martella, G., & Peppe, A. Beyond the microbiota: Understanding the role of the enteric nervous system in Parkinson's disease from mice to human. Biomedicines 2023; 11(6):1560. https://doi.org/10.3390/biomedicines11061560
55. Mischley, L. K., Lau, R. C., & Bennett, R. D. Role of diet and nutritional supplements in Parkinson's disease progression. Oxid Med Cell Longev 2017; 1:6405278. https://doi.org/10.1155/2017/6405278
56. Park, H. A., & Ellis, A. C. Dietary antioxidants and Parkinson's disease. Antioxidants 2020; 9(7):570. https://doi.org/10.3390/antiox9070570
57. Hodges, R. E., & Minich, D. M. Modulation of metabolic detoxification pathways using foods and food-derived components: A scientific review with clinical application. J Nutr Metab 2015; 1:760689. https://doi.org/10.1155/2015/760689
58. Monti, D. A., Zabrecky, G., Kremens, D., et al. N-Acetyl cysteine may support dopamine neurons in Parkinson's disease: Preliminary clinical and cell line data. PLoS One 2016; 11(6):0157602. https://doi.org/10.1371/journal.pone.0157602
59. Costantini, A., Pala, M. I., Grossi, E., et al. Long-term treatment with high-dose thiamine in Parkinson disease: An open-label pilot study. J Altern Complement Med 2015; 21(12):740–747. https://doi.org/10.1089/acm.2014.0353
60. Kwok, J. Y. Y., Kwan, J. C. Y., Auyeung, M., et al. Effects of mindfulness yoga vs stretching and resistance training exercises on anxiety and depression for people with Parkinson disease: A randomized clinical trial. JAMA Neurol 2019; 76(7):755–763. https://doi.org/10.1001/jamaneurol.2019.0534
61. Ramig, L., Halpern, A., Spielman, J., Fox, C., & Freeman, K. Speech treatment in Parkinson's disease: Randomized controlled trial (RCT). Mov Disord 2018; 33(11):1777–1791. https://doi.org/10.1002/mds.27460
62. Ramig, L. O., Countryman, S., Thompson, L. L., & Horii, Y. Comparison of two forms of intensive speech treatment for Parkinson disease. J Speech Hear Res 1995; 38(6):1232–1251. https://doi.org/10.1044/jshr.3806.1232
63. Ramig, L. O., Sapir, S., Countryman, S., et al. Intensive voice treatment (LSVT®) for patients with Parkinson's disease: A 2 year follow up. J Neurol Neurosurg Psychiatry 2001; 71(4):493–498. https://doi.org/10.1136/jnnp.71.4.493
64. Bega, D., Gonzalez-Latapi, P., Zadikoff, C., & Simuni, T. A review of the clinical evidence for complementary and alternative therapies in Parkinson's disease. Curr Treat Options Neurol 2014; 16(10):1–19. https://doi.org/10.1007/s11940-014-0314-5
65. Zhang M., Li F., Wang D., Ba X., & Liu Z. Exercise sustains motor function in Parkinson's disease: evidence from 109 randomized controlled trials on over 4,600 patients. Front Aging Neurosci. 2023;15:1071803.
66. Wei, L., Chai, Q., Chen, J., et al. The impact of tai chi on cognitive rehabilitation of elder adults with mild cognitive impairment: A systematic review and meta-analysis. Disabil Rehabil 2022; 44(11):2197–2206. https://doi.org/10.1080/09638288.2020.1830311

67. Yang, Y., Wang, G., Zhang, S., et al. Efficacy and evaluation of therapeutic exercises on adults with Parkinson's disease: A systematic review and network meta-analysis. BMC Geriatr 2022; 22(1):813. https://doi.org/10.1186/s12877-022-03510-9
68. Chen, S., Zhang, Y., Wang, Y. T., Liu, X., Song, W., & Du, X. The effect of qigong-based therapy on patients with Parkinson's disease: A systematic review and meta-analysis. Clin Rehabil 2020; 34(12):1436–1448. https://doi.org/10.1177/0269215520946695
69. Yin, R., Xue, J., Tan, Y., et al. The positive role and mechanism of herbal medicine in Parkinson's disease. Oxid Med Cell Longev 2021; 2021. https://doi.org/10.1155/2021/9923331
70. Padmanabhan S, Polinski NK, Menalled LB, Baptista MAS, Fiske BK. The Michael J. Fox Foundation for Parkinson's Research strategy to advance therapeutic development of PINK1 and Parkin. Biomolecules. 2019;9(8):296. doi:10.3390/biom9080296
71. Sadural, A., MacDonald, J., Johnson, J., Gohil, K., & Rafferty, M. Occupational therapy for people with early Parkinson's disease: A retrospective program evaluation. Parkinsons Dis 2022; 1:1931468. https://doi.org/10.1155/2022/1931468
72. Marques, O., & Outeiro, T. F. Alpha-synuclein: From secretion to dysfunction and death. Cell Death Dis 2012; 3(7):350–350. https://doi.org/10.1038/cddis.2012.94
73. Sardi, S. P., Cedarbaum, J. M., & Brundin, P. Targeted therapies for Parkinson's disease: From genetics to the clinic. Mov Disord 2018; 33(5):684–696. https://doi.org/10.1002/mds.27414
74. Masliah, E., Rockenstein, E., Adame, A., et al. Effects of α-synuclein immunization in a mouse model of Parkinson's disease. Neuron 2005; 46(6):857–868. https://doi.org/10.1016/j.neuron.2005.05.010
75. Ghochikyan, A., Petrushina, I., Davtyan, H., et al. Immunogenicity of epitope vaccines targeting different B cell antigenic determinants of human α-synuclein: Feasibility study. Neurosci Lett 2014; 560: 86–91. https://doi.org/10.1016/j.neulet.2013.12.028
76. Affiris, A. G. Affiris Announces Encouraging Long-Term Data from a Series of First-in-Human Studies Using Affitope® PD01A Targeting Oligomeric Alpha-Synuclein in Early Parkinson's Disease Patients. 2018. Retrieved from https://affiris.com/investors/
77. Weihofen, A., Patel, H., Huy, C., et al. Human-derived α-synuclein antibody BIIB054 binds pathologic forms of α-synuclein and attenuates transmission of α-synuclein In Vitro and In Vivo [Abstract]. Mov Disord 2016; 31(suppl 2). Retrieved from http://www.mdsabstracts.org/abstract/human-derived-synuclein-antibody-biib054-binds-pathologic-forms-of-synuclein-and-attenuates-transmission-of-synuclein-in-vitro-and-in-vivo/
78. Tronci E., Fidalgo C., & Carta M. Foetal cell transplantation for Parkinson's disease: focus on graft-induced dyskinesia. Parkinsons Dis. 2015;2015(1):563820.
79. Cha, Y., Park, T. Y., Leblanc, P., & Kim, K. S. Current status and future perspectives on stem cell-based therapies for Parkinson's disease. J Mov Disord 2023; 16(1):22. https://doi.org/10.14802/jmd.22141
80. Shalaby, K. E., & El-Agnaf, O. M. A. Gene-based therapeutics for Parkinson's disease. Biomedicines 2022; 10(8):1790. https://doi.org/10.3390/biomedicines10081790
81. Dumbhare, O., & Gaurkar, S. S. A review of genetic and gene therapy for Parkinson's disease. Cureus 2023. https://doi.org/10.7759/cureus.34657
82. Zharikov, A. D., Cannon, J. R., Tapias, V., et al. ShRNA targeting α-synuclein prevents neurodegeneration in a Parkinson's disease model. J Clin Invest 2015; 125(7):2721–2735. https://doi.org/10.1172/JCI64502
83. Heman-Ackah, S. M., Bassett, A. R., & Wood, M. J. A. Precision modulation of neurodegenerative disease-related gene expression in human iPSC-derived neurons. Sci Rep 2016; 6(1):28420. https://doi.org/10.1038/srep28420
84. Zhao, H. T., John, N. Delic, V., et al. LRRK2 antisense oligonucleotides ameliorate α-synuclein inclusion formation in a Parkinson's disease mouse model. Mol Ther Nucleic Acids 2017; 8(1):508–519. https://doi.org/10.1016/j.omtn.2017.08.002
85. Sardi, S. P., Clarke, J., Viel, C., et al. Augmenting CNS glucocerebrosidase activity as a therapeutic strategy for Parkinsonism and other Gaucher-related synucleinopathies. Proc Nat Acad Sci US Amer 2013; 110(9):3537–3542. https://doi.org/10.1073/pnas.1220464110
86. Zhou, H., Su, J., Hu, X., et al. Glia-to-neuron conversion by CRISPR-CasRx alleviates symptoms of neurological disease in mice. Cell 2020; 181(3):590–603. https://doi.org/10.1016/j.cell.2020.03.024
87. Gold, A., Turkalp, Z. T., & Munoz, D. G. Enteric alpha-synuclein expression is increased in Parkinson's disease but not Alzheimer's disease. Mov Disord 2013; 28(2):237–241. https://doi.org/10.1002/mds.25292
88. Holmqvist, S., Chutna, O., Bousset, L., Aldrin-Kirk, P., Li, W., Björklund, T., Wang, Z. Y., Roybon, L., Melki, R., & Li, J. Y. Direct evidence of Parkinson pathology spread from the gastrointestinal tract to the brain in rats. Acta Neuropathol 2014; 128:805–820. https://doi.org/10.1007/s00401-014-1343-6

89. Segal, A., Zlotnik, Y., Moyal-Atias, K., Abuhasira, R., & Ifergane, G. Fecal microbiota transplant as a potential treatment for Parkinson's disease – A case series. Clin Neurol Neurosurg 2021; 207:106791. https://doi.org/10.1016/j.clineuro.2021.106791
90. Tamtaji, O. R., Taghizadeh, M., Kakhaki, R. D., Kouchaki, E., Bahmani, F., Borzabadi, S., Oryan, S., Mafi, A., & Asemi, Z. Clinical and metabolic response to probiotic administration in people with Parkinson's disease: A randomized, double-blind, placebo-controlled trial. Clin Nutr 2019; 38(3):1031–1035. https://doi.org/10.1016/j.clnu.2018.04.010
91. Dumitrescu, L., Marta, D., Dănău, A., Lefter, A., Tulbă, D., Cozma, L., Manole, E., Gherghiceanu, M., Ceafalan, L. C., & Popescu, B. O. Serum and fecal markers of intestinal inflammation and intestinal barrier permeability are elevated in Parkinson's disease. Front Neurosci 2021; 15:689723. https://doi.org/10.3389/fnins.2021.689723
92. Srivastav, S., Neupane, S., Bhurtel, S., Katila, N., Maharjan, S., Choi, H., Hong, J. T., & Choi, D. Y. Probiotics mixture increases butyrate, and subsequently rescues the nigral dopaminergic neurons from MPTP and rotenone-induced neurotoxicity. J Nutr Biochem 2019; 69:73–86. https://doi.org/10.1016/j.jnutbio.2019.03.011
93. van Kessel, S. P., Frye, A. K., El-Gendy, A. O., Castejon, M., Keshavarzian, A., van Dijk, G., & El Aidy, S. Gut bacterial tyrosine decarboxylases restrict levels of levodopa in the treatment of Parkinson's disease. Nat Commun 2019; 10(1):310. DOI: 10.1038/s41467-019-08294-y
94. Dinan, T. G., Stanton, C., & Cryan, J. F. Psychobiotics: A novel class of psychotropic. Biol Psychiatry 2013; 74(10):720–726. https://doi.org/10.1016/j.biopsych.2013.05.001
95. Zheng, H. B., Liu, B., Shen, J., Xie, F., Ji, Q. M., & Zhu, X. Y. Non-invasive brain stimulation for treating psychiatric symptoms in Parkinson's disease: A systematic review and meta-analysis. J Clin Neurosci 2022; 106:83–90. https://doi.org/10.1016/j.jocn.2022.01.029
96. Cardoso, E. F., Fregni, F., Martins Maia, F., Boggio, P. S., Luis Myczkowski, M., Coracini, K., Lopes Vieira, A., Melo, L. M., Sato, J. R., Antonio Marcolin, M., & Rigonatti, S. P. rTMS treatment for depression in Parkinson's disease increases BOLD responses in the left prefrontal cortex. Int J Neuropsychopharmacol 2008; 11(2):173–183. DOI: 10.1017/S1461145707007961
97. França, C., Carra, R. B., Diniz, J. M., Munhoz, R. P., & Cury, R. G. Deep brain stimulation in Parkinson's disease: State of the art and future perspectives. Arq Neuropsiquiatr 2022; 80(5 Suppl 1):105–115. https://doi.org/10.1590/0004-282X-anp-2021-0275
98. Huang, H. T., Huang, T. W., & Hong, C. T. Bright light therapy for Parkinson disease: A literature review and meta-analysis of randomized controlled trials. Biol 2021; 10(11):1205. https://doi.org/10.3390/biology10111205
99. Ning, B., Wang, Z., Wu, Q., Deng, Q., Yang, Q., Gao, J., Fu, W., Deng, Y., Wu, B., Huang, X., & Mei, J. Acupuncture inhibits autophagy and repairs synapses by activating the mTOR pathway in Parkinson's disease depression model rats. Brain Res 2023; 1808:148320. https://doi.org/10.1016/j.brainres.2023.148320
100. Takamiya, A., Seki, M., Kudo, S., Yoshizaki, T., Nakahara, J., Mimura, M., & Kishimoto, T. Electroconvulsive therapy for Parkinson's disease: A systematic review and meta-analysis. Mov Disord 2021; 36(1):50–58. https://doi.org/10.1002/mds.28328

7 Mechanism of Neurodegeneration in Other Neurodegenerative Diseases

Harikrishna Reddy Dontiboina, Srikanth Yadava, Sri Lekha Chintha, Naresh Dumala, and Guntupalli Chakravarthy
KL College of Pharmacy, Koneru Lakshmaiah Education Foundation, Vaddeswaram, Guntur, Andhra Pradesh, India

Hasniza Zaman Huri
KL College of Pharmacy, Koneru Lakshmaiah Education Foundation, Vaddeswaram, Guntur, Andhra Pradesh, India
Department of Clinical Pharmacy & Pharmacy Practice, Faculty of Pharmacy, Universiti Malaya, Kuala Lumpur, Malaysia

Tan Hui Yin
KL College of Pharmacy, Koneru Lakshmaiah Education Foundation, Vaddeswaram, Guntur, Andhra Pradesh, India
Faculty of Applied Sciences, Tunku Abdul Rahman University of Management and Technology, Jalan Genting Kelang, Setapak, Kuala Lumpur, Malaysia

*Kakarla Ramakrishna**
KL College of Pharmacy, Koneru Lakshmaiah Education Foundation, Vaddeswaram, Guntur, Andhra Pradesh, India

7.1 INTRODUCTION

Neurodegenerative diseases exhibit a gradual degeneration of both the structure and function of the nervous system, specifically the neurons located in the brain and spinal cord. These are slowly advancing degenerative processes that result in a decline in cognitive and physical abilities, thereby severely diminishing the quality of life for those affected. Often, these diseases are untreatable and lead to considerable disability, necessitating long-term care and support. Globally, Alzheimer's disease (AD) and Parkinson's disease (PD) represent the most common neurodegenerative disorders, responsible for 60–80% of dementia instances, while other neurodegenerative conditions encompass Huntington's disease (HD), motor neuron disease, amyotrophic lateral sclerosis (ALS), multiple system atrophy (MSA), PSP, Friedreich ataxia (FA), Lewy body disease (LBD), spinal muscular atrophy (SMA), and multiple sclerosis (MS).

Neurodegenerative disease frequently results in cognitive decline, motor deficits, difficulties in speech and swallowing, and muscle weakness, which are common among the neurodegenerative

*Corresponding Author. kakarlaramakrishna@kluniversity.in

 DOI: 10.1201/9781003530510-7

diseases. Cognitive decline resulting in memory impairment, disorientation, and alterations in personality, and motor symptoms like tremors, rigidity, and impaired coordination can render everyday tasks progressively more challenging. Further, challenges with speech and swallowing may arise, impacting both communication and dietary intake and muscle weakness and paralysis can advance with time, resulting in considerable mobility difficulties. Additionally, the majority of neurodegenerative disease exhibit anxiety, depression, and other mood changes that critically influence the patient's ability to function independently. The symptoms associated with the neurodegenerative diseases vary according to the type of neurodegenerative diseases, and these symptoms are independent of disease pathophysiology. In the current chapter, we have discussed the various pathophysiology of various neurodegenerative diseases, except AD and PD.

7.2 HUNTINGTON'S DISEASE

HD is a neurodegenerative disease that exhibits motor, cognitive, and psychiatric dysfunction. Motor symptoms include involuntary jerking (chorea), muscle rigidity (dystonia), abnormal eye movements, balance issues, and speech/swallowing difficulties. Cognitive decline affects focus, organization, learning, and adaptability, often leading to perseveration and slowed thinking. Psychiatric symptoms include irritability, depression, impulsivity, and risky behavior. Individuals may struggle with self-awareness and experience frequent thoughts of death or suicide. HD progressively worsens, significantly impacting daily life and independence.

7.2.1 Pathophysiology of HD

HD is characterized by neuronal loss in the striatum and cortex, also affecting the globus pallidus, thalamus, hypothalamus, subthalamic nucleus, substantia nigra, and cerebellum. It results from a huntingtin (HTT) gene mutation, causing abnormal polyglutamine (polyQ) expansion in the N-terminal region of the Htt protein. This expansion alters Htt function, making it toxic. Mutant Htt interacts abnormally with neuronal proteins, triggering various pathological events that contribute to neurodegeneration. HD arises from an abnormal increase in CAG (cytosine–adenine–guanine) repeats within the HTT gene located on chromosome 4. This expansion of trinucleotide repeats leads to the creation of a mutated form of the huntingtin protein, which causes progressive neurodegeneration that predominantly impacts the basal ganglia. The quantity of CAG repeats is a key factor in determining both the risk and severity of the disease. Individuals possessing 26 or fewer repeats are within the normal range and face no risk of developing HD. Those with 27–35 repeats do not exhibit symptoms but may transmit an expanded repeat to their offspring. When the repeat count falls between 36 and 39, the disease demonstrates reduced penetrance, indicating that some individuals may experience symptoms, while others remain symptom-free. If the count of CAG repeats is 40 or greater, the individual is guaranteed to develop HD, with higher repeat counts correlating to an earlier onset and a more severe progression. In instances where the repeats exceed 60, the condition manifests as juvenile HD, which occurs before the age of 20 and progresses swiftly, often accompanied by significant motor impairment, seizures, and cognitive decline. The instability of CAG repeats is a vital factor in the hereditary nature of the disease. Paternal inheritance is more prone to result in repeat expansion, as mistakes during sperm formation can lead to longer repeat sequences. This condition heightens the probability of earlier onset in subsequent generations, a phenomenon referred to as genetic anticipation. Generally, the greater the number of CAG repeats an individual possesses, the earlier the disease manifests and the more severe its progression. HD presents with a combination of motor, cognitive, and psychiatric symptoms due to shrinkage (atrophy) of the striatum and cerebral cortex caused by mutated huntingtin (mHTT). Patients with HD experience involuntary dance-like movements (chorea) and abnormal body postures, along with challenges in behavior, emotions, cognition, and personality.

Apart from mutations in HTT gene, various factors contribute to the development of HD, such as oxidative stress, inflammation, excitotoxicity, mitochondrial dysfunction, reduced energy supply, and inadequate neurotrophic support (decreased brain-derived neurotrophic factor, BDNF). Excitotoxicity is a key feature of many neurological disorders, including HD, caused by overstimulation of glutamate receptors, particularly N-methyl-D-aspartate (NMDA) subtypes. Preclinical studies suggest that NMDA receptor agonists can mimic HD pathology, indicating excitotoxicity's role in neuronal death. Dopamine and glutamate signaling pathways interact to enhance excitotoxicity, with elevated striatal dopamine levels increasing susceptibility to mHTT via D2 receptor mechanisms.

Mitochondrial dysfunction is a critical factor in HD, playing a significant role in neuronal degeneration and the progression of the disease. The presence of mHTT interferes with mitochondrial function, resulting in energy deficits, increased oxidative stress, and disrupted cellular processes. It impacts the mitochondrial respiratory chain complexes, leading to a decrease in adenosine triphosphate (ATP) production, thereby rendering neurons particularly susceptible to degeneration. Dysfunctional mitochondria produce an excess of reactive oxygen species (ROS), inflicting oxidative damage on proteins, lipids, and DNA, which hastens neuronal death. Furthermore, mHTT disturbs calcium homeostasis, resulting in mitochondrial overload, decreased ATP synthesis, and the triggering of apoptotic pathways. Abnormalities in mitochondrial fission and fusion further impede energy distribution, while transport defects compromise neuronal function and survival. Ultimately, these dysfunctions activate pro-apoptotic pathways, expediting neurodegeneration. Possible therapeutic strategies encompass antioxidants such as Coenzyme Q10 and creatine to mitigate oxidative stress, mitochondrial-targeted treatments to restore energy production, and calcium regulators to avert mitochondrial overload. Given that mitochondrial dysfunction plays a significant role in the pathology of HD, it continues to be a primary focus for prospective therapies.

Oxidative stress plays a key role in HD, contributing to neuronal damage and progression. It occurs when ROS production exceeds the body's ability to eliminate them. mHTT disrupts mitochondria, increasing ROS and reducing ATP synthesis, weakening cellular defenses. Lower antioxidant enzyme levels further heighten neuronal vulnerability. Excitotoxicity worsens oxidative damage in HD. NMDA receptor overactivation causes calcium influx, triggering ROS buildup, mitochondrial dysfunction, and neuronal death. ROS also damage membranes, proteins, and DNA, accelerating neurodegeneration. As oxidative stress drives HD pathology, it remains a key therapeutic target.

Inflammation plays a key role in HD, contributing to neurodegeneration and disease progression. mHTT triggers a persistent inflammatory response in both the central nervous system (CNS) and peripheral immune system, worsening neuronal damage. Microglial and astrocyte activation is a major factor in HD-related inflammation. mHTT leads to sustained microglial activity, causing the release of pro-inflammatory cytokines, such as tumor necrosis factor (TNF)-α, interleukin (IL)-1β, and IL-6, which increase neurotoxicity. Chronic neuroinflammation accelerates oxidative stress, mitochondrial dysfunction, and neuronal death, while prolonged cytokine release disrupts synaptic function, further advancing HD. Potential therapeutic approaches include anti-inflammatory drugs to reduce cytokine levels and microglial activation and astrocyte-targeted treatments to lower inflammation. Since inflammation is a key driver of HD, it remains a promising target for therapy.

Neurotrophins are essential for neuronal survival, growth, and synaptic plasticity, and their dysfunction contributes to HD progression. A key issue in HD is the reduction of BDNF, which is crucial for striatal neuron survival. mHTT disrupts BDNF production, transport, and signaling, leading to neuronal dysfunction and death. In healthy neurons, BDNF is transported from the cortex to the striatum, but in HD, this process is impaired, reducing neurotrophic support. Additionally, BDNF signaling through TrkB receptors is disrupted, weakening neuroprotection and increasing susceptibility to apoptosis. The loss of neurotrophic support contributes to motor, cognitive, and psychiatric symptoms in HD. Potential therapeutic strategies include enhancing BDNF production through gene therapy, exercise, or pharmacological agents; improving BDNF transport; and activating TrkB receptors to restore neurotrophic support. Since neurotrophin dysfunction plays a crucial role in HD, targeting BDNF deficits remains a promising treatment approach.

Apart from mHTT, HD is accompanied with various pathological events, including oxidative stress, inflammation, mitochondrial dysfunction, and decreased neurotrophins, which are the additional events that critically contribute to the HD progress. Therefore, targeting these pathways would yield beneficial outcomes in preserving further neuronal function in HD.

7.3 MULTIPLE SCLEROSIS

MS is a chronic neurological, autoimmune, inflammatory disorder that predominantly affects the myelinated axons of CNS, which is characterized by inflammation, demyelination of nerve fibers, and the formation of scar tissue. MS is distinguished from other neurodegenerative diseases by its characteristic of causing damage to myelin sheath, the protective insulation of neurons. MS is a common neurological disorder and a leading cause of non-traumatic impairment in young adults. It affects millions worldwide, with higher prevalence in women. Smoking increases MS risk, and while onset typically occurs between 20 and 40, it can develop at any age. Genetic and environmental factors contribute to its development, and relocating to high-risk areas in childhood increases susceptibility. MS is classified into four major types based on disease progression: relapsing-remitting, secondary-progressive MS, primary-progressive MS, and progressive-relapsing MS. MS symptoms include, but are not limited to, fatigue, difficulty walking, numbness or weakness in one or more limbs, problems with coordination and balance, visual acuity loss, and sphincter incontinence. Additionally, individuals with MS may experience cognitive changes such as memory problems or difficulty concentrating.

7.3.1 Pathophysiology of MS

MS is distinguished by the development of plaques within the CNS, accompanied by inflammation, loss of myelin, axonal injury, and neurodegeneration. These plaques manifest in both white and gray matter, affecting regions such as the brain, spinal cord, optic nerves, corpus callosum, and cerebellar peduncles. MS is classified as an autoimmune condition wherein autoreactive immune cells traverse the blood–brain barrier (BBB) to target the CNS. Typically, these cells are removed through central tolerance mechanisms within the thymus and bone marrow, but in MS, breakdowns in peripheral tolerance contribute to the onset of the disease. Dysregulation of regulatory T cells and the resilience of autoreactive T cells against suppression are pivotal mechanisms underlying this breakdown. A blend of genetic susceptibility and environmental factors is instrumental in activating these immune cells. $CD8^+$ T cells, along with $CD4^+$ Th1 and Th17 cells, are involved, releasing cytokines such as interferon-gamma, IL-17, and granulocyte-macrophage colony-stimulating factors that promote inflammation and demyelination.

B cells also play a significant role in the pathogenesis of MS. Increased levels of immunoglobulins in cerebrospinal fluid (CSF) point to their involvement, with oligoclonal bands (OCBs) acting as a diagnostic indicator. Memory B cells, which are clonally expanded in the CSF and brain tissue, produce excessive immunoglobulins that further stimulate inflammation. In individuals with MS, meningeal inflammation characterized by B cell infiltrates is associated with more severe cortical damage and neurodegeneration. The Epstein–Barr virus (EBV) infection has been associated with MS, as B cells serve as reservoirs for the virus, enhancing antigen presentation and immune activation. B cells infected with EBV cross-present myelin antigens, eliciting cytotoxic $CD8^+$ T-cell responses. However, as the disease advances, these T cells become exhausted and lose their ability to regulate EBV-infected cells, resulting in a self-perpetuating cycle of immune dysregulation and neurodegeneration. Microglia and macrophages are also involved by secreting pro-inflammatory cytokines such as TNF-α and IL-1β, which contribute to the injury of oligodendrocytes and neurodegeneration. They produce glutamate, which may lead to excitotoxicity. In addition, reactive oxygen and nitrogen species generated by these cells worsen oxidative stress and mitochondrial dysfunction. Despite their detrimental effects, microglia have the potential to adopt an anti-inflammatory phenotype that facilitates remyelination.

As MS advances, chronic neuroinflammation, oxidative stress, and mitochondrial dysfunction lead to increased neuronal damage and loss. The deterioration of axons results in irreversible disability, especially in the progressive variants of the disease. While the precise cause is still unclear, genetic factors and environmental influences such as viral infections, vitamin D deficiency, and smoking are thought to play a role in the onset and progression of the disease. Current therapies aim to modulate the immune response, minimize inflammation, and safeguard nerve function to decelerate disease progression and reduce symptoms. The degree and site of demyelination influence the severity of symptoms, which may encompass motor weakness, sensory abnormalities, cognitive impairment, and visual issues. Cumulatively, inflammation and immune cells together demyelinate the neurons, thereby causing MS. Hence, targeting these would be promising to combat the MS and its associated pathology.

7.4 AMYOTROPHIC LATERAL SCLEROSIS (ALS)

ALS, also referred to as Lou Gehrig's disease, constitutes a progressive neurodegenerative condition that impacts neuronal cells within the CNS, specifically in the cerebral cortex and spinal cord. The disorder predominantly affects motor neurons, resulting in muscular weakness, diminished motor coordination, and eventually paralysis. As the disease advances, it severely impairs verbal communication, ingestion, locomotion, and respiration, depending on the person. Initial ALS patients exhibit muscular weakness, particularly localized in the hands, upper limbs, lower limbs, or feet. As ALS progresses, muscular weakness and atrophy become increasingly evident. Patients may struggle with ambulation, endure recurrent falls, experience a loss of coordination and equilibrium, or exhibit episodes of uncontrollable laughter or weeping, clinically referred to as pseudobulbar affect. In the advanced phases of ALS, individuals lose nearly all voluntary muscular control, rendering them incapable of ambulation, verbal expression, or deglutition. Respiratory difficulties intensify, often necessitating mechanical ventilation assistance. The precise etiology of ALS remains elusive; however, several risk factors have been elucidated. Genetic predispositions contribute to 5–10% of cases, indicating a hereditary component to the disease. Age further serves as a significant factor, with ALS being most frequently diagnosed in individuals aged between 40 and 70 years. Environmental influences, including exposure to neurotoxins, traumatic brain injuries, or rigorous physical exertion, may also play a role in the pathogenesis of this disorder.

7.4.1 Pathophysiology of ALS

In ALS, both upper motor neurons (UMNs) located in the motor cortex and lower motor neurons (LMNs) situated in the brainstem and spinal cord are degenerated. The impairment of UMNs manifests as spasticity and hyperreflexia, whereas the degeneration of LMNs results in muscle weakness, atrophy, and fasciculations. A principal pathological mechanism underlying ALS is excitotoxicity, which arises from an excessive accumulation of glutamate. Typically, the excitatory amino acid transporter 2 (EAAT2) is responsible for the clearance of surplus glutamate from synapses. However, in the context of ALS, diminished function of EAAT2 leads to glutamate overstimulation, ultimately culminating in the demise of motor neurons.

Oxidative stress and mitochondrial dysfunction are also critical factors contributing to the progression of ALS. Motor neurons affected by ALS exhibit heightened susceptibility to oxidative damage due to an imbalance between free radicals and the organism's antioxidant defenses. Genetic mutations in the SOD1 gene, which encodes the antioxidant enzyme superoxide dismutase 1, result in amplified oxidative stress, mitochondrial dysfunction, and eventual cellular apoptosis. Another notable characteristic of ALS is the aberrant accumulation of misfolded proteins within motor neurons. The protein TDP-43 (TAR DNA-binding protein 43) frequently exhibits misfolding and accumulates in neurons affected by ALS, leading to perturbations in RNA processing and cellular functionality. Similarly, mutations in FUS and SOD1 facilitate protein aggregation, which is detrimental to neuronal health. Impaired axonal transport represents an additional contributing factor to the pathophysiology of ALS. Neurons depend on axonal transport to facilitate the movement of essential nutrients, organelles, and signaling molecules between the cell body and synaptic terminals. In ALS, defects in microtubule-associated proteins and motor proteins, such as dynein and kinesin, disrupt this transport system, resulting in neuronal dysfunction and subsequent cell death.

Neuroinflammation is also a significant contributor to the progression of ALS. Microglia, the resident immune cells of the CNS, become activated in ALS and secrete pro-inflammatory cytokines, including TNF-α, IL-1β, and IL-6. This results in a chronic inflammatory milieu that exacerbates the degeneration of motor neurons. Furthermore, astrocytes, which typically provide support to motor neurons, become dysfunctional in ALS and fail to deliver the requisite metabolic and structural assistance. Another critical pathological mechanism implicated in ALS is RNA dysregulation. Mutations in RNA-binding proteins such as TDP-43 and FUS disrupt RNA metabolism, which adversely affects the proper splicing, translation, and transport of mRNA. This, in turn, impedes the synthesis of essential proteins necessary for the survival of motor neurons.

Consequently, as a result of these pathological processes, programmed cell death, or apoptosis, is initiated within motor neurons. The interplay of excitotoxicity, oxidative stress, protein misfolding, and inflammation collectively activates apoptotic pathways, culminating in progressive neuronal attrition. The continuous loss of motor neurons leads to a gradual weakening of muscles, diminished mobility, and ultimately, respiratory failure. Despite ongoing research efforts, no definitive cure for ALS currently exists; however, elucidating these mechanisms aids in the formulation of targeted therapeutic strategies aimed at decelerating disease progression.

7.5 MULTIPLE SYSTEM ATROPHY (MSA)

MSA is a progressive neurodegenerative disorder with autonomic dysfunction, parkinsonism, and cerebellar ataxia. MSA is classified into two main subtypes based on predominant features, including Parkinsonian subtype (MSA-P) and cerebellar subtype (MSA-C). MSA-P presents core symptoms like PD, including bradykinesia, rigidity, and tremors, and exhibits urinary disturbances, irregular sweating, and digestive challenges. On the other hand, MSA-C is primarily characterized by ataxia, leading to impairments in balance and coordination. Individuals with MSA-C may also struggle with swallowing, speech difficulties, and irregular eye movements. Common symptoms across both subtypes include hand or limb stiffness, Pisa syndrome, and anterocollis. Additionally, psychological symptoms such as anxiety and depression are frequently observed in MSA patients. Rapid eye movement (REM) sleep disturbances are another notable symptom associated with MSA. Both MSA-P and MSA-C significantly impact daily life, requiring comprehensive management strategies.

MSA represents a progressive neurodegenerative condition that significantly impacts the autonomic nervous system, motor functions, and coordination. The precise etiology of MSA remains elusive; however, its pathophysiological characteristics are marked by extensive neurodegeneration and abnormal protein deposition. A salient characteristic of MSA is the deposition of the protein α-synuclein within oligodendrocytes, which are the supportive cells in the CNS tasked with the maintenance of myelin. In contrast to PD, where α-synuclein aggregates within neurons, in MSA, this protein accumulates to form glial cytoplasmic inclusions (GCIs) within oligodendrocytes. The emergence of GCIs precipitates oligodendrocyte dysfunction, thereby contributing to demyelination and pervasive neurodegeneration. MSA encompasses multiple regions of the brain, including the striatonigral pathway, olivopontocerebellar system, and autonomic nervous system. Impairment of the striatonigral pathway is associated with the manifestation of Parkinson-like symptoms such as muscular rigidity and bradykinesia, whereas degeneration of the olivopontocerebellar system culminates in cerebellar ataxia, which induces difficulties in coordination and balance. Furthermore, dysfunction within the autonomic nervous system gives rise to symptoms including orthostatic hypotension, urinary incontinence, and various autonomic dysfunctions.

Neuroinflammation also significantly contributes to the progression of MSA. Activation of microglia and astrocytosis are frequently documented in the regions of the brain affected, resulting in enhanced oxidative stress and neuronal injury. Mitochondrial dysfunction further exacerbates the condition by hindering energy production and augmenting cellular stress. Although MSA is predominantly regarded as a sporadic disorder, certain genetic mutations, such as variants in COQ2, have been recognized in infrequent familial instances. Nonetheless, the environmental factors that may play a role in the pathogenesis of the disease remain ambiguous. Despite the continuous pursuit of research, there is presently no curative treatment for MSA, and the therapeutic approach is centered on the alleviation of symptoms and enhancement of the patient's quality of life.

Overall, these pathological processes primarily affect the basal ganglia, cerebellum, brainstem, and autonomic nervous system, leading to the motor and autonomic dysfunction characteristic of MSA. The combined effects of alpha-synuclein toxicity, inflammation, oxidative stress, and impaired protein degradation contribute to the rapid progression of neurodegeneration in MSA.

7.6 PROGRESSIVE SUPRANUCLEAR PALSY (PSP)

PSP is a rare, progressive, and ultimately fatal neurodegenerative brain disease resulting from the deposition of misfolded and neurotoxic forms of tau protein in specific areas of the midbrain, basal ganglia, and cortex. It is one of the most representative forms of tauopathy (DeRosier, 2024). The prevalence of PSP has been studied over the years, and their cases are increasing day by day. Life expectancy after disease onset is approximately seven to eight years, and PSP is currently incurable (Nysetvold et al., 2024). PSP affects a person's movements and can lead to loss of balance, difficulty walking or swallowing, slurred speech, problems with eye movements and exhibits alterations in mood, behavior, and thinking.

7.6.1 Pathophysiology of PSP

PSP is a neurodegenerative disorder primarily characterized by the accumulation of tau protein in the brain. The key mechanism of neurodegeneration in PSP involves the aggregation of tau protein into neurofibrillary tangles, which disrupt neuronal function. These tangles primarily affect areas of the brain responsible for motor control, particularly the brainstem and basal ganglia, leading to progressive motor symptoms, including balance problems, difficulty with eye movements, and muscle rigidity. The brains of patients with PSP have tufted astrocytes and other neuropathological findings, including globose neurofibrillary tangles in the gray matter and coiled tau bodies in oligodendrocytes in the white matter. Further neuronal findings in PSP include tau threads and pretangles. The area of tau deposition depends on the specific phenotype of PSP, but the most common brain areas affected are the brainstem, globus pallidus, subthalamic nucleus, substantia nigra, and dentate nucleus of the cerebellum. Tau sometimes deposits in the dentate nucleus and frontal, temporal, and parietal cortices. The exact pattern of tau deposition varies with the disease phenotype. For instance, the total tau load is generally less in the cortex in PSP parkinsonism (PSP-P) and PSP-progressive gait freezing, while it is more prominent in the PSP speech language subtype. Significant differences in the tau burden and cytopathology are present among the various clinical phenotypes of PSP. Tau accumulation usually begins in the subcortical regions and proceeds to the neocortical regions and the cerebellum (Dunning et al., 2024a). Tau aggregates impair axonal transport and mitochondrial function and disrupt synaptic activity, leading to neuronal death. Additionally, the dysfunction of the tau protein may cause abnormal interactions with microtubules, which play a critical role in maintaining cell structure and intracellular transport. The damage to neurons in PSP also activates neuroinflammatory responses, further contributing to neurodegeneration. As the disease progresses, these tau aggregates spread to other regions of the brain, including the cortex, affecting cognitive and behavioral functions. This widespread accumulation of tau, coupled with neuroinflammation and neuronal death, ultimately leads to the motor and cognitive impairments seen in PSP (Ichikawa-Escamilla et al., 2024).

7.7 FRIEDREICH ATAXIA (FA)

FA is an uncommon, inherited neurodegenerative disorder that mainly affects the cooperation (ataxia) in the arms and legs, and it mainly impacts the spinal cord and causes progressive damage to the peripheral nervous system. Patient with FA develops motor weakness and sensory loss (Keita et al., 2022). FA is the most common hereditary ataxia; furthermore, patients who are younger than 25 years make up about 75% of the ataxia cases, with 1 in 50,000 people suffering ataxia in western European states. Globally the prevalence of FA is around 1 in 40,000, frequently observed in Europe, the Middle East, South Asia, and North America. FA causes ataxia, muscle weakness, slurred speech, and foot abnormalities like high arches, hammer toes, scoliosis, sensory impairments, hearing and vision problems, and an increased risk of diabetes (Pandolfo, 2008). Severe complications involve cardiac issues such as hypertrophic cardiomyopathy, arrhythmias, and congestive heart failure (Williams & De Jesus, 2024).

7.7.1 Pathophysiology of FA

FA is caused by mutations in the *FXN* gene, resulting in reduced expression or absence of the protein frataxin, which is crucial for mitochondrial function. This deficiency disrupts iron metabolism, leading to excessive iron accumulation in the mitochondria. The buildup of iron causes oxidative stress, generating ROS that damage proteins, lipids, and DNA, particularly in neurons and glial cells. This oxidative stress is a key factor driving neurodegeneration in FA. Mitochondrial dysfunction impairs energy production, contributing to symptoms such as muscle weakness and ataxia, and compromises the ability of cells to repair damage. As neurons progressively degenerate, particularly in the spinal cord, cerebellum, and peripheral nervous system, patients experience motor impairments like loss of coordination and balance. FA also affects the heart, particularly the left ventricle, causing hypertrophic cardiomyopathy and other cardiac issues. The combination of oxidative stress and iron accumulation further exacerbates neuronal damage, leading to the progressive decline in function seen in FA.

7.8 DEMENTIA OF LEWY BODIES

LBD is a progressive neurodegenerative disorder caused by the abnormal accumulation of alpha-synuclein in the brain, affecting regions related to motor, cognitive, and autonomic functioning. It includes disorders like PD, PD dementia (PDD), and dementia with Lewy bodies (DLB), all classified as synucleinopathies. Symptoms overlap with PD, including muscle tremors, rigidity, and bradykinesia, as well as visual hallucinations, fluctuating cognitive function, and REM sleep behavior disorder. Diagnosing and distinguishing these diseases can be challenging (Armstrong, 2019; Jellinger, 2003; Outeiro, Koss et al., 2019).

7.8.1 Pathophysiology

DLB is characterized by the abnormal accumulation of alpha-synuclein protein in the neurons of the cerebral cortex and subcortical regions, forming Lewy bodies. These protein aggregates disrupt normal brain function, particularly in areas responsible for cognition, memory, motor control, and autonomic regulation. The accumulation of Lewy bodies impairs the dopaminergic system, leading to motor symptoms similar to PD. The precise mechanism of neurodegeneration in DLB is not fully understood, but it is known to be conducted majorly due to the neurotoxic effects of alpha-synuclein aggregation. Apart from the α-synuclein accumulation, DLB is also accompanied with neuroinflammation caused by microglial activation and oxidative stress; mitochondrial dysfunction further accelerates neurodegeneration. The symptoms and pathophysiologies of these neurodegenerative diseases are depicted in Table 7.1.

TABLE 7.1
Symptoms, Pathophysiology, and Genetic Causes of Neurodegenerative Diseases

S. No.	Disease	Genetic Cause/Mutation	Symptoms	Pathophysiology	Mechanism of Neurodegeneration	Key Proteins/Biomarkers
1	HD	CAG repeat expansion in HTT gene	Chorea, cognitive decline, and personality changes	Neuronal loss in striatum and cortex	mHTT protein aggregation, transcriptional dysregulation, mitochondrial dysfunction, and excitotoxicity	Mutant huntingtin (mHTT) and BDNF
2	ALS	Mutations in SOD1, TDP-43, and C9ORF72	Muscle weakness and speech/breathing difficulties	Motor neuron degeneration in brain and spinal cord	Oxidative stress, protein aggregation, excitotoxicity, mitochondrial and DNA damage	TDP-43, SOD1, and C9ORF72
3	MSA	Sporadic	Parkinsonism, cerebellar ataxia, and autonomic failure	Alpha-synuclein inclusions in oligodendrocytes	α-Synuclein toxicity, mitochondrial dysfunction, neuroinflammation, and impaired clearance	α-Synuclein and P25α
4	PSP	Tauopathy involving abnormal tau protein	Gait instability, vertical gaze palsy, and cognitive issues	Tau deposition in midbrain, basal ganglia, cortex	Tau aggregates, impaired axonal transport, mitochondrial dysfunction, and inflammation	Tau and tufted astrocytes
5	FA	GAA repeat in FXN gene	Ataxia, muscle weakness, scoliosis, and cardiac issues	Iron accumulation in mitochondria	Oxidative stress, mitochondrial dysfunction, and energy metabolism disruption	Frataxin, ROS, and iron
6	DLB	Alpha-synuclein aggregation	Hallucinations, fluctuating cognition, and Parkinsonism	Lewy body deposits in cortex and subcortical regions	α-Synuclein toxicity, impaired autophagy, mitochondrial dysfunction, and inflammation	α-Synuclein, neurofilament light chain
7	SMA	Mutation in SMN1 gene	Muscle weakness and respiratory difficulty	Loss of motor neurons in spinal cord	SMN deficiency, impaired RNA processing, axonal transport, and mitochondrial dysfunction	SMN protein
8	MS	Autoimmune, HLA-DRB1, EBV, low vitamin D	Fatigue, numbness, coordination, and vision issues	Immune attack on CNS myelin	T-/B-cell inflammation, demyelination, oxidative stress, and excitotoxicity	Myelin basic protein, oligoclonal bands, and neurofilament

7.9 SPINAL MUSCULAR ATROPHY

SMA is a genetic disorder that is caused by the mutation of survival motor neuron (SMN1) gene on chromosome 5 that effects the CNS and PNS, thereby causing dysfunction of involuntary muscle movements (skeletal muscles). The SMA shows primary action on muscles, which do not receive signals from the nerve cells. SMN is produced by the SMN1 and SMN2 genes, which are essential for motor activity (Boyer et al., 2014). Clinically, the SMA is classified into four types, which often do not exhibit cognitional defects (Zerres & Rudnik-Schöneborn, 1995). Globally people effected by SMA are in the ratio of 1:7000–11,000 live births (Vill et al., 2019).

7.9.1 Pathophysiology of SMA

SMA is an autosomal neuromuscular disorder characterized by the degeneration of motor neurons in the spinal cord that leads to progressive muscle weakness and respiratory difficulties in the body **(TL, 1992)**. It is mainly caused by the mutations in the SMN1 located on chromosome 5. It is responsible for the production of SMN protein, which is essential for the survival of motor neurons and functions **(Clermont et al., 2004)**. The deficiency of the SMN protein, caused by mutations in the SMN1 gene, leads to neurodegeneration. SMN protein is essential for RNA splicing, axonal transport, and mitochondrial function. Its deficiency disrupts RNA processing, impairing protein synthesis and cellular function. Additionally, axonal transport is compromised, causing energy depletion and neurodegeneration. Mitochondrial dysfunction and increased oxidative stress further contribute to cell death, while neuroinflammation accelerates neuronal damage. These combined effects result in the progressive loss of motor neurons, leading to muscle weakness and atrophy. The pathophysiology of the above-discussed neurodegenerative diseases is depicted in Figure 7.1.

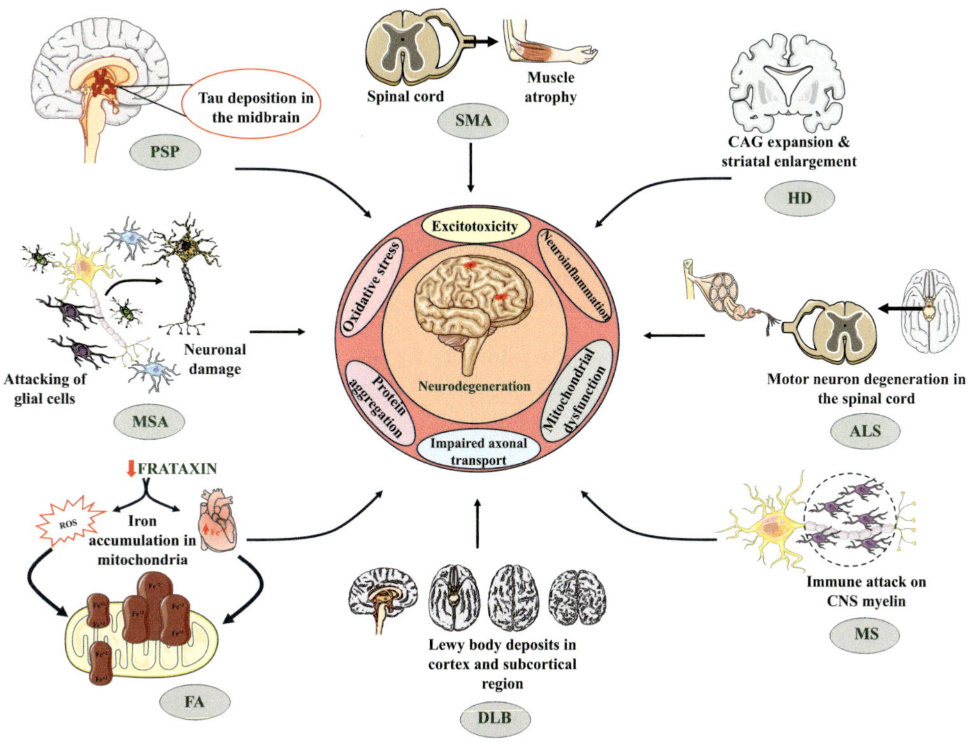

FIGURE 7.1 Pathophysiology of neurodegenerative diseases.

7.10 CONCLUSION

Pathophysiological mechanisms for each neurodegenerative disease are unique, but common mechanisms for neurodegenerative disease include protein misfolding and aggregation, oxidative stress, mitochondrial dysfunction, and inflammation in the brain. These mechanisms for neurodegenerative disease ultimately led to progressive neuronal degeneration and a subsequent increase in the disease symptoms. Indeed, the exact pathophysiology of each of the above-discussed neurodegenerative diseases is not known. Therefore, future studies must investigate the in-depth understanding of this disease pathophysiology.

ACKNOWLEDGMENTS

Figure 7.1 is made partially using Servier Medical Art (http://smart.servier.com/), licensed under a Creative Commons Attribution 4.0 Generic License.

REFERENCES

Armstrong, M. J. (2019). Lewy body dementias. *Continuum (Minneap Minn)*, 25(1), 128–146.

Boyer, J. G., Deguise, M.-O., Murray, L. M., Yazdani, A., De Repentigny, Y., Boudreau-Larivière, C., & Kothary, R. (2014). Myogenic program dysregulation is contributory to disease pathogenesis in spinal muscular atrophy. *Human Molecular Genetics*, *23*(16), 4249–4259.

Clermont, O., Burlet, P., Benit, P., Chanterau, D., Saugier-Veber, P., Munnich, A., & Cusin, V. (2004). Molecular analysis of SMA patients without homozygous SMN1 deletions using a new strategy for identification of SMN1 subtle mutations. *Human Mutation*, *24*(5), 417–427.

DeRosier, F., Hibbs, C., Alessi, K., Padda, I., Rodriguez, J., Pradeep, S., & Parmar, M. S. (2024). Progressive supranuclear palsy: Neuropathology, clinical presentation, diagnostic challenges, management, and emerging therapies. *Disease-a-month : DM*, 70(8), 101753. https://doi.org/10.1016/j.disamonth.2024.101753

Ichikawa-Escamilla, E., Velasco-Martínez, R. A., & Adalid-Peralta, L. (2024). Progressive supranuclear palsy syndrome: An overview. *IBRO Neuroscience Reports*, *16*, 598–608. https://doi.org/10.1016/j.ibneur.2024.04.008

Jellinger, K. A. (2003). Neuropathological spectrum of synucleinopathies. *Movement Disorders*. *18*(Suppl 6), S2–S12.

Keita, M., McIntyre, K., Rodden, L. N., Schadt, K., & Lynch, D. R. (2022). Friedreich ataxia: clinical features and new developments. *Neurodegenerative Disease Management*, *12*(5), 267–283. https://doi.org/10.2217/nmt-2022-0011

Nysetvold, E., Lopez, L. N., Cogell, A. N., Fryk, H., Pace, N. D., Taylor, S. S., Rhoden, J., Nichols, C. A., Pillas, D., Klein, A., Gasalla, T., & Scowcroft, A. (2024). Progressive Supranuclear palsy (PSP) disease progression, management, and healthcare resource utilization: a retrospective observational study in the US and Canada. *Orphanet Journal of Rare Diseases*, *19*(1), 215. https://doi.org/10.1186/s13023-024-03168-z

Outeiro, T. F., Koss, D. J., Erskine, D., Walker, L., Kurzawa-Akanbi, M., Burn, D., Donaghy, P., Morris, C., Taylor, J.-P., Thomas, A., Attems, J., & McKeith, I. (2019). Dementia with Lewy bodies: An update and outlook. *Molecular Neurodegeneration*, 14, Article 5.

Pandolfo, M. (2008). Friedreich ataxia. *Archives of Neurology*, *65*(10), 1296–1303. https://doi.org/10.1001/archneur.65.10.1296

TL, M. (1992). International SMA consortium meeting (26–28 June 1992, Bonn, Germany). *Neuromuscul Disord*, *2*, 423–428.

Vill, K., Kölbel, H., Schwartz, O., Blaschek, A., Olgemöller, B., Harms, E., Burggraf, S., Röschinger, W., Durner, J., & Gläser, D. (2019). One year of newborn screening for SMA–results of a German pilot project. *Journal of Neuromuscular Diseases*, *6*(4), 503–515.

Williams, C. T., & De Jesus, O. (2024). Friedreich ataxia. In *StatPearls*. StatPearls Publishing. https://www.ncbi.nlm.nih.gov/books/NBK563199/

Zerres, K., & Rudnik-Schöneborn, S. (1995). Natural history in proximal spinal muscular atrophy: clinical analysis of 445 patients and suggestions for a modification of existing classifications. *Archives of Neurology*, *52*(5), 518–523.

8 Pharmacological, Non-Pharmacological, and Emerging Therapies for Other Neurodegenerative Diseases

*Srikanth Yadava, Harikrishna Reddy Dontiboina, Lalitha Bhaskarini Dyvapu, Manikanta Murahari, and Kakarla Ramakrishna**
KL College of Pharmacy, Koneru Lakshmaiah Education Foundation, Vaddeswaram, Guntur, Andhra Pradesh, India

Tan Hui Yin
Faculty of Applied Sciences, Tunku Abdul Rahman University of Management and Technology, Jalan Genting Kelang, Setapak, Kuala Lumpur, Malaysia

Yvonne Liew Jing Mei
Department of Pharmaceutical Life Sciences, Faculty of Pharmacy, Universiti Malaya, Kuala Lumpur, Malaysia

8.1 INTRODUCTION

Neurodegenerative diseases such as Alzheimer's disease (AD), Parkinson's disease (PD), Huntington's disease (HD), and amyotrophic lateral sclerosis (ALS) prevalence rates are increasing day by day. In the majority of the instances, these diseases impair the quality of life of the patients. Indeed, these diseases are accompanied by neurological impairments such as motor deficits, cognitive and memory dysfunction, and psychological disturbances. These neurodegenerative diseases are caused by progressive loss of neuronal cells in a specific location of the brain [1], resulting in neurological dysfunction and eventually leading to fatal consequences. For several decades, there has been continuous growth in understanding the pathophysiology of these neurodegenerative diseases. However, there is a lacuna in knowing the exact pathophysiology of these diseases. Currently, there is no treatment available in the clinics to combat these diseases, except for symptomatic relief [2]. Amidst these therapeutic limitations, ongoing research is increasingly directed toward the use of natural compounds [3] and the discovery of novel molecular targets, with the goal of enhancing brain health and developing effective neuroprotective strategies [4]. In the present work, we have comprehensively reviewed the pharmacological and non-pharmacological treatments available for various neurological diseases along with recent advancements.

*Corresponding Author. kakarlaramakrishna@kluniversity.in

DOI: 10.1201/9781003530510-8

8.2 HUNTINGTON'S DISEASE

HD is a genetic disease where neurons in the brain are progressively damaged and die, leading to severe motor, cognitive, and psychological disturbances. HD is attributable to a mutation in the gene encoding a protein known as huntingtin. This genetic aberration results in the excessive repetition of the DNA building blocks cytosine, adenine, and guanine (CAG), far exceeding their typical occurrence. The majority of individuals possess fewer than 27 CAG repeats within their HD gene, thereby rendering them not susceptible to the disorder. Individuals exhibiting CAG repeats within the intermediate range (27–35) are generally unlikely to manifest the disease; however, they retain the potential to transmit it to subsequent generations. Those afflicted with HD typically exhibit 36 or more CAG repeats. Typically, HD patients exhibit involuntary, dance-like movements (chorea) and atypical bodily postures, in addition to experiencing difficulties with behavior, emotional regulation, cognitive processes, and personality traits. As of now, in the clinics, various drugs are used to control these symptoms. HD patients exhibit numerous mental and psychological issues, including social withdrawal, sleep deprivation, fatigue, suicidal thoughts, sadness, apathy, irritability, mania, bipolar disorder, and obsessive–compulsive disorder. Antipsychotic drugs are commonly recommended for this illness in HD patients.

8.2.1 Pharmacological Management

There is no definitive treatment for HD. However, certain medications have been symptomatically reduced and are approved by the FDA, primarily aimed at chorea. It is imperative to underscore the pathophysiology of HD, and the majority of these agents are employed principally to alleviate the symptomatic manifestations of HD. These include psychiatric medications, sedatives, cognitive enhancers, and various neuroprotective agents.

HD patients manifest a diverse motor impairment, including involuntary jerking or writhing movements (chorea), muscle rigidity or contractures, diminished or atypical ocular movements, difficulties in ambulation or maintaining postural stability and balance, challenges in articulation or deglutition, as well as the inability to regulate voluntary motor functions. Tetrabenazine (TBZ) is the only medication approved for the treatment of chorea in HD. It selectively reduces central monoamines by binding to the Type 2 vesicular monoamine transporter (VMAT-2). Since VMAT-2 contains and transports serotonin, dopamine, and norepinephrine, its inhibition leads to the early degradation of these monoamines. The reduction of dopamine improves chorea, while depletion of serotonin and norepinephrine may result in worsening depression and anxiety, and increase the risk of suicide [5]. Despite its beneficial effects, it also possesses several adverse effects, including drowsiness and bradykinesia. Deutetrabenazine was approved for the treatment of chorea. It is a deuterated TBZ with a longer half-life (two daily dosing) and better tolerability than TBZ. Valbenazine, a prodrug of an isomer of TBZ, with a longer half-life allowing once-a-day dosing, previously approved for tardive dyskinesia, is under FDA submission for chorea [6].

Antipsychotic drugs are used to treat schizophrenia. These drugs are typical and atypical. Antipsychotic drugs majorly block the D_2 receptors and serotonin receptors. Atypical drugs such as risperidone, olanzapine, quetiapine, clozapine, and ziprasidone are considered to be a second generation of antipsychotics, which are commonly recommended to HD patients. These have a high affinity for D_2 and 5-HT_2 receptors and have a low risk of adverse effects when compared to typical antipsychotics. For example, clozapine mitigates psychotic symptoms, and its combination with risperidone controls both psychiatric and severe choreoathetotic movements [7]. Typical antipsychotics include haloperidol, phenothiazine, pimozide, sulpiride, amisulpiride, tiapride, and aripiprazole, which are also recommended when needed. Indeed, by using antipsychotics, several adverse events including extrapyramidal symptoms such as tremors, rigidity, bradykinesia and restlessness, weight gain, increased appetite, diabetes mellitus, and dyslipidemia are experienced. Therefore, while using these drugs, greater care must be given to avoid the adverse effects. Mirtazapine is recommended to manage the depression in HD patients.

HD patients often exhibit cognitive deficits, including a lack of organization and awareness and memory impairment. Rivastigmine and memantine are generally recommended to mitigate memory impairment. SAGE-817, an NMDA receptor regulator, is currently under investigation to treat the cognitive dysfunction associated with HD [8].

8.2.2 Emerging Therapies and Clinical Trials in HD

Gene therapy is the process of delivering genes to cells that previously did not express a particular protein. It is mainly used to treat/prevent the disease by modifying genetic material within a person's cell to compensate for abnormal genes and make beneficial proteins. Most of the gene therapies employ viral vectors to express genes of interest. Gene therapy mainly aims to modulate immune responses, modulate gene expressions, and correct genetic defects. In HD, one of the most important strategies is to ameliorate the disease by modulating the mutated huntingtin gene (mHtt) level through gene transfer [9]. There are currently active programs using antisense oligonucleotides (ASOs), RNA interference, small molecule splicing modulators, and zinc-finger protein transcription factors. ASOs reduce mutant HTT protein levels. Except for ASOs and RNA interference approaches, the remaining therapeutic strategies are at a preclinical stage of development. ASO therapies, which bind to pre-RNA or mRNA and can alter protein expression via RNA degradation, blocking translation, or splice modulation [10].

Several therapeutic strategies for HD are in various stages of development. Tominersen (Hoffmann-La Roche) is an allele non-specific therapy previously in Phase 3 trials, though its current status is unclear. WVE-120102 and WVE-120101 (Wave Life Sciences) are allele-specific therapies targeting single-nucleotide polymorphisms (SNPs) rs362307 and rs362331, in Phase 1b/2a trials, but updates on their progress are limited. AMT-130 (UniQure) is an allele non-specific gene therapy in Phase 1b/2a trials, delivered via an adeno-associated virus serotype 5 (AAV5) vector. Recent data show significant reductions in neurofilament light chain (NfL) and slower disease progression. PTC518 (PTC Therapeutics), an oral allele non-specific therapy, has shown up to a 43% reduction in mHTT levels. Branaplam (Novartis), another oral allelic non-specific therapy, was discontinued due to safety concerns. AK-686 and TAK-686 (Takeda/Sangamo) are allele-specific therapies targeting expanded CAG repeats. Among these, AMT-130 and PTC518 emerged as the most promising candidates. Others include AAV.shHD2.1a (Spark), VY-HTT01 (Voyager), and ZF-KOX1 (European Research Council/Imperial College London/Fingers4Cure), which currently under preclinical studies [10].

8.3 AMYOTROPHIC LATERAL SCLEROSIS (ALS)

ALS is a progressive neurodegenerative disease causing motor neuron loss in the brain and spinal cord, leading to impaired voluntary muscle control. As ALS progresses, muscle fasciculations and atrophy result in impaired walking, chewing, swallowing, and talking, and difficulties in breathing. While no cure exists, certain treatments can slow progression, enhance quality of life, and extend survival.

8.3.1 Pharmacological Management

Disease-modifying therapies like riluzole have been the only drug proven to extend survival in ALS, reducing motor neuron loss and increasing one-year survival by 9%. It decreases excitotoxicity by inhibiting glutamate release, enhancing GABA activity, and preventing TDP-43 mislocalization [11]. Edaravone, approved in 2017, is an antioxidant that slows ALS progression by scavenging free radicals and protecting motor neurons from oxidative stress. Edaravone's reported adverse effects include contusion, headache, and rare severe reactions like anaphylaxis [12]. Baclofen, tizanidine, and benzodiazepines can be used to alleviate spasticity in patients with ALS. Masitinib reduces neuroinflammation by inhibiting proinflammatory cytokines, mast cells, macrophages, and microglia, slowing motor neuron degeneration and prolonging survival [13].

Relyvrio (sodium phenylbutyrate/taurursodiol) was designed to reduce neuronal cell death and oxidative injury by diminishing endoplasmic reticulum (ER) stress and mitochondrial dysfunction, slowing ALS progression. Initially, it was approved in 2022, and later withdrawn from the market in 2024 after larger studies failed to confirm its effectiveness [14]. Tofersen, a SOD1 ASO, reduces SOD1 protein synthesis and plasma NfL levels, indicating neurodegeneration. It is administered intrathecally in a structured dosing regimen [15]. Lithium carbonate has been investigated for its potential benefits in ALS patients with mutations in UNC13A. Though the precise mechanism remains unclear, lithium is believed to promote neuronal sprouting, enhance synaptogenesis, regulate intracellular calcium homeostasis, and stimulate autophagy, which may mitigate neurodegeneration [16].

8.3.2 Emerging Therapies and Clinical Trials in ALS

Gene therapy silences harmful genes by replacing mutated genes with regular copies or editing the mutant genome. It also adds genes that introduce protective factors or knockout effects to decrease the expression of causative genes. Cellular therapy for ALS uses various human and nonhuman cells to address multiple disease mechanisms, including replacing damaged motor neurons, modulating the immune system, and reducing neuroinflammation. Key cell types include mesenchymal stem cells (MSCs), neural stem cells (NSCs), and induced pluripotent stem cells. These cells can be delivered intrathecally, intravenously, or intramuscularly, aiming to repair neuromuscular damage and improve ALS symptoms.

MicroRNAs (miRNAs) regulate gene expression by forming the RNA-induced silencing complex (RISC) with Argonaute 2, binding to the 3′-UTR of target mRNA to suppress translation or trigger degradation. Histone phosphorylation on serine, threonine, and tyrosine residues regulates gene expression, cell cycle, and DNA repair, with enzyme dysregulation linked to ALS and other neurodegenerative diseases. It raised the proportion of B cells and monocytes while lowering IL-18 protein and mRNA relative expression of the pro-inflammatory cytokine IL-18. Another emerging advancement is stem cell therapy, which aims to repair nerve tissue, protect neurons, and reduce inflammation using NSCs and MSCs, delivered intravenously or intraspinally. Robotic and assistive technologies enhance mobility and independence. Exoskeletons aid gait via EEG/EMG control, while electric wheelchairs, adjustable beds, and brain–computer interfaces (BCIs) improve movement and communication [17].

NurOwn (MSC-NTF cell therapy) improved cerebrospinal fluid (CSF) biomarkers but lacked primary efficacy in Phase III. Tofersen (ASO for SOD1) reduced CSF SOD1 levels in Phase I/II, with positive secondary outcomes in Phase III. AMX0035 (sodium phenylbutyrate + taurursodiol) slowed decline and increased survival in Phase II, earning approval in the U.S. and Canada. Aldesleukin (IL-2) boosted Treg cells and reduced inflammation in Phase II; Phase II/III is ongoing. Ravulizumab (anti-C5 complement inhibitor) failed in Phase III, and Verdiperstat (MPO inhibitor) showed no significant efficacy in Phase II/III [18, 19].

8.4 MULTIPLE SYSTEM ATROPHY

Multiple system atrophy (MSA) is a rare neurodegenerative disorder affecting the central and autonomic nervous systems, causing motor impairment, autonomic dysfunction, and cerebellar ataxia. It results from the accumulation of misfolded α-synuclein in oligodendrocytes, causing widespread neurodegeneration in the basal ganglia, cerebellum, brainstem, and autonomic nervous system. Symptoms include blood pressure instability, urinary and breathing issues, poor motor control, and coordination loss. Unlike PD, MSA progresses rapidly (6–10 years), resists levodopa, and has no cure, with treatment focused on symptom management.

8.4.1 Pharmacological Management of ALS

Curative therapies are absent at present; instead, the current therapeutic interventions aim to alleviate symptoms and improve quality of life. They include anti-PD-drug treatment, levodopa, carbidopa, and amantadine, which are commonly used to mitigate PD-like symptoms. MSA exhibits a limited or temporary improvement in comparison to those with PD.

Fludrocortisone and droxidopa are used to increase blood volume in the management of orthostatic hypotension. Midodrine helps raise blood pressure by constricting blood vessels, while pyridostigmine acts as an acetylcholinesterase inhibitor, which can also be used to improve orthostatic hypotension. For postprandial hypotension, octreotide is commonly used. Bladder dysfunction can be managed with desmopressin, which helps control nocturia and assists in managing urinary urgency and incontinence. Erectile dysfunction can be treated with sildenafil, tadalafil, and vardenafil. For rapid eye moment (REM) sleep behavior disorder, melatonin and clonazepam are commonly prescribed. Pain and muscle spasms may be treated with baclofen and tizanidine. Constipation can be managed with stool softeners such as senna, as well as laxatives, enemas, polyethylene glycol 3350, and magnesium citrate. Additionally, linaclotide and lubiprostone are used to relieve constipation. Indeed, many of these drugs have potential side effects; therefore, careful monitoring and appropriate dosage adjustments are essential to minimize risks [20].

8.4.2 Emerging Therapies and Clinical Trials in MSA

Emerging therapies for MSA focus on neuroprotection, symptom relief, and disease modification. Gene therapy (AB-1005) enhances dopamine production through GDNF delivery, potentially improving motor function. Immunotherapies (TAK-341, AMULET, UB-312) use monoclonal antibodies and vaccines to clear α-synuclein aggregates, a key factor in MSA pathology. ASOs are designed to suppress α-synuclein production, aiming to slow neurodegeneration. Ampreloxetine, a norepinephrine reuptake inhibitor, may help stabilize blood pressure and alleviate orthostatic hypotension [20].

AFFITOPE® PD01A and PD03A, vaccines targeting α-synuclein, have shown promise in reducing its accumulation and improving motor function in transgenic MSA mouse models. Phase 1 trials in PD and early-stage MSA patients confirmed their safety and tolerability. Passive immunization approaches like Lu AF82422, a monoclonal antibody targeting toxic α-synuclein, showed safety in PD patients and are now in a Phase 2 trial for MSA [21]. Similarly, BIIBO101 and Anle 138b, designed to prevent α-synuclein aggregation, are in Phase 2 and Phase 1 trials, respectively. In neuroinflammation research, minocycline failed to meet efficacy goals in Phase 3 trials, while intravenous immunoglobulin (IVIG) showed positive results in Phase 2. Verdiperstat is undergoing further evaluation for its effects on microglial activation. In neuroprotection, riluzole, aimed at inhibiting excitotoxicity in MSA, did not achieve its intended goals in Phase 3 trials [20].

8.5 PROGRESSIVE SUPRANUCLEAR PALSY

Progressive supranuclear palsy (PSP) is a rare neurodegenerative condition that impacts motor function, equilibrium, verbal communication, visual perception, and cognitive processes. The etiology of this disorder is attributed to the pathological accumulation of aberrant tau protein within the cerebral tissue, resulting in the degeneration of particular neural substrates, notably the brainstem and basal ganglia. The clinical manifestations of PSP encompass difficulties in balance and ambulation, impairments in ocular motility, challenges in speech and deglutition, alterations in cognitive functioning and behavior, increased muscular rigidity and bradykinesia, as well as facial stiffness and apathy. As of now, there is no treatment available for PSP, except for symptomatic management.

8.5.1 Pharmacological Management of PSP

There is currently no treatment that effectively stops or slows the progression of PSP. The motor deficits associated with PSP are managed with levodopa and amantadine. Zolpidem is a GABA agonist that is recommended to improve motor function, sleep disturbances, speech, and eye problems. Selective serotonin reuptake inhibitors (SSRIs) are used to treat depression, anxiety, obsessive–compulsive behavior, and emotional lability. However, they may worsen apathy. Memantine is used to help people manage symptoms in a specific subtype of PSP, which is characterized by speech difficulties (progressive non-fluent aphasia, or PNFA). Botulinum toxin injections to the pretarsal muscles can be effective in treating eyelid spasms and problems with keeping the eyes open, known as eyelid-opening apraxia. It is also used for significant neck pain and stiffness known as cervical dystonia. Clonazepam is a benzodiazepine sometimes used for sleep problems and muscle stiffness [22]. Artificial tears/eye drops are used to reduce dryness and eye irritation.

8.5.2 Emerging Therapies and Clinical Trials for PSP

Emerging therapies for PSP focus on targeting tau protein accumulation, neuroprotection, and genetic interventions. Tau-targeting approaches include ASOs to reduce tau production, tau aggregation inhibitors to prevent toxic clumping, and immunotherapy using monoclonal antibodies to clear tau aggregates. Neuroprotective agents like AZP2006 have shown promise in promoting neuron survival, improving synaptic function, and reducing tau hyperphosphorylation and neuroinflammation. Gene therapy research explores manipulating genetic targets such as discoidin domain receptor tyrosine kinase 2 (DDR2), KN motif and ankyrin repeat domain-containing Protein 2 (KANK2), and stomatin (STOM) to slow disease progression.

Currently, several chemicals are under clinical investigation to manage PSP. TPN-101 has shown reductions in NfL and interleukin-6 (IL-6) levels, biomarkers linked to neurodegeneration and inflammation. Sodium selenate, a protein phosphatase 2A (PP2A) activator, enhances tau dephosphorylation to counteract tau buildup. Tideglusib, a GSK-3β inhibitor, was tested for its potential to slow brain atrophy but failed to demonstrate clinical efficacy in PSP trials [23, 24].

8.6 FRIEDREICH'S ATAXIA

Friedreich's ataxia (FRDA) is a rare genetic neurodegenerative condition that predominantly impacts the nervous system and muscular structures. The etiology of this disorder is attributed to a mutation within the FXN gene, which culminates in insufficient production of frataxin, a protein that is critical for mitochondrial functionality. Consequently, this deficiency precipitates a progressive deterioration of both the nervous system and cardiac tissues. FA, also known as spinocerebellar ataxia, is a rare genetic autosomal-recessive disorder that damages the neurons of the cerebellum, spinal cord, and peripheral nerves. FA exhibits a variety of symptoms, including loss of coordination, muscle weakness, and spasticity, speech and swallowing complications, cardiovascular disorders, diabetes mellitus or impaired glucose tolerance, auditory and visual impairments, as well as scoliosis and foot deformities. Currently, there is no cure for FA, but various pharmacological and non-pharmacological therapies are available for the management of FA.

8.6.1 Pharmacological Management of FRDA

Omaveloxolone is an antioxidant and the only approved drug for the treatment of FA, which acts by nuclear factor erythroid 2-related factor 2 (Nrf2) activation, natural antioxidants such as coenzyme Q10, vitamin E, along synthetic analogs such as idebenone are used as antioxidants for the management of oxidative stress in FA. Selegiline, N-acetylcysteine, dehydroepiandrosterone, and the anti-diabetic drug pioglitazone are also shown to improve antioxidant defenses. Deferiprone is an iron chelator; it works by binding to excess iron and helps in redistributing the overloaded iron from the mitochondria to the cytosol. Agents such as nicotinamide and 2-aminobenzamide are used in the FRDA as histone deacetylase inhibitors (HDACis). They work by reversing the gene silencing of the frataxin gene, by increasing the expression of frataxin, leading to increased levels of frataxin [25–28].

Botulinum toxin, baclofen, and tizanidine are commonly used muscle relaxants for treating the ataxia-related muscle spasms, cramps, and stiffness. Gabapentin and pregabalin are used to reduce neuropathic pain. Cardiac myopathy is one of the frequently associated comorbidities with FRDA. Beta blockers and ACE inhibitors are used for treating the cardiomyopathy associated with FRDA [29]. Diabetes mellitus is often associated with FRDA, as comorbidity can be treated with various oral hypoglycemic agents and insulin. Pioglitazone, a peroxisome proliferator-activated receptor gamma (PPARY) molecule, is considered for drug repurposing to treat FRDA as it also contains antioxidant properties [28].

8.6.2 Emerging Therapies and Clinical Trials in FRDA

Emerging therapies for FRDA are focused on increasing frataxin levels and mitigating the consequences of its deficiency. Gene therapy, using AAV-mediated frataxin delivery, has shown promise in preclinical models, while CRISPR/Cas9 gene editing is being explored for FXN mutation correction. TAT-frataxin (CTI-1601) delivers frataxin directly to mitochondria and is in clinical trials. Nrf2 activators to enhance mitochondrial function, deuterated polyunsaturated fatty acids (dPUFAs), catalytic antioxidants to reduce oxidative stress, iron chelators to prevent iron-induced damage, and histone deacetylase (HDAC) inhibitors to boost frataxin expression. Oligonucleotide-based therapies, including ASOs and anti-gene oligonucleotides, aim to disrupt GAA repeat-induced R-loop formation in the FXN gene, enhancing frataxin expression in FA [30].

Several therapies for FA are in clinical trials. Vatiquinone (PTC-743) reduces oxidative stress (Phase II, PTC Therapeutics). RT001 (dPUFAs) limits lipid peroxidation (Phase III, Retrotope). Leriglitazone (MIN-102), a PPAR-γ agonist, lowers oxidative stress (Phase II, Minoryx Therapeutics). Micronized resveratrol provides neuroprotection (Phase II, Murdoch Children's Research Institute). Etravirine (FAEST1) increases FXN protein (Phase II, IRCCS Eugenio Medea). AAVrh.10hFXN and LX2006 deliver FXN via gene therapy (Phase I/II, Weill Medical College & Lexeo Therapeutics). Elamipretide, a mitochondrial antioxidant, is in Phase I/II (Children's Hospital of Philadelphia & Stealth BioTherapeutics). Calcitriol (vitamin D) boosts FXN (Phase IV, Institut de Recerca Biomèdica de Lleida). DT-216, targeting FXN gene GAA expansions, completed Phase I (Design Therapeutics) [30].

8.7 DEMENTIA WITH LEWY BODIES

Dementia with Lewy bodies (DLB) is the second most common form of dementia after AD. Globally, around 1.4 million people are affected by DLB. It is more prevalent in males than females and is characterized by cognitive decline, visual hallucinations, fluctuating alertness, Parkinsonian motor symptoms, and autonomic dysfunction. The disease results from abnormal deposits of α-synuclein in the brain, leading to widespread neurodegeneration.

8.7.1 Pharmacological Management of DLB

Currently, there are no preventive medicines available to prevent DLB. Drugs such as donepezil, rivastigmine, and galantamine are used as cholinesterase inhibitors to treat the cognitive function and behavioral symptoms caused by the loss of cholinergic neurons in the brain. Side effects such as nausea, vomiting, diarrhea, insomnia, and potential exacerbation of tremors and other Parkinsonian symptoms are common. Drugs such as memantine act as NMDA receptor blockers, thereby reducing excitotoxicity by blocking glutamate from binding to NMDA receptors. Used to reduce excitotoxicity and improve cognitive symptoms in DLB. Side effects include dizziness, headache, constipation, confusion, and hallucinations. Anti-PD drugs such as levodopa and carbidopa help in increasing dopamine levels, which were reduced in DLB due to the loss of dopaminergic neurons in the brain. This reduces motor symptoms such as bradykinesia, tremors, and rigidity. Common side effects include dyskinesias, nausea, dizziness, orthostatic hypotension, and hallucinations. SSRIs such as sertraline and citalopram are used to treat depression associated with DLB. Anticonvulsants such as valproic acid control the fluctuations and agitation, neuronal hyperactivity, and excitability, which are common to DLB. They are used as an adjuvant therapy alongside atypical antipsychotics such as Quetiapine. Benzodiazepines such as clonazepam and hormones such as melatonin are used to treat the REM sleep behavior associated with DLB [31].

8.7.2 Emerging Therapies and Clinical Trials in DLB

Recent research related to DLB has helped to develop certain emerging therapies that might be able to target various pathological mechanisms underlying DLB. Some of the early-stage developing drugs for the treatment of DLB include nilotinib, bosutinib, and E2027, and some drugs such as ambroxol are being explored as potential drug repurposing [32].

Several promising clinical trials are investigating novel approaches for PD and lewy body dementia (LBD). The anti-diabetic medications to fight PD and LBD trial, led by Mayo Clinic, is in Phase IV and explores whether DPP4 inhibitors (sitagliptin) and SGLT2 inhibitors (dapagliflozin) offer neurological benefits beyond glucose control. The dementias and microbiota composition (DEM-BIOTA) study examines the role of gut microbiota in dementia, investigating whether modifying microbiota composition can reverse cognitive decline. In targeted pharmacological interventions, the RewinD-LB trial (Phase II) is evaluating neflamapimod for its potential to enhance cognitive functions like verbal learning, memory, and attention in DLB patients. Additionally, the COG1201 trial is assessing the safety, tolerability, and efficacy of CT1812 in mild to moderate DLB cases [10]. The overall pharmacological and non-pharmacological management of other neurodegenerative diseases is illustrated in Table 8.1.

8.8 SPINAL MUSCULAR ATROPHY

Spinal muscular atrophy (SMA) is a rare genetic neuromuscular disorder caused by mutations in the SMN1 gene, resulting in progressive degeneration of motor neurons in the spinal cord, causing muscle weakness, atrophy, and impaired movement. The prevalence of SMA can vary by population. The severity of the disease ranges from Type 0 (the most severe, with prenatal onset and early mortality) to Type 4 (adult-onset with mild symptoms).

8.8.1 Pharmacological Management of SMA

Currently, there is no cure for SMA, but disease-modifying treatments help slow disease progression and improve motor function. **Nusinersen**, an ASO approved in 2016, is administered via repeated intrathecal injections due to its inability to cross the BBB. It enhances **SMN2** pre-mRNA splicing, promoting the production of full-length **SMN** protein necessary for motor neuron survival. **Risdiplam**, an oral small molecule approved in 2020, increases systemic SMN protein levels and can cross the BBB, offering a less invasive alternative to nusinersen. However, it is contraindicated in pregnancy due to potential fetal harm. Another investigational therapy, **Olesoxime**, protects motor neurons by stabilizing mitochondrial membranes, reducing apoptosis, and preserving cellular energy production, potentially improving neuronal survival. These therapies represent significant advancements in SMA treatment, improving quality of life and survival, though continued research is needed for more effective and accessible treatments [33].

TABLE 8.1
Pharmacological, Non-Pharmacological Management, Emerging Therapies, and Clinical Trials of Neurodegenerative Diseases

Disease	Pharmacological Therapies	Non-Pharmacological Therapies	Emerging Therapies and Clinical Trials
Huntington's disease (HD)	i. Tetrabenazine, deutetrabenazine (for chorea) ii. Antipsychotics: clozapine, risperidone, olanzapine, quetiapine iii. Cognitive enhancers: rivastigmine, memantine	i. Physical therapy for motor symptoms ii. Occupational therapy iii. Cognitive rehabilitation iv. Speech therapy	i. Gene therapy (ASOs, RNA interference, CRISPR-based approaches) ii. NMDA antagonists like memantine
Amyotrophic lateral sclerosis (ALS)	i. Riluzole (extends survival) ii. Edaravone (reduces oxidative stress) iii. Baclofen, tizanidine (spasticity management)	i. Assistive technology (wheelchairs, voice synthesizers) ii. Respiratory support (ventilation therapy) iii. Nutritional interventions (high-caloric diet)	i. NurOwn (MSC-NTF cell therapy, Phase III) ii. Tofersen (SOD1-targeting ASO) iii. Masitinib (anti-inflammatory, phase III)
Multiple system atrophy (MSA)	i. Levodopa, amantadine (for motor symptoms) ii. Fludrocortisone, midodrine (for orthostatic hypotension) iii. Sildenafil, tadalafil (for erectile dysfunction)	i. Physical therapy (to maintain mobility) ii. Speech therapy (for dysphagia and speech difficulties) iii. Cognitive training	i. Ampreloxetine (norepinephrine reuptake inhibitor) ii. Immunotherapies targeting α-synuclein iii. Gene therapy approaches
Progressive supranuclear palsy (PSP)	i. Levodopa, amantadine (for motor symptoms) ii. SSRIs (for mood disorders) iii. Memantine (for cognitive decline) iv. Botulinum toxin (for dystonia)	i. Gait and balance training ii. Speech and swallowing therapy iii. Psychological support for depression and anxiety	i. Tau-targeting therapies (antisense oligonucleotides, aggregation inhibitors) ii. Sodium selenate (PP2A activator, Phase II) iii. TPN-101 (neurofilament light chain modulator)
Friedreich's ataxia (FA/FRDA)	i. Omaveloxolone (antioxidant, FDA-approved) ii. Idebenone, coenzyme Q10 (antioxidants) iii. Deferiprone (iron chelator)	i. Rehabilitation (physical and occupational therapy) ii. Mobility aids (walkers, wheelchairs) iii. Surgical correction of scoliosis	i. Gene therapy (AAV-based FXN gene delivery) ii. Etravirine (FAEST1 trial) iii. Elamipretide (mitochondrial therapy)
Dementia with Lewy bodies (DLB)	i. Cholinesterase inhibitors (donepezil, rivastigmine) ii. NMDA receptor antagonists (memantine) iii. Dopaminergic drugs (levodopa/carbidopa)	i. Occupational therapy (to maintain independence) ii. Cognitive behavioral therapy (CBT) iii. Speech therapy (for dysphagia) iv. Light therapy (for sleep disturbances)	i. Nilotinib, bosutinib (kinase inhibitors) ii. RewinD-LB trial (neflamapimod for cognition) iii. COG1201 trial (CT1812, Phase II)
Spinal muscular atrophy (SMA)	i. Nusinersen (ASO; intrathecal; enhances SMN2 splicing) ii. Risdiplam (oral; BBB-permeable; increases SMN protein) iii. Olesoxime (mitochondrial membrane stabilizer; neuroprotection)	i. Respiratory support (mechanical ventilation, airway clearance therapy) ii. Nutritional management (tube feeding for oral intake difficulties) iii. Physical therapy and motor training (to improve strength and delay atrophy)	i. Onasemnogene abeparvovec (zolgensma; AAV9-SMN1 gene therapy) ii. Apitegromab (myostatin inhibitor; muscle enhancement, Phase III) iii. Combination therapies (gene therapy + neuroprotection + SMN2 splicing correction)
Multiple sclerosis (MS)	i. Interferon beta-1a/1b, pegylated interferon beta-1a (Plegridy) ii. Glatiramer acetate, fingolimod, teriflunomide, dimethyl fumarate, cladribine iii. Monoclonal antibodies: natalizumab, alemtuzumab, ocrelizumab iv. Symptomatic: methylprednisolone, baclofen, tizanidine, gabapentin, modafinil, SSRIs, anticholinergics	i. Physical rehabilitation and occupational therapy ii. Psychological counseling (for mood disorders, fatigue) iii. Dietary management (to support immune and metabolic health) iv. Bladder training and pelvic floor therapy	i. Ublituximab-xiiy (CD20-targeted monoclonal antibody) ii. Opicinumab (LINGO-1 inhibitor; remyelination promoter) iii. Fecal microbiota transplantation and probiotics (gut-immune modulation) iv. CAR-T cell therapy (targeting autoreactive B-cells) v. BTK inhibitors: tolebrutinib (ongoing) vi. Ibudilast (neuroprotection; reduces brain atrophy)

8.8.1.1 Supportive Therapies

Supportive treatments for SMA focus on managing symptoms and improving quality of life. Pharmacological interventions include muscle relaxants, neuroprotective agents, respiratory support, gastrointestinal management, and pain relief. Respiratory care is essential, as SMA patients are prone to infections such as pneumococcal disease, influenza, and severe respiratory syndromes, which can be managed with monoclonal antibodies. Airway clearance therapy helps prevent infections, and mechanical insufflation-exsufflation has been shown to improve vital capacity. Depending on SMA severity, 61% of patients require intubation within a year, and long-term ventilator support is needed for respiratory failure. Physical exercise (PE), including stretching and motor training, has shown benefits in improving motor function and stimulating SMN expression in preclinical studies. Chewing difficulties, tongue dysfunction, and muscle fatigue affect oral intake, making tube feeding necessary in some cases to ensure proper nutrition. These multidisciplinary approaches help enhance motor function, respiratory health, and overall well-being in SMA patients [34].

8.8.2 Emerging Therapies and Clinical Trials for SMA

Emerging therapies for SMA focus on gene therapy, SMN2 splicing modifiers, neuroprotective agents, and muscle-targeted treatments. Onasemnogene abeparvovec (Zolgensma), an AAV9-mediated gene therapy, delivers a functional SMN1 gene, offering a one-time treatment to halt disease progression. Risdiplam and Nusinersen, SMN2 splicing modifiers, enhance SMN protein production and are effective across SMA types. Neuroprotective agents, such as Olesoxime, aim to preserve motor neurons by stabilizing mitochondrial function. Muscle-directed therapies, including myostatin inhibitors, seek to boost muscle strength and compensate for motor neuron loss. Combination approaches integrating gene therapy, splicing correction, and neuroprotection may further optimize SMA management.

Clinical trials have led to significant advancements in SMA treatments. Nusinersen (Spinraza), an ASO, demonstrated improved motor function in infants with SMA Type 1 during the ENDEAR trial, leading to its FDA approval. Onasemnogene abeparvovec (Zolgensma), a gene therapy, has shown promise in treating SMA Type 2, potentially expanding its use beyond infants. Risdiplam (Evrysdi), an oral SMN2 splicing modifier, is under investigation for early intervention in presymptomatic SMA. Additionally, Apitegromab, a myostatin inhibitor, demonstrated notable motor function gains in a Phase 3 trial, offering a novel approach to muscle preservation in SMA patients [35].

8.9 MULTIPLE SCLEROSIS

Multiple sclerosis (MS) is a chronic autoimmune disorder that selectively causes the demyelination of neurons in the central nervous system (CNS), disrupting the axonal signal transmission. Currently, there is no preventive medicine for MS, but the risk of onset can be reduced or delayed with some supplements such as vitamin D and some lifestyle adjustments such as quitting smoking and alcohol.

8.9.1 Pharmacological Therapy for MS

Pharmacological treatment of MS consists of the use of disease-modifying drugs (DMTs) as well as symptomatic treatment. DMTs aim to alter the course of MS by modulating the immune response. Interferon beta-1a and interferon beta-1b reduce inflammation by decreasing pro-inflammatory cytokines and immune cell activity, and adverse effects include flu-like symptoms and liver enzyme abnormalities. Pegylated interferon beta-1a (Plegridy) has an extended half-life, allowing for less frequent dosing. Glatiramer acetate (Copaxone) mimics myelin proteins, shifting the immune response to a less inflammatory state, with side effects like injection site reactions and transient flushing. Other immunomodulators include fingolimod, which traps lymphocytes in lymph nodes, preventing them from reaching the CNS, though it may cause headaches and bradycardia. Teriflunomide inhibits pyrimidine synthesis, reducing activated lymphocytes, but has the side effect profile of liver enzyme abnormalities and gastrointestinal issues. Dimethyl fumarate (Tecfidera) activates the Nrf2 pathway and reduces inflammation. Cladribine (Mavenclad) selectively depletes B and T cells and is reserved for highly active MS cases; major side effects include increased risk of infection.

Monoclonal antibodies provide targeted immune suppression. Natalizumab prevents immune cells from crossing the blood–brain barrier but carries a risk of progressive multifocal leukoencephalopathy (PML). Alemtuzumab depletes lymphocytes, offering long-term disease control but increasing the risk of autoimmune conditions. Ocrelizumab targets CD20-positive B cells and is approved for both relapsing and primary progressive MS, though it may lead to infusion reactions and infections. Symptom management focuses on alleviating MS-related complications. Corticosteroids, such as methylprednisolone, reduce inflammation during relapses with potential side effects of mood swings and elevated blood sugar. Muscle spasticity is managed with GABA receptor agonists such as baclofen and tizanidine, an alpha-2 adrenergic agonist, with side effects such as drowsiness. Neuropathic pain is treated with anticonvulsants like gabapentin and pregabalin, which modulate neurotransmitter release but may lead to dizziness and peripheral edema. Fatigue, a common MS symptom, is treated with stimulants such as modafinil and amantadine, though they can cause headaches and insomnia. Depression, a frequent comorbidity, is managed with SSRIs (e.g., sertraline, fluoxetine) or SNRIs (e.g., venlafaxine, duloxetine). Bladder dysfunction is treated with anticholinergics like oxybutynin and tolterodine.

8.9.2 Emerging Therapies and Clinical Trials for MS

Ublituximab-xiiy, a recently approved monoclonal antibody, works by depleting the CD-20 B-cells. It can be used in various phases of MS, such as the relapsing phase and secondary progressive phase of MS. Potential adverse effects that may arise are upper respiratory tract infections, urinary tract infections, herpes reactivations, and injection site reactions. Remyelination promoters such as opicinumab inhibit LINGO-1 protein and negatively affect the process of myelination. By blocking this, it will help the body in the natural remyelination process. The gut microbiome plays an integral role in maintaining the Immune system of individuals. The gut microbiota of MS patients is shown to be altered, especially during the relapse phase of the disease. In such individuals, treatment with fecal microbiota transfer and use of probiotics might improve the disease condition, as they can promote a good microbiome environment. These approaches may complement existing disease-modifying therapies by promoting immune homeostasis [36].

Chimeric antigen receptor T-cell (CAR-T) therapy, originally developed for B-cell lymphomas, is being tested in clinical trials for MS by engineering T-cells to target autoreactive B-cells, the primary drivers of MS [37]. Bruton's tyrosine kinase (BTK) inhibitors, evobrutinib, and tolebrutinib reduce inflammation in MS by suppressing B-cell and microglial activation. Evobrutinib failed in Phase III and was discontinued, while tolebrutinib showed a 31% reduction in disability progression in progressive MS [38, 39]. Ibudilast, initially developed for asthma and stroke, is undergoing trials for its neuroprotective properties in MS, with Phase II data showing reduced brain volume loss in progressive MS [40]. The various therapies for the management of the above-mentioned neurological diseases are illustrated in Figure 8.1.

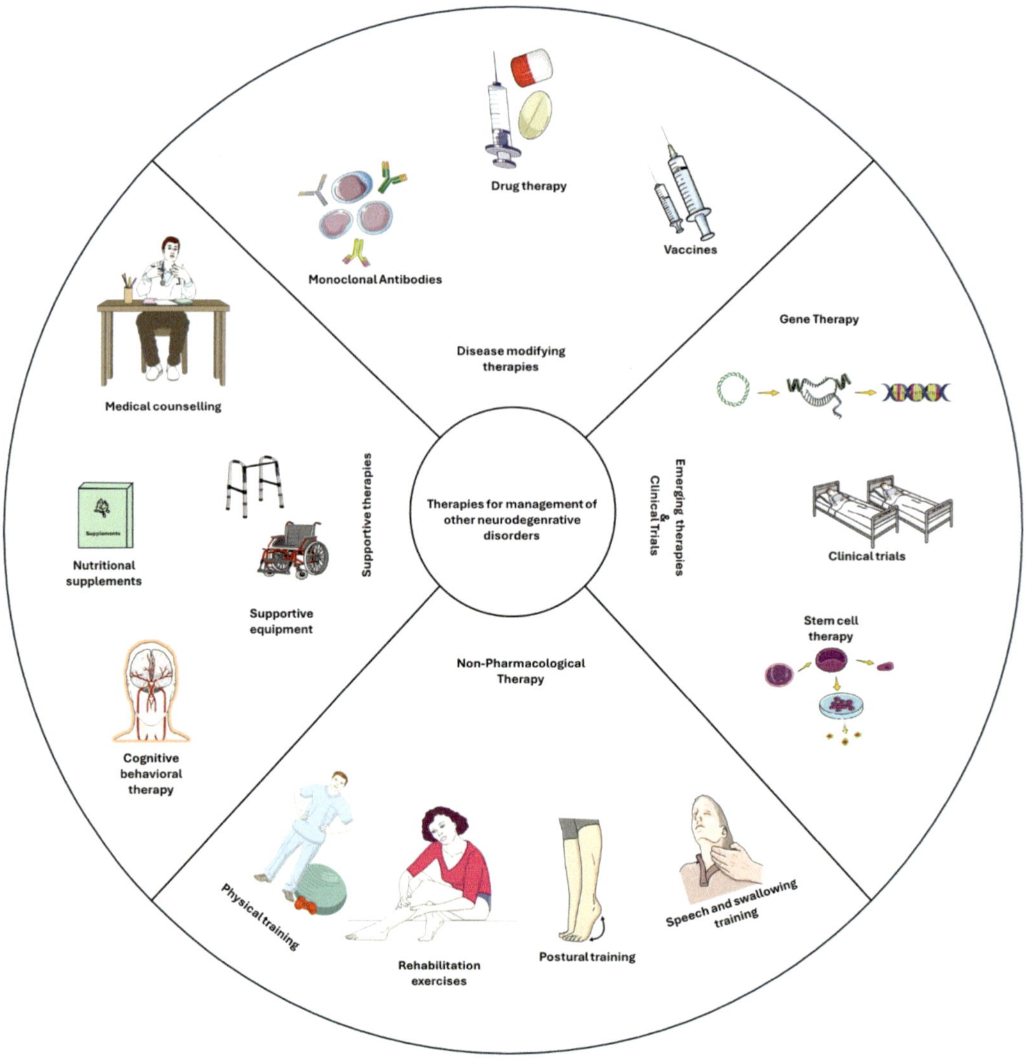

FIGURE 8.1 Classification of various strategies for the management of neurodegenerative disorders. Common pharmacological, non-pharmacological, supportive, and emerging therapies for nuerodegenerative diseases used or under investigation to improve patient outcomes and disease management.

8.10 NON-PHARMACOLOGICAL MANAGEMENT OF NEURODEGENERATIVE DISEASES

Neurodegenerative diseases, including HD, ALS, MSA, PSP, MS, FRDA, and DLB, demand a comprehensive approach to care that extends beyond pharmacological interventions. Medication may provide symptomatic relief, but it often fails to address the full spectrum of functional decline and quality of life concerns. In this context, non-pharmacological interventions represent crucial treatment approaches that promote independence, alleviate symptoms, and improve quality of life. A wide range of therapies, including physical rehabilitation, occupational therapy, speech and swallowing therapy, cognitive rehabilitation, nutritional support, psychological interventions, sleep, and lifestyle modifications, are considered for improving the lives of affected individuals. Many of these interventions share mutual benefits across multiple neurodegenerative disorders, but specific techniques and adaptations are required to address the unique challenges posed by each disease.

One of the key non-pharmacological management techniques is physical therapy, which plays a fundamental role in preserving motor function and preventing secondary complications associated with immobility. Regular PE has been widely recognized for its neuroprotective effects, as it promotes neurogenesis, increases the levels of brain-derived neurotrophic factor (BDNF), and enhances cognitive function [41]. Patients with conditions such as HD, ALS, MS, and MSA benefit from structured exercise programs that focus on maintaining muscular strength, flexibility, coordination, and balance. Strength training, low-impact aerobic exercises such as walking and swimming, and gait training techniques have been shown to improve mobility and reduce the risk of falls, a concern that is particularly pronounced in PSP and DLB due to the prevalence of postural instability [42, 43]. For conditions marked by progressive motor impairment, such as ALS and FRDA, physical therapy also incorporates the use of assistive devices, including braces, walkers, and wheelchairs, to maximize independence and facilitate safe movement [44].

Occupational therapy complements physical rehabilitation by addressing the functional limitations that affect daily living activities. The progressive nature of neurodegenerative diseases often results in a loss of dexterity, coordination, and strength, making tasks such as dressing, cooking, and writing increasingly difficult. Occupational therapy focuses on teaching patients adaptive techniques that compensate for motor deficits while promoting efficiency and energy conservation [45]. Installing grab bars, ramps, and adaptive furniture are examples of house adaptations that make living spaces safer and meet the changing needs of people with MSA and PSP [46].

Speech and swallowing difficulties are common across various neurodegenerative diseases, necessitating the involvement of speech-language pathologists in patient care. In disorders such as ALS, MS, and HD, progressive weakening of the oral and pharyngeal muscles leads to dysarthria, which affects speech clarity and intelligibility. Speech therapy aims to strengthen these muscles through targeted exercises while also introducing alternative communication methods, including augmentative and assistive communication devices, to preserve the ability to interact with others. Swallowing dysfunction, or dysphagia, is another major concern, particularly in PSP, MSA, and DLB, where patients face an increased risk of aspiration pneumonia [47]. Techniques such as modifying food textures, incorporating specific swallowing maneuvers, and adjusting head positioning during meals are crucial in preventing choking episodes and ensuring adequate nutritional intake. In more advanced cases, as seen in ALS, patients may require enteral feeding via gastrostomy tubes to maintain sufficient caloric intake and prevent severe weight loss [48].

Cognitive rehabilitation is another important aspect of non-pharmacological management, particularly for conditions associated with memory impairment, executive dysfunction, and attention deficits. Patients with MS and DLB frequently experience cognitive decline, which can significantly impact their ability to perform routine activities and maintain social engagement. Structured cognitive training exercises, memory aids such as planners and digital reminders, help individuals retain functional independence for as long as possible [49]. In the case of HD, cognitive rehabilitation is also aimed at managing impulsivity and impaired judgment, which are hallmarks of the disease's psychiatric manifestations. Additionally, behavioral interventions, including cognitive-behavioral therapy (CBT), play a role in addressing the emotional and psychological challenges that accompany neurodegenerative conditions. Depression, anxiety, and mood disturbances are frequently observed in patients with MS, HD, and DLB, making psychological support a critical component of care.

Proper nutritional support is equally important in managing the systemic effects of neurodegenerative disorders. Malnutrition, weight loss, and dehydration are prevalent concerns, particularly in HD, ALS, and MSA, where increased metabolic demands and swallowing difficulties contribute to a decline in overall health [50, 51]. A carefully tailored diet that incorporates high-caloric foods, adequate protein intake, and fiber-rich meals can help maintain energy levels, support muscle function, and alleviate gastrointestinal complications such as constipation, which is a common issue in MSA and DLB [52]. In conditions like ALS, where respiratory function is progressively compromised, dietary modifications must also account for the increased difficulty in coordinating

TABLE 8.2
Mutidisciplinary Members and Their Role and Responsibilities in Implementing Non-Pharmacological Management of Neurodegenerative Diseases

Multidisciplinary Team Members	Roles and Responsibilities
Caretakers	Organizes and supervises practice; welcomes and orients the patient; assists with tests (e.g., weight, lung function); oversees clinical data collection.
Dietician	Assesses nutritional status and dietary needs; advises on maintaining caloric balance.
Occupational therapist	Evaluates dexterity and functional independence in daily activities.
Physical therapist	Evaluates motor function, mobility, and safety; assesses the need for adaptive equipment.
Psychologist	Screens for anxiety and depressive symptoms; provides coping strategies; evaluates cognitive functions (memory, attention, concentration); offers emotional support.
Social worker	Assists with disability benefits, insurance, home care, and advance directives; offers psychosocial support.
Speech therapist	Assesses speech, voice, and swallowing function; recommends alternative communication devices (e.g., speech boards, synthesizers).

breathing and swallowing, necessitating early interventions such as thickened liquids or, in severe cases, feeding tube placement [53].

Rehabilitative approaches, sleep, and lifestyle modifications play a crucial role in managing symptoms and improving patient comfort. Many neurodegenerative diseases are associated with sleep disturbances, including insomnia, fragmented sleep, and excessive daytime sleepiness. In DLB and PSP, for example, disturbances in circadian rhythms and REM sleep behavior disorder contribute to significant sleep-related issues, necessitating interventions such as bright light therapy to regulate sleep-wake cycles [54]. Similarly, lifestyle adaptations such as maintaining a consistent sleep schedule, avoiding large meals before bedtime, and limiting caffeine and alcohol intake can help alleviate sleep dysfunction in MS, HD, and MSA [55, 56]. The non-pharmacological management of neurological disorders necessitates an integrated multidisciplinary approach. This involves collaboration among healthcare professionals from various fields, each responsible for distinct yet complementary aspects of patient care. It is through coordinated assessments and interventions ranging from physical rehabilitation and cognitive support to nutritional planning and psychosocial assistance. Table 8.2 outlines the core members of the multidisciplinary team and their specialized contributions to the non-pharmacological evaluation and management of neurological disorders.

8.11 CONCLUSION

The management of neurodegenerative diseases requires a multifaceted approach that combines pharmacological treatments, nutritional support, and medical or psychiatric counseling. Non-pharmacological strategies such as assistive devices, CBT, speech and swallowing therapies, PE, and rehabilitation play a key role in maintaining function and improving quality of life. Emerging therapies like gene and stem cell treatments are under investigation and offer hope for future disease-modifying options. As research advances, integrating traditional and innovative therapies will be essential for more effective, personalized care.

ACKNOWLEDGMENTS

Figure 8.1 is made partially using Servier Medical Art (http://smart.servier.com/), licensed under a Creative Common Attribution 4.0 Generic License.

REFERENCES

1. Brown, R.C., A.H. Lockwood, and B.R. Sonawane, *Neurodegenerative diseases: An overview of environmental risk factors*. Environmental Health Perspectives, 2005. **113**(9): p. 1250–1256.
2. Moraes, W., *Current pharmacological and non-pharmacological therapies for neurodegenerative diseases*. Young Perspectives for Old Diseases, 2015. p. 23–48.
3. Kakarla, R., et al., *Current understanding and future directions of cruciferous vegetables and their phytochemicals to combat neurological diseases*. Phytotherapy Research, 2024. **38**(3): p. 1381–1399.
4. Ramakrishna, K., et al., *WNT-β catenin signaling as a potential therapeutic target for neurodegenerative diseases: Current status and future perspective*. Diseases, 2023. **11**(3): p. 89.
5. Hayden, M.R., et al., *Tetrabenazine*. Nature Reviews Drug Discovery, 2009. **8**(1).
6. Bashir, H., and J. Jankovic, *Treatment options for chorea*. Expert Review of Neurotherapeutics, 2018. **18**(1): p. 51–63.
7. Unti, E., et al., *Antipsychotic drugs in Huntington's disease*. Expert Review of Neurotherapeutics, 2017. **17**(3): p. 227–237.
8. Koenig, A., et al., *Dalzanemdor (SAGE-718), a novel, investigational N-methyl-D-aspartate receptor positive allosteric modulator: Safety, tolerability, and clinical pharmacology in randomized dose-finding studies in healthy participants and an open-label study in participants with Huntington's disease*. Clinical and Translational Science, 2024. **17**(7): p. e13852.
9. Byun, S., M. Lee, and M. Kim, *Gene therapy for Huntington's disease: The final strategy for a cure?* Journal of Movement Disorders, 2022. **15**(1): p. 15.
10. Tolea, M.I., et al., *Emerging drugs for dementia with Lewy bodies: A review of Phase II & III trials*. Expert Opinion on Emerging Drugs, 2023. **28**(3): p. 167–180.
11. Andrews, J.A., et al., *Real-world evidence of riluzole effectiveness in treating amyotrophic lateral sclerosis*. Amyotrophic Lateral Sclerosis and Frontotemporal Degeneration, 2020. **21**(7–8): p. 509–518.
12. Jiang, J., Y. Wang, and M. Deng, *New developments and opportunities in drugs being trialed for amyotrophic lateral sclerosis from 2020 to 2022*. Frontiers in Pharmacology, 2022. **13**.
13. Ketabforoush, A.H.M.E., et al., *Masitinib: The promising actor in the next season of the amyotrophic lateral sclerosis treatment series*. Biomedicine & Pharmacotherapy, 2023. **160**: p. 114378.
14. Bowser R, An J, Mehta L, Chen J, Timmons J, Cudkowicz M, and Paganoni S. *Effect of sodium phenylbutyrate and taurursodiol on plasma concentrations of neuroinflammatory biomarkers in amyotrophic lateral sclerosis: results from the CENTAUR trial*. Journal of Neurology, Neurosurgery & Psychiatry, 2024 Jul 1;95(7): p. 605–608.
15. US Food and Drug Administration, US Food and Drug Administration. *FDA approves treatment of amyotrophic lateral sclerosis associated with a mutation in the SOD1 gene [Internet]*. 2023.
16. Willemse, S.W., et al., *UNC13A in amyotrophic lateral sclerosis: From genetic association to therapeutic target*. Journal of Neurology, Neurosurgery & Psychiatry, 2023. **94**(8): p. 649–656.
17. Gupta, D., et al., *Advances in understanding and treating amyotrophic lateral sclerosis (ALS): A comprehensive review*. Cureus, 2024. **15**(11): p. e48691.
18. Berry, J.D., et al., *NurOwn, phase 2, randomized, clinical trial in patients with ALS*. Neurology, 2019. **93**(24): p. e2294–e2305.
19. Mead, R.J., et al., *Amyotrophic lateral sclerosis: A neurodegenerative disorder poised for successful therapeutic translation*. Nature Reviews Drug Discovery, 2023. **22**(3): p. 185–212.
20. Liu, M., Z. Wang, and H. Shang, *Multiple system atrophy: An update and emerging directions of biomarkers and clinical trials*. Journal of Neurology, 2024. **271**(5): p. 2324–2344.
21. Mészáros, L., et al., *Current symptomatic and disease-modifying treatments in multiple system atrophy*. International Journal of Molecular Sciences, 2020. **21**(8): p. 2775.
22. Giagkou, N., G.U. Höglinger, and M. Stamelou, *Chapter three – Progressive supranuclear palsy*, in *International Review of Neurobiology*, M. Stamelou and G.U. Höglinger, Editors. 2019. Academic Press. p. 49–86.
23. Giagkou, N., and M. Stamelou, *Emerging drugs for progressive supranuclear palsy*. Expert Opinion on Emerging Drugs, 2019. **24**(2): p. 83–92.
24. Vivash, L., et al., *Sodium selenate as a disease-modifying treatment for progressive supranuclear palsy: Protocol for a phase 2, randomised, double-blind, placebo-controlled trial*. BMJ Open, 2021. **11**(12): p. e055019.
25. Ghanekar, S.D., et al., *Orphan drugs in development for the treatment of Friedreich's Ataxia: Focus on omaveloxolone*. Degenerative Neurological and Neuromuscular Disease, 2019. **9**: p. 103–107.

26. Profeta, V., et al., *Omaveloxolone: An activator of Nrf2 for the treatment of Friedreich Ataxia.* Expert Opinion on Investigational Drugs, 2023. **32**(1): p. 5–16.
27. Herman, D., et al., *Histone deacetylase inhibitors reverse gene silencing in Friedreich's Ataxia.* Nature Chemical Biology, 2006. **2**(10): p. 551–558.
28. Kearney, M., et al., *Pharmacological treatments for Friedreich Ataxia.* Cochrane Database of Systematic Reviews, 2016(8).
29. Cook, A., and P. Giunti, *Friedreich's Ataxia: Clinical features, pathogenesis and management.* British Medical Bulletin, 2017. **124**(1): p. 19–30.
30. Scott, V., et al., *New and emerging drug and gene therapies for Friedreich Ataxia.* CNS Drugs, 2024. **38**(10): p. 791–805.
31. McKeith, I.G., et al., *Diagnosis and management of dementia with Lewy bodies.* Neurology, 2017. **89**(1): p. 88–100.
32. Skylar-Scott, I.A., and S.J. Sha, *Lewy body dementia: An overview of promising therapeutics.* Current Neurology and Neuroscience Reports, 2023. **23**(10): p. 581–592.
33. Darras, B.T., et al., *Risdiplam-treated infants with type 1 spinal muscular atrophy versus historical controls.* New England Journal of Medicine, 2021. **385**(5): p. 427–435.
34. Schroth, M.K., et al., *Spinal muscular atrophy update in best practices: Recommendations for treatment considerations.* Neurology: Clinical Practice, 2025. **15**(1): p. e200374.
35. Ponomarev, A.S., et al., *Emerging gene therapy approaches in the management of spinal muscular atrophy (SMA): An overview of clinical trials and patent landscape.* International Journal of Molecular Sciences, 2023. **24**(18): p. 13743.
36. Olejnik, P., et al., *Multiple sclerosis: A narrative overview of current pharmacotherapies and emerging treatment prospects.* Pharmacological Reports, 2024. **76**(5): p. 926–943.
37. Yu, W.-L., and Z.-C. Hua, *Chimeric antigen receptor T-cell (CAR T) therapy for hematologic and solid malignancies: Efficacy and safety—A systematic review with meta-analysis.* Cancers, 2019. **11**(1): p. 47.
38. Carnero Contentti, E., and J. Correale, *Bruton's tyrosine kinase inhibitors: A promising emerging treatment option for multiple sclerosis.* Expert Opinion on Emerging Drugs, 2020. **25**(4): p. 377–381.
39. Krämer, J., et al., *Bruton tyrosine kinase inhibitors for multiple sclerosis.* Nature Reviews Neurology, 2023. **19**(5): p. 289–304.
40. Fox, R.J., et al., *Phase 2 trial of ibudilast in progressive multiple sclerosis.* New England Journal of Medicine, 2018. **379**(9): p. 846–855.
41. Baker, L.D., et al., *Effects of aerobic exercise on mild cognitive impairment: A controlled trial.* Archives of Neurology, 2010. **67**(1): p. 71–9.
42. Papamichail, P., et al., *The effectiveness of exercise programs on balance, functional ability, quality of life, and depression in progressive supranuclear palsy: A case study.* Applied Sciences, 2024. **14**(18): p. 8368.
43. Gupta, A., N.B. Prakash, and G. Sannyasi, *Rehabilitation in dementia.* Indian Journal of Psychological Medicine, 2021. **43**(5 Suppl): p. S37–s47.
44. Gruis, K.L., P.A. Wren, and J.E. Huggins, *Amyotrophic lateral sclerosis patients' self-reported satisfaction with assistive technology.* Muscle Nerve, 2011. **43**(5): p. 643–7.
45. Matuska, K., V. Mathiowetz, and M. Finlayson, *Use and perceived effectiveness of energy conservation strategies for managing multiple sclerosis fatigue.* The American Journal of Occupational Therapy, 2007. **61**(1): p. 62–69.
46. Samuelson, M., (2021). *Degenerative diseases of the central nervous system.* Early's Physical Dysfunction Practice Skills for the Occupational Therapy Assistant E-Book, 2021. 461.
47. De Pauw, A., et al., *Dysphagia in multiple sclerosis.* Clinical Neurology and Neurosurgery, 2002. **104**(4): p. 345–351.
48. Greenwood, D.I., *Nutrition management of amyotrophic lateral sclerosis.* Nutrition in Clinical Practice, 2013. **28**(3): p. 392–399.
49. O'Brien, A.R., et al., *Evidenced-based cognitive rehabilitation for persons with multiple sclerosis: A review of the literature.* Archives of Physical Medicine and Rehabilitation, 2008. **89**(4): p. 761–769.
50. Li, S., et al., *Prevalence and prognostic significance of malnutrition in early-stage multiple system atrophy.* Frontiers in Nutrition, 2023. **10**.
51. Zarco-Martín, M.T., et al., *Malnutrition in amyotrophic lateral sclerosis: Insights from morphofunctional assessment and global leadership initiative on malnutrition criteria.* Nutrients, 2024. **16**(16).
52. Goh, Y.Y., et al., *Multiple system atrophy.* Practical Neurology, 2023. **23**(3): p. 208–221.

53. Greenwood, D.I., *Nutrition management of amyotrophic lateral sclerosis.* Nutrition in Clinical Practice, 2013. **28**(3): p. 392–9.
54. Gooley, J.J., *Treatment of circadian rhythm sleep disorders with light.* Annals of the Academy of Medicine Singapore, 2008. **37**(8): p. 669–76.
55. Faraclas, E., *Interventions to improve quality of life in multiple sclerosis: New opportunities and key talking points.* Degenerative Neurological and Neuromuscular Disease, 2023. **13**: p. 55–68.
56. Santiago, J.A., and J.A. Potashkin, *Physical activity and lifestyle modifications in the treatment of neurodegenerative diseases.* Frontiers in Aging Neuroscience, 2023. **15**.

9 Brain Biomarkers and Diagnosis of Neurodegenerative Diseases

Prasada Chowdari Gurram, Chandra Teja Uppuluri, Kavuma Asadi, Radha Kasaraneni, Guntupalli Chakravarthi, and Buchi N Nalluri*
KL College of Pharmacy, Koneru Lakshmaiah Education Foundation, Green Fields, Vaddeswaram, A.P, India

Dimple Mohan Kundla
Chennupati Indo American School of Pharmacy, Jonnalagadda, Narasaraopet, Palnadu, A.P, India

Sairaj Satarker
Goa University, Taleigao Plateau, Goa, India

9.1 INTRODUCTION

The progressive loss of brain function has become a hallmark of neurodegenerative diseases. The magnitude of the occurrence of these diseases in humans has contributed to a greater global concern (Gurram et al., 2022). Every year, the rise of several neurodegenerative diseases increases and affects millions of individuals globally. To adequately understand the disease initiation and progression, accurate diagnostic biomarkers specific to neurodegenerative diseases need to be discovered. By definition, biomarkers are evaluable, objectively detectable signs that can be used to characterize diseased and normal biological processes as well as the pharmacological reactions to therapeutic intervention. Biomarkers are mainly responsible for enhancing the accuracy of clinical diagnosis. In addition to reflecting treatment response rates, biomarkers ought to assist in determining an illness's stage and rate of progression. The biomarkers vary in their sensitivity and specificity. The effectiveness of a test in identifying the existence of conditions being seen in a person is reflected in its sensitivity. The capacity to precisely identify instances in which the studied qualities do not exist is reflected in specificity; novel biomarkers are needed for better diagnostic and therapy options for people. A biomarker serves as a quantifiable indicator depicting a biological state or condition, measured and assessed to examine normal and pathogenic processes and pharmacologic responses to a therapeutic intervention (Shusharina et al., 2023). In clinical medicine, biomarkers are employed as diagnostic tools to forecast illness risk or prognosis and to track the effectiveness of therapies.

For therapeutic strategies to be developed in order to halt disease progression and enhance patient outcomes, early and precise diagnosis of these disorders is essential. Unfortunately, many of the biomarkers for these disorders that are now on the market have low sensitivity and specificity, thus making their use unsuitable for the initial manifestations of the disease

*Corresponding Author.

 DOI: 10.1201/9781003530510-9

(Shusharina et al., 2023). With difficulties arising in evaluating patient brains, diagnosing and treating neurodegenerative disorders can be challenging. Accurate diagnosis for the majority of these disorders requires postmortem brain tissue examination. Significant progress has been made in imaging, cerebrospinal fluid (CSF), and blood-based biomarker evaluation for various neurodegenerative illnesses (Shusharina et al., 2023).

Our understanding of immunoassays can be divided into four categories: first generation, which is the immunoblot; second generation, which is the enzyme-linked immunosorbent assay (ELISA); third generation, which uses electrochemiluminescence (ECL) technology; and fourth generation, which is based on single molecule array technology (Simoa). These encouraging results created a favorable environment for the continued development of immunoassays (Khalil et al., 2018). The fourth generation of tests was especially successful in translating CSF assays into blood-based testing. Over the past few years, this category has grown, aside from the Simoa (Khalil et al., 2018).

9.1.1 Image-Based Biomarkers

Regional neurodegeneration is assessed by studying patterns of brain shrinkage using structural magnetic resonance imaging (MRI) (Table 9.5), which includes T1- and T2-weighted imaging. Since MRI parameters such as cortical thickness and gray matter volume (Table 9.3) frequently show minimal intra-individual variability over time, longitudinal clinical trials can identify relatively small changes in atrophy rates (Wankhede et al., 2025). However, specific cell populations and cellular morphology that have been lost cannot be determined by MRI. For example, in the basal ganglia, the density of dopaminergic nerve terminals can be assessed via molecular imaging, such as positron emission tomography (PET) (Table 9.3), and it is found to be lower in Parkinson's disease (PD) as well as in progressive supranuclear palsy (PSP), multiple systemic atrophy (MSA), and corticobasal degeneration (CBD) (Wankhede et al., 2025).

9.1.2 CSF Biomarkers

CSF biomarkers for diagnosing neurological illnesses are widely employed in clinical practice and translational research. These biomarkers are reflective of brain pathology. Nonetheless, inter-individual variability unrelated to illness may have an impact on CSF biomarker concentrations. The CSF markers have the potential to significantly enhance the disease's diagnosis process, which is becoming more crucial as powerful disease-modifying therapies for neurodegenerative diseases have recently been developed. The neurofilament heavy chain (NfH) forms the heavily phosphorylated protein in the central nervous system (CNS) and serves as an integral biomarker for dementia, which affects axonal transport, homeostasis, and cell shape. CSF neurofilament light chain (NfL) and phosphorylated NfH (pNfH) concentrations in individuals afflicted with neurodegenerative disorders are positively associated and exhibit comparable variations (Olsson et al., 2016) (Figure 9.1).

9.1.3 Blood-Based Biomarkers

Blood-based biomarkers aid significant advancement in the clinical assessment of neurodegeneration. Blood markers like neurofilament light, ubiquitin-C-terminal-hydrolase-L1, β-synuclein for neuronal degeneration, and glial fibrillary acidic protein (GFAP) pathologies, along with tau pathologies specific to Alzheimer's disease (AD)-like P-tau and amyloid-β peptides, suggest glial degeneration. The use of these indicators in the future for disease detection, diagnosis, screening, and therapy response tracking will be crucial. Such markers have been rapidly employed in research and clinical contexts; they may soon be used in clinical practice (Olsson et al., 2016).

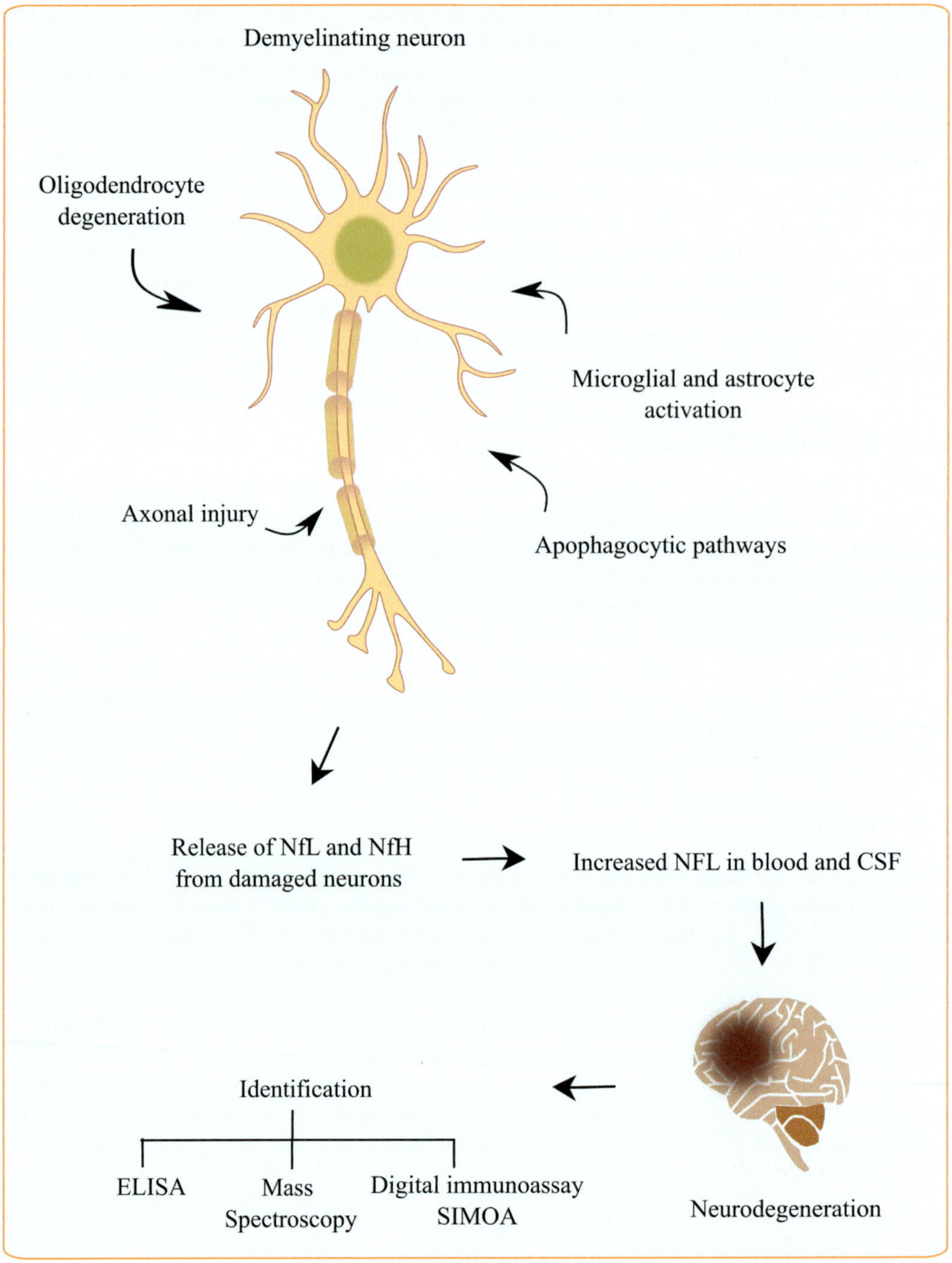

FIGURE 9.1 Demyelinated neurons get injured by axonal injury, and glial cell activation leads to the release of NFL. The accumulation of neuroinflammation in the brain tissue causes neuroinflammation followed by neurodegeneration. NFL is detected by ELISA, mass spectroscopy, and digital immune assay.

9.1.4 Biomarkers of Blood–Brain Barrier Function and Glial Activation

Reactive astrogliosis is indicated by GFAP, which is mostly associated with astrocytes. Neurodegenerative diseases like AD and PD markedly upregulate GFAP. In early AD/PD, the levels of GFAP in the CSF may be a useful marker to forecast the degree and course of cognitive decline together with alterations in AD-associated pathology markers (Gurram et al., 2022). Pereira et al. found that amyloid-β PET (Aβ-PET) showed longitudinal changes, where even CSF and plasma GFAP predicted such cognitive decline in the entire sample of AD patients (Wankhede et al., 2025). Diagnoses of untreated PD patients revealed a potential relationship between CSF GFAP levels at baseline and Montreal Cognitive Assessment (MoCA) score at six-year follow-up. The levels of GFAP in plasma at baseline showed a statistically significant negative correlation with Mini-Mental State Examination scores. It also functioned as a prognostic factor for the development of mild cognitive impairment (MCI) in PD patients (Gurram et al., 2022). However, further findings are needed to establish a link between the postmortem number of reactive astrocytes identified by immunohistochemistry and antemortem levels of plasma GFAP.

Several neurodegenerative illnesses affect the integrity of the blood–brain barrier (BBB), which can be assessed using specific MRI techniques. These approaches include dynamic contrast-enhanced and dynamic susceptibility contrast MRI, which are sensitive to the leakage of gadolinium-based contrasting agents from the brain (Wankhede et al., 2025). Additionally, the CSF-to-plasma ratio of albumin, demonstrated to be raised in the majority of dementia disorders without AD pathologies, reflects BBB integrity. Soluble platelet-derived growth factor receptor beta (sPDGFR-β) levels in the CSF denote the destruction of capillary pericytes, which is considered to be linked to a decline in cognitive performance (Karlsson et al., 2024).

Discovering biomarkers that are sensitive enough to forecast the course of a disease is vital. In addition to being valuable in clinical trial results, an ideal biomarker should respond to treatment, have diagnostic and prognostic value, and correlate with particular disease activity like relapse or progression. A perfect biomarker should also be non-invasive, secure, reliable, repeatable, affordable, and simple to identify in patients.

9.2 ALZHEIMER'S DISEASE (AD)

AD forms the most prevalent type of dementia. AD is primarily identified by progressive dementia, which includes executive dysfunction, memory loss, cognitive impairment, personality and behavior abnormalities, and, in most cases, mental disorder symptoms (Nampoothiri et al., 2023). While careful attention and treatment can temporarily alleviate these symptoms, prevention and treatment of AD remain elusive (Nampoothiri et al., 2023). The majority of cases of dementia have been identified in the geriatric population, mainly in low- and middle-income nations. It has caused a global health concern, placing significant financial strain on affected individuals and their families. In 2016, 43.8 million people were affected by dementia, which saw a rise from 20.3 million in 1990. There will be 152 million people with AD and other dementias by 2050, according to estimates.

In the past, AD was typically diagnosed in the later stages of the illness and has always been a challenge to diagnose. However, the illness can persist for years, taking a heavy toll on the affected individuals, their caretakers, and even the medical system. Early and precise identification of AD-related symptoms clinically is difficult. Patients with symptoms of early AD will find it difficult to be seen in a primary healthcare setup. Some patients may exhibit subtle behavioral or cognitive changes during a normal wellness check or during a consultation to address additional comorbidities (Gurram et al., 2022). The Aβ plaques and neurofibrillary tangles (NFTs), the neuropathologic markers of AD, can be found decades before symptoms appear. Biomarkers that reflect this underlying pathology offer a valuable chance to screen high-risk patients predisposed to developing MCI as a result of early AD. The US Food and Drug Administration (FDA) has established guidelines for the use of biomarkers in the AD population to aid AD diagnosis (Hansson, 2021).

Numerous PET tracers are available to measure the amount of Aβ fibrillar content present in the CNS. Clinically, three licensed Aβ PET tracers are used, namely flumetamol, florbetapir, and florbetaben. Multiple studies showed a correlation between the density of Aβ plaques in postmortem samples and the in vivo mediated uptake of PET tracers (Hansson, 2021). In the majority of patients with cognitive symptoms, a normal Aβ PET scan suggests the absence of AD as a causative agent. Young patients with cognitive symptoms having an abnormal Aβ PET scan are indicative of AD.

On the contrary, such findings in older patients should be interpreted cautiously since 40% of people over the age of 90 years have brain Aβ plaques (Jansen et al., 2022). In the USA, only one tau PET tracer, flortaucipir, is authorized for therapeutic use. This tracer can accurately measure the density of neuropil threads and NFTs because it has been verified against neuropathology (Jansen et al., 2022), even though its sensitivity is insufficient to identify the earliest stages of tau. When separating Alzheimer's dementia from the majority of other neurodegenerative illnesses, tau PET has demonstrated exceptional diagnostic accuracy. Identification of tau in neocortical regions for high specificity in AD PET retention, having high specificity for AD patients, can be utilized to rule out AD in patients with cognitive impairment, even in the geriatric population (Wankhede et al., 2025). In a recent study, the odds of developing dementia and MCI were nearly 40 times and 20 times greater, respectively, in patients having normal cognition and a positive tau PET, and positive Aβ PET scans than in those with normal PET scans. Individuals with normal cognition, positive Aβ PET, but negative tau PET were less likely to become cognitively impaired. All of these findings are consistent with the study framework for AD developed by the National Institute on Aging-Alzheimer's Association (NIA-AA), which specifies that people having both tau and Aβ pathology should be classified as having AD regardless of their cognitive state (Wankhede et al., 2025).

Crucial inferences obtained from prior investigations of CSF-based biomarkers have been instrumental in the advancement of blood-based biomarkers for brain (Table 9.1) diseases, primarily AD. Evidence states that Aβ42, total tau (T-tau), and phosphorylated tau (P-tau) levels in CSF have been used to differentiate AD patients from healthy controls (Olsson et al., 2016). These fundamental AD biomarkers form the integral elements of the AD diagnosis because the CSF Aβ42 to Aβ40 ratio is unaffected by variations in amyloidogenic amyloid precursor protein (APP) processing measured within individuals. It has consistently been demonstrated to be more preferred than CSF Aβ42 alone. Typically, non-AD dementia patients have normal CSF concentrations of the three primary AD biomarkers: Aβ42, T-tau, and P-tau (Olsson et al., 2016).

The smallest of the neurofilament triplet proteins, forming the axonal structural elements, is called the NfL. It is released from axons and paves its way into the extracellular space during healthy aging, leading to higher quantities of NfL in the CSF. Nevertheless, axonal damage, which can result from neurodegeneration, inflammation, trauma, or stroke, accelerates the release of NfL. CSF NfL concentrations spike in brain diseases associated with subcortical pathology, including normal pressure hydrocephalus and vascular dementia, according to several investigations (Abukuri, 2024). Interestingly, CSF NfL levels were greater in frontotemporal dementia (FTD) patients than in AD patients, indicating that this biomarker may be used to distinguish between these two conditions (Abukuri, 2024). While CSF NfL concentrations in people with PD are generally normal, a number of studies have found that, when compared to healthy individuals, people with atypical parkinsonian conditions like corticobasal syndrome, multiple system atrophy, or PSP have significantly higher CSF NfL concentrations (Abukuri, 2024) (Figure 9.1).

In clinical trials, biomarkers that accurately reflect the pathology of AD during life are a crucial component of the inclusion criteria. They are also crucial for diagnosis, especially with the advent of disease-modifying treatments. The classic pathophysiological markers of AD, namely tau and neurodegeneration, can now be identified through imaging or CSF methods. Tau- and amyloid-PET scans serve as the gold standard for determining tau and amyloid pathology in clinical studies. Nevertheless, these techniques are costly, intrusive, or both. Finding affordable biomarkers that

TABLE 9.1
CSF/Blood Biomarkers in AD

Biomarkers	Disease Indicators	Detection Methods	References
Aβ42, T-tau, P-tau, YKL-40	Elevated levels of β-amyloid 42, P-tau, T-tau, and YKL-40 in CSF due to astrocyte and microglial activation	ELISA/WB, mass spectrometry, and advanced immunoassays	Olsson et al. (2016)
BDNF	Low levels of BDNF modulated by Aβ42	ELISA, WB	Ng et al. (2019)
sTREM2	High levels of sTREM2 that are involved in neuronal injury, microglial activation, neuroinflammation, and plaque formation	ELISA	Suárez-Calvet et al. (2016)
mtDNA	Peripheral lymphocytes in late-stage AD patients are higher than in AD patients in the early stage	Quantitative polymerase chain reaction (PCR)	Gao and Ma (2022)
GFAP	Astrogliosis and microgliosis upregulated	ELISA, IHC quartz-crystal-microbalance, surface-acoustic-wave lab-on-chip, recombinant GFAP, functionalization, and biosensing experiment	Olsson et al. (2016)
Apolipoprotein E	ApoE 4 carriers show higher T-tau, Aβ aggregation, and deposition with faster disease progression	ELISA, WB technique	Leoni (2011)
S100B	Elevated levels observed in correlation with Aβ and tau biomarkers	HPLC	Michetti et al. (2019)
Cell-free DNA (cfDNA)/RNA	Elevated TLR3 gene expression in AD patients increases with disease progression. Ribosomal cfRNA/cfDNA fragments derived from the damaged tissue and organs can circulate alone or can be associated with proteins and EVs. miR-21-5p, miR-126-3p, and miR-128 are upregulated	High-sensitivity next-generation sequencing (hsNGS), digital PCR	Bivona et al. (2024)
Clusterin	Clusterin levels in CSF elevated in individuals with AD. Clusterin dramatically increases Aβ clearance, an action that LRP2 mediates	ELISA	Foster et al. (2019)
α-2-macroglobulin (A2M)	A2M upregulated in the preclinical AD brain and blood in response to tau phosphorylation in the brain via the RCAN1-calcineurin pathway is linked to preclinical AD and represents early neuronal damage	ELISA, immunoturbidimetry	Varma et al. (2017)
Neurogranin	High CSF levels of neurogranin in AD brains, which binds to calmodulin exclusively in the brain	ELISA	Agnello et al. (2021)

can be serially evaluated and more readily collected in a less invasive manner is, therefore, crucial from a medical standpoint. Blood-based biomarkers (Table 9.1) are probably going to meet this need (Qiu et al., 2025). It is now possible to find blood-based biomarkers for dementias such as AD. The utilization of blood indicators of neurodegeneration, such as NfL, as well as the core pathological biomarkers, such as Aβ and P-tau, has great potential, according to the results obtained from well-defined cohorts (Abukuri, 2024). However, once more, the intrusive character of CSF biomarkers raises concerns that keep them out of large-cohort examinations. However, numerous studies have highlighted the viability of using the newly developed capacity to measure plasma tau, namely plasma-tau181 and plasma-tau217, to evaluate tau pathology in AD.

9.2.1 miRNA and Circulating miRNA as Putative AD Biomarkers

A class of non-coding RNAs known as "circulating microRNAs" (cimiRNA) is suggested as a viable biomarker, according to research findings. It belongs to a class of single-stranded RNAs that are spliced backward or from head to tail. Upon nuclear production, cimiRNA is subsequently moved into the cytoplasm, which is then actively or passively passed into body fluids. Because they have a half-life of more than 48 hours and no loose ends, cimiRNAs have incredible stability and resistance to lysis by endogenous RNA activity (Doxakis, 2022). They can also tolerate harsh environmental circumstances like boiling, high pH levels, prolonged storage, and repeated defrost cycles. The advantages include their ability to be identified using semiquantitative techniques like next-generation sequencing (NGS) and quantitative techniques like real-time quantitative reverse transcription PCR (qRT-PCR). The cimiRNAs associated with genes linked to AD, like APP, PSEN1 (presenilin 1), and PSEN2 (presenilin 2), are important in the pathogenesis of AD. On the other hand, it has been demonstrated that patients with certain cimiRNA groups, including miR-9-5p, miR-125-5p, miR-146a-5p, and miR-155-5p, were upregulated prior to a significant increase in the link between Aβ levels and development of AD pathogenesis. This highlights miRNA's potential value as a biomarker for an early AD diagnosis (van den Berg et al., 2020).

MiRNAs seem to control processes related to synaptic plasticity and homeostasis. This suggests that they could play a role in the early synaptic disruption that occurs throughout AD. As a result, miRNAs offer trustworthy, inexpensive, and non-invasive AD biomarkers that can identify the disease's asymptomatic phases (Nassar et al., 2022).

9.3 PARKINSON'S DISEASE (PD)

PD is a type of neurodegenerative disorder that involves loss of dopaminergic neurons in the region of the substantia nigra (SN). It is generally chronic and progressive and mainly causes motor and non-motor symptoms. PD is known to be the most commonly prevailing neurodegenerative disease after AD and affects 1% of adults over 60 and 3% of those over 80 (Erkkinen et al., 2018). The striking feature includes a reduction in dopamine levels as a result of significant degeneration of dopaminergic neurons. These neurons mainly exist in the SN pars compacta. Certain neurons in the SN have intracytoplasmic accumulations named Lewy bodies (Nampoothiri et al., 2023). At present, the primary method for diagnosing PD is the detection of cardinal motor symptoms, such as rigidity, bradykinesia, and resting tremor. Unfortunately, around 70% of nigral neurons are destroyed when motor symptoms occur (Nampoothiri et al., 2023). Additionally, there are significant clinical similarities to atypical parkinsonian syndromes (APSs), which belong to neurodegenerative disorders distinct from PD but manifest common symptoms like akinetic rigidity. Examples of APSs include CBD, dementia with Lewy bodies (DLB), PSP, and MSA. As a result, events occur wherein there is an underdiagnosis of APSs and also the overdiagnosis of PD (Nampoothiri et al., 2023).

The identification of the gene instrumental for an uncommon autosomal dominant variant of PD in the late 1990s sparked a resurgence of interest in the hunt for pathogenic processes. α-synuclein, a synaptic protein with an unclear function, is encoded by the mutant gene. It has been proposed that it contributes to the recycling of vesicles into the synaptic cleft that have released their neurotransmitter contents (Tsao et al., 2022). For example, the DNA samples of G2019S carriers showing SN hyperechogenicity might not evolve to clinical PD during life; the mutation, where glycine (G) is replaced by serine (S) at position 2019 of the LRRK2 protein, is observed. Midbrain-like organoids were produced from sporadic or familial PD patients carrying the LRRK2-G2019S mutation that, respectively, displayed α-syn accumulation and dopaminergic neuron degeneration observed by DaT-SPECT and transcriptional sonography (TCS) in pluripotent stem cells (Rideout et al., 2020).

Recently, Xiang et al. reviewed CSF biomarkers (Table 9.2) that distinguish PD patients from those with other APSs or controls. To be more precise, we found that NFL, YKL-40, T-tau, and C-reactive proteins were able to differentiate better between PD and MSA, whereas DJ-1, t-α-syn,

TABLE 9.2
CSF and Blood Biomarkers in PD

Biomarkers	Samples	Diagnostic Applications	Finding Methods	References
Alpha-synuclein	CSF, plasma, saliva, blood	Elevated levels of alpha-synuclein in CSF or blood.	Western blot, ELISA, mass spectrometry	Tsao et al. (2022)
LRRK2	Blood, CSF exosomes, urine exosomes	LRRK2 protein levels are elevated in the prefrontal cortex of PD patients.	Western immunoblot, ELISA, mass spectroscopy	Rideout et al. (2020)
PARK7/DJ-1	CSF/plasma	DJ-1 is oxidized in response to PARK7 mutations. Elevated DJ-1 levels are seen in PD patients.	ELISA	Lind-Holm Mogensen et al. (2023)
PTEN-induced kinase 1 (PINK1)	CSF/serum	In PD patients with PINK1 mutations, PINK1 levels are often reduced.	ELISA, western blotting, and IHC	Iwaki et al. (2019)
Circulating microRNAs	Plasma	k-TSP1 (miR-1826/miR-450b-3p), miR-626, miR-505, and miR-1826 have been found to be upregulated in plasma.	Microarrays, qt-PCR, SAM analysis	Khoo et al. (2012)
Altered glycosylation patterns (Ig-G)	Plasma samples	Deviations in the peripheral IgG glycome were seen in the PD cases.	Nano-LC-ESI-MS/MS, ultra-performance liquid chromatography (UPLC)	Russell et al. (2017)
Short-chain fatty acids (SCFAs)-AA/PA/BA	Plasma/feces	Marked reduction in SCFAs associated with progression in PD.	ELISA, western blotting	Duan et al. (2024)
Neurotransmitter metabolites	CSF samples	CSF levels of DOPAC and HVA were reduced in early PD.	DaT-SPECT	Kremer et al. (2021)
Matrix metalloproteinases (MMPs)	Serum	Altered levels of MMP3 and MMP9 denote PD.	ELISA	Liu et al. (2022)

p-α-syn, o-α-syn, Zn, and YKL-40 in CSF were unable to differentiate between PD and control samples. Additionally, Aβ42 and NFL were able to differentiate between PSP and PD. Moreover, NfL could distinguish PD from CBD, whereas Aβ42, T-tau, and P-tau were able to distinguish PD from DLB. Thus, it was suggested that markers like t-α-syn, NFL, and Aβ42 could be instrumental in PD diagnostics (Xiang et al., 2022).

It has been suggested that the reduced presence of total monomeric α-synuclein in CSF is a biomarker (Table 9.2) for both DLB and PD. The majority of research, however, has only discovered a small (though statistically significant) drop in this marker's concentrations in PD or DLB patients in comparison to healthy persons (Xiang et al., 2022). PD patients showed a significant reduction in total α-synuclein concentrations in the CSF. They also showed greater oligomer formation and phosphorylated α-synuclein levels than healthy subjects, according to a meta-analysis that included almost 3,000 participants from 17 individual research studies (Tsao et al., 2022). The reason for this negative result could be that blood contains 10,000 times more α-synuclein than CSF does, indicating that the presence of blood in CSF could introduce peripherally derived α-synuclein.

In addition to being expressed in neurons, α-syn is also found in blood, primarily in erythrocytes (Tsao et al., 2022). Studies on total α-syn, however, produced contradictory findings. Serum and plasma levels of total α-syn levels in PD were either higher than in controls, not notably different, or

TABLE 9.3
Image Biomarkers in PD

Biomarkers	Samples	Diagnostic Applications	Finding Methods	References
C-RT132	Dopamine transporter imaging	C-RT132 PET imaging helps visualize and quantify the reduction in dopamine transporters. PD patients have been reported to have lower 11C-RT132.	PET	Brooks and Pavese (2011)
F-deoxyglucose (FDG)	Dopamine transporter imaging (midbrain and striata)	The regional cerebral glucose metabolic rate is impacted in PD and is often associated with deteriorating nigral cells and hypometabolism.	PET SCAN (detects metabolic activity of tissues)	Brooks and Pavese (2011)
Electroencephalography (EEG)	QEEG	Identification of specific patterns or abnormalities in brain wave activity QEEG parameters can be useful in defining the risk of incident dementia in PD.	Spectral analysis, topographic mapping	Klassen et al. (2011)
Digital biomarker assessment	Detection and recording of potential parameters related to PD	(1) Rest-activity circadian rhythm. (2) Gold-standard Epworth sleepiness scale (ESS) positive in the progression of PD. (3) HRV parameters combined with respiration rate or skin impedance. (4) Alterations in high-frequency (HF) power, total frequency (TF) power, and the low-frequency/high-frequency power (LF/HF) ratio between early PD, advanced PD, and healthy controls. (5) Distill proxies for hallucinations in PD.	(1) Wrist-worn multimodal device (2) Wrist-worn photoplethysmography (3) Smartphone-measured gaze to estimate fatigue levels (4) Photoplethysmography (5) Wrist-worn accelerometer	Janssen Daalen et al. (2024)
Electromyography (EMG)	Electromyography technique	The prolonged motor unit potential (MUP) duration and polyphasic MUPs usually indicate collateral reinnervation, as seen in early PD. Higher %DET, lower SampEn, and higher theta, alpha, and beta band intermuscular coherences are observed in PD.	External anal sphincter electromyography (EAS-EMG) Surface electromyography (sEMG)	Hu (2022)
Voxel-based morphometry (VBM)	Analyzing brain images through segmenting into different tissue types (e.g., gray matter)	Reductions in gray matter volume. Gray matter atrophy in the affected area includes the substantia nigra, putamen, and other parts of the basal ganglia.	MRI	Zheng et al. (2019)

may even be lower. Notably, a strong association was found between motor severity and greater levels of α-syn in serum and plasma (Tsao et al., 2022). Extracellular vesicles (EVs) include exosomes, which are capable of transferring α-syn across cells and from the CNS to peripheral circulation. Exosomal α-syn levels in PD patients have been found to be much greater than in controls (Tsao et al., 2022). Recently, a study with a limited sample size and a 22-month follow-up period indicated that a longitudinal rise in α-syn level in exosomes was linked to a greater likelihood of motor progression in individuals with early-stage PD (Niu et al., 2020).

The most recurring post-translational modification of α-syn is phosphorylation at serine 129. Phosphorylation of α-syn in RBC and plasma was higher in PD patients (Tsao et al., 2022). Plasma samples from approximately 189 newly diagnosed PD patients were used in a rare longitudinal experiment to measure total and phosphorylated α-syn at intervals of four to six months over three to four years (Foulds et al., 2013). Increased α-syn phosphorylation was discovered in the plasma of PD patients (Table 9.2), which was stable in contrast to total plasma α-syn levels in PD patients, which increased with time. There is a higher rate of clearance of α-syn pathology into the blood circulation, which explains the dynamics of α-syn. Another long-term study found that among people at risk for PD, baseline and longitudinal upregulation in total plasma α-syn was predictive of the development of memory impairment, as indicated by hyposmia and reduced dopamine transporter (DAT) binding (Tsao et al., 2022).

Overall, we find that while the total α-syn levels may serve as a proxy marker to denote the advancement of PD, the plasma level of oligomeric and phosphorylated α-syn may be useful as a diagnostic tool. Further validation is required to determine the usefulness of α-syn incorporating exosomes. Standardizing preanalytical factors will be required to enable comparability (Table 9.3).

9.4 MULTIPLE SCLEROSIS (MS)

Axonal damage and demyelination are hallmarks of multiple sclerosis (MS), a chronic neuroinflammatory disorder that can result in neurological abnormalities as well as severe physical and cognitive impairment. Finding lesions in at least two different regions of the CNS, such as the brain, spinal cord, and optic nerves, is the most commonly used diagnostic procedure for MS. The number of MS treatments available has increased significantly in recent years. A tailored strategy based on each patient's prognosis and treatment risks typically determines the best course of action. Making ideal decisions about the management of MS requires the use of biomarkers that can track the evolution of the disease, predict impairment, and evaluate treatment response. Chitinase-3-like-1 protein (CHI3L1), sometimes called YKL-40, has been shown to be a viable biomarker in understanding the advancement of MS. Microglia, monocytes, and activated astrocytes secrete the glycoprotein CHI3L1. It is associated with numerous inflammatory illnesses, including MS, which is considered important in controlling CNS inflammation. Because CHI3L1 is detectable in serum and CSF, it could be used as a biomarker in MS patients. Numerous other cell types, such as neutrophils, chondrocytes, osteoblasts, synovial cells, and vascular smooth muscle cells, also generate CHI3L1. It is associated with chronic inflammatory diseases and is expressed at an increased level following an inflammatory trigger (Talaat et al., 2023). Research has demonstrated that CHI3L1 concentration in CSF is elevated in MS patients in comparison to the levels in healthy subjects and that the severity and course of the disease are correlated with these levels. For example, a study found that in MS patients, CHI3L1 levels were greater in CSF and serum than those of healthy individuals. Such levels represented higher disability and disease occurrence (Talaat et al., 2023).

In addition to its many extra-skeletal benefits, vitamin D probably has other non-immunomodulatory functions. Specifically, this vitamin penetrates the many CNS cell types, including microglia, oligodendrocytes, astrocytes, and neurons, all of which have vitamin D receptors. Additionally, it may have other effects on CNS disease, such as neuroprotective, neurotrophic, and remyelinating effects (Boltjes et al., 2021). Numerous positive immunological effects were observed with vitamin D administration, including the activation of regulatory T cells (Tregs), the helpful cytokine IL-10, the reduction of pro-inflammatory Th17 cells and the harmful cytokine IL17, and the attenuation of B-cell immunoreactivity (Boltjes et al., 2021).

Neurofilaments are cytoskeleton proteins specific to neurons that can be quantified in different parts of the body. A novel MS biomarker that may be evaluated in serum samples is called NfL; it causes neuroaxonal injury (Karikari et al., 2020). A sensitive and highly selective serum marker of neuromyelitis optica (NMO) which can help distinguish between NMO and classical MS is aquaporin-4 antibody (also known as NMO-IgG or AQP4-Aβ) (Pittock et al., 2014).

Using PCR methods, non-coding single-stranded miRNA can be quantified in CSF and serum of MS patients. MS patients have dysregulated miRNAs in their immune systems and CNS, which means that individual miRNAs can be up- or downregulated in MS patients, changing the expression of different mRNA transcripts (Pérez & Eisele, 2022). In particular, there is an upregulation of miR-19a, miR-21, miR-22, miR-142-3p, miR-146a, miR-146b, miR-155, miR-210, miR-233, and miR-326 in peripheral blood mononuclear cells (PBMCs), white matter lesions, plasma, and blood cells in the brains of MS patients (Table 9.4). On the other hand, miR-328, miR-181c, miR-15a, and miR-15b are downregulated (Pérez & Eisele, 2022).

TABLE 9.4
CSF/Blood Biomarkers in MS

Biomarkers	Samples	Diagnostic Applications	Investigation Techniques	References
IL-12p40, TNF-α, TNFβ, and chitinase-3-like-1	CSF and serum samples	Elevated in active MS patients due to astrocytic activation.	Spectrophotometric analysis, ELISA	Ciubotaru et al. (2025)
Anti-VZV and anti-aquaporin-4 antibodies	CSF, serum, and plasma	Elevated in MS patients due to astrocyte damage, demyelination, and necrosis.	ELISA, western blot	Pittock et al. (2014)
Tubulin beta (TUBβ)	CSF	Higher in MS patients compared to patients with other neurological diseases.	PCR, western blotting	Ciubotaru et al. (2025)
Vitamin D	CSF	Low vitamin D with elevated EBNA-1 antibodies seen prior to MS onset.	MRI and optical coherence tomography (OCT)	Boltjes et al. (2021)
Serum gamma-glutamyltransferase and CRP	CSF, venous blood samples for serum	GGT and CRP levels were elevated.	Auto Analyzer 7180-ISE	Shu et al. (2017)
Myelin basic protein	Serum	Elevated levels in active MS.	ELISA	Ciubotaru et al. (2025)
CLEC16A gene	Whole blood/ thymic tissue samples	Astrocytic activation causes single nucleotide polymorphism (SNP) in the C-type lectin associated with MS in the prospect of an autoimmune disorder.	Microarray expression, reverse-transcriptase polymerase chain reaction analyses	Berge et al. (2013)
Interleukin 7 receptor (IL-7R)	CSF (zebra fish model)	Myelin sheath's size and structural integrity altered significantly in MS and have a major impact on immunological dysregulation and T-cell activation.	Microarray analysis, MRI	Lei et al. (2017)

MS patients had upregulated levels of serum miRNAs that support pro-regenerative and anti-inflammatory polarization. On the other hand, they discovered that pro-inflammatory miR-155 was reduced in RRMS and primary progressive multiple sclerosis (PPMS). The possible explanation for these results is that monocytes are trying to reduce CNS inflammation. However, miR-124, which stimulates an anti-inflammatory condition, is downregulated in many MS patients, particularly those with progressive MS. This suggests that homeostatic monocyte regulation is absent in progressive MS. Additionally, this study discovered that whereas miR-181a was only elevated in RRMS patients, miR-23a, miR-30c, miR-125a, miR-146a, and miR-223 were all upregulated in both RRMS and PPMS patients (Pérez & Eisele, 2022). Additionally, research has revealed that miR-181c levels in CSF and miR-150 are connected to the clinically isolated syndrome (CIS) to MS conversion. Furthermore, it has been discovered that MS patients have higher levels of miR-150 than controls (Pérez & Eisele, 2022).

9.4.1 Biomarkers 14-3-3 Protein

It is possible to assess the neuronal protein 14-3-3 in the CSF samples of MS patients and patients with Creutzfeldt–Jakob disease. The 14-3-3 protein has a mixed function in MS. The occurrence of 14-3-3 protein in the CSF is linked to a greater degree of disability, greater association of the spinal cord, and a faster development of MS or other diseases. The early accumulation of 14-3-3 protein in CSF could be associated with slower healing rates (Figure 9.2). However, other studies have reported low detection of the 14-3-3 protein in the CSF; 2 of 22 MS patients showed the protein in a study (Ciubotaru et al., 2025), while in another study, the protein was evident in 1 out of 21 CIS patients (Hein (née Maier) et al., 2008).

9.4.2 Neuron-Specific Enolase (NSE)

Patients with cerebral hemorrhage, hypoxic brain injury, or trauma have higher concentrations of neuron-specific enolase (NSE). This is a neuronal and axonal enzyme that is used to evaluate neuronal density in both their CSF and serum. According to recent study, patients with CIS lower serum and CSF NSE levels than healthy controls, showing a negative relationship between the MS severity score (MSSS), extended disability status scale (EDSS), and plasma NSE levels (Ciubotaru et al., 2025).

9.4.3 Protein S100β

There have been reports of elevated amounts of S100β protein, a glial cell-derived subunit of S100 protein, in serum and plasma in MS patients. People with "primary progressive multiple sclerosis" (PPMS) or secondary progressive multiple sclerosis (SPMS) have the highest levels. S100β is associated with a number of processes, such as oligodendrocyte differentiation, neuronal proliferation, and maintenance of astrocyte integrity. S100β levels have also been demonstrated to rise during acute exacerbations in RRMS patients; however, this window of opportunity is narrow because S100β levels were not elevated in subjects who experienced acute exacerbations more than seven days (Figure 9.2). S100β alterations are seen in individuals with amyotrophic lateral sclerosis (ALS) cerebral ischemia. When samples were collected more than seven days after an acute exacerbation, the investigation failed to note a statistically significant change in CSF and serum S100β protein content between CIS patients and healthy controls. Additionally, the EDSS score and S100β protein content were not shown to be significantly correlated in the same study (Hein (née Maier) et al., 2008). There have also not been any reported variations in the plasma S100β protein levels among the different clinical subtypes of MS (Ciubotaru et al., 2025).

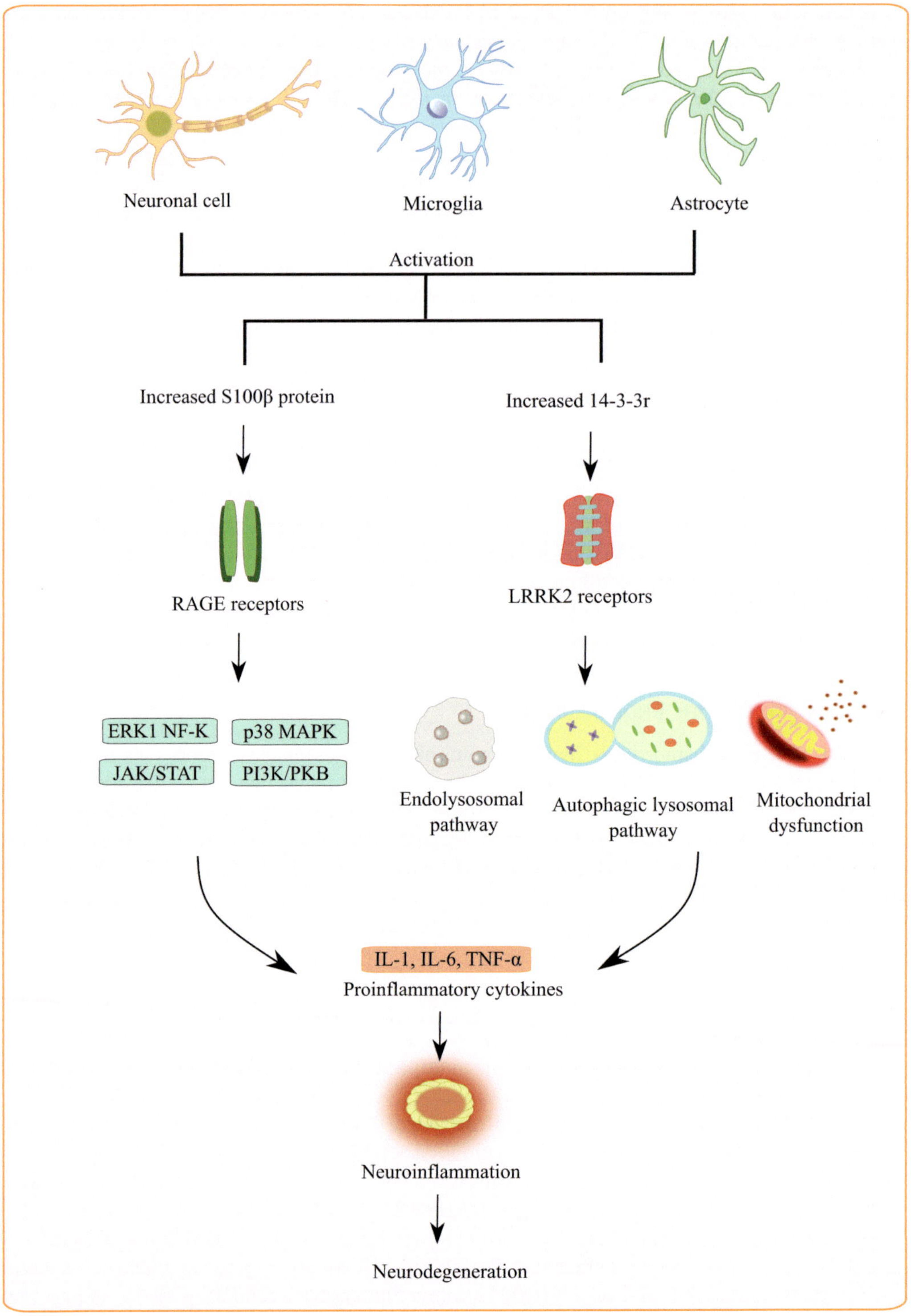

FIGURE 9.2 Damaged neurons, astrocytes, and microglia produce increased S100β and 14-3-3 proteins. These bind to RAGE and LRRK2 receptors, activate cytoplasmic inflammatory pathways, including NF-KB, p38, and MAPK, and cause mitochondrial and lysosomal dysfunction, leading to neuroinflammation, finally causing neurodegeneration.

9.4.4 Antibodies Against Aquaporin-4

Aquaporin-4 (AQP4) is expressed by astrocytes to facilitate the passage of water across cell membranes and support CNS homeostasis. On the other hand, research indicates that AQP4 is not present in MS patients. This test is intended to assist in the challenging process of distinguishing MS from NMO. This rare disease also manifests as optic nerve and spinal cord demyelination (Pittock et al., 2014).

9.4.5 Myelin Basic Protein (MBP)

Myelin basic protein (MBP), which is produced by CNS oligodendrocytes, is discovered to be elevated in MS patients' CSF and linked with QWQAS scores. MBPs in the CSF have the potential to remyelinate demyelinated lesions, which makes their use difficult (Ciubotaru et al., 2025) (Table 9.5).

9.5 HUNTINGTON'S DISEASE (HD)

Huntington's disease (HD) is a monogenic neurodegenerative condition that progresses according to age, including motor impairment, mental symptoms, persistent progressive involuntary chorea, and cognitive decline. The first exon of the huntingtin protein (HTT) gene on chromosome 4 codes for polyglutamine in the HTT and has an extension of the cytosine–adenine–guanine (CAG) repeat. This is the cause of HD. It mostly impacts the brain and striatum with a neuropathological advancement. Early intervention in the disease course can be facilitated by identifying bearers of mutations

TABLE 9.5
Imaging Biomarkers in MS

Samples/Techniques	Imaging	Diagnostic Applications	Investigation Techniques	References
Optical coherence tomography	Retinal imaging	Non-invasive measures of axonal damage and neurodegeneration that correlate with brain lesions (pRNFL) thickness are significantly reduced.	Thinner retinal nerve fiber layer (RNFL), thinner peripapillary RNFL (pRNFL) atrophy findings	Satue et al. (2016)
Sphingolipids	Human autopsy brain tissues	Sphingoids and their derivatives (dhSph and Sph, as well as dhS1P and S1P) are quantified in particular phases in the disease progression.	HPLC	Podbielska et al. (2020)
Optical coherence tomography	Retinal imaging	Reduced macular volume, retinal ganglion layer thinning in patients, and atrophy with relapsing-remitting and progressive MS.	Logistic regression analysis	Satue et al. (2016)
Arterial spin labeling (ASL)	Labeling water molecules in the blood (imaging)	Perfusion changes, lower CBF (cerebral blood flow), and increased BAT (brown adipose tissue) due to altered metabolism.	DSC-MRI, EDSS, and SDMT evaluation associated with cognitive dysfunction and disability	Satue et al. (2016)
N-acetyl-aspartate (NAA)	H-MRSI (imaging)	Reduction of N-acetyl-aspartate in MS.	MRI and spectroscopy	Vafaeyan et al. (2015)
Myo-inositol (Myo)	H-MRSI (imaging)	Increased Myo levels in MS.	MRI and spectroscopy	Vafaeyan et al. (2015)

TABLE 9.6
Biomarkers in HD

Biomarkers/Samples	Diagnostic Applications	Finding Methods	References
24-hydroxycholesterol (24OHC)/CSF	Elevated 24OHC levels correlate with disease progression and severity. Higher levels are often found in patients with more advanced stages of HD—alterations in the activity of CYP46A1, the enzyme that turns cholesterol into 24OHC.	Microarray studies	Aqel et al. (2025)
TNF-α, IL-6, IL-1β, tau, YKL-40, chitotriosidase/CSF	IL-6, chitotriosidase, tau, and YKL-40 show disease-related elevations in CSF in HD. CSF YKL-40 can independently predict clinical severity and neuronal death.	ELISA, Meso Scale Discovery antibody-based tetra-plex array with electrochemiluminescence detection.	Aqel et al. (2025)
Dopamine/CSF	Increased DA metabolite levels in the CSF of HD patients. Extra-striatal changes in the DA systems in patients have shown reductions in D2R in the amygdala and cortical areas in HD patients.	ELISA	Aqel et al. (2025)
Huntingtin/CSF	Mutant HTT can be quantified, and measures of neuronal damage and elevated levels of mHTT are noted.	Femtomolar-sensitive "single molecule counting" (SMC) immunoassay	Aqel et al. (2025)
Fatty acid amide hydrolase (FAAH)/ blood	Lymphocytes of HD patients show a dramatic decrease in FAAH activity. Anandamide hydrolase is defective in the peripheral cells of human HD patients.	Western blot analysis and ELISA	Battista et al. (2007)
HPA axis (cortisol)/ blood	Cortisol levels are higher in HD patients, who have an autosomal dominant neurodegenerative disorder caused by an expanded CAG repeat size in the gene encoding the protein huntingtin.	RIA measured plasma cortisol	Aziz et al. (2009)
HPG axis (GnRH)/ serum/plasma samples	Decrease in levels of testosterone. GnRH neurons in the hypothalamus can be detected, suggesting that testicular pathology results from a direct toxic effect of mutant huntingtin in the testes. There was no significant change in females for LH FSH.	ELISA, FT analysis	Aziz et al. (2009)
8-OHdG, TGFβ1 factor/blood-plasma	8-hydroxy-2-deoxyguanosine (8-OHdG) and TGFβ1 factor levels were found to be elevated in HD.	ELISA, HPLC, IHC, LC–MS	Killoran and Biglan (2012); Plinta et al. (2021)

prior to the onset of symptoms. The best biomarkers (Table 9.6) are those that are sensitive to prospective disease-modifying therapies and accurately represent neuropathological and clinical development. The advantages of blood-based biomarkers are their low cost, ease of use, safety, and little invasiveness.

Finding the best biochemical indicators holds promise for monitoring the neuropathological and clinical development of HD and assessing how well therapies work. Many cross-sectional and longitudinal investigations, such as the large-cohort studies PREDICT-HD and TRACK-HD, have reported biochemical biomarkers in great detail (Aqel et al., 2025).

9.5.1 Biomarkers—TDP-43

In agreement with ubiquitylated inclusions, the primary pathogenic protein in ALS and the most prevalent subtype of FTD is the transactive response (TAR) DNA binding protein of 43 kDa (TDP-43) (Nguyen et al., 2023). TDP-43 is a widely distributed RNA/DNA-binding protein that is a member of the heterogeneous nuclear ribonucleoprotein (hnRNP) family. It controls a number of processes, including autophagy, stress granule formation, microRNA maturation, mRNA transport, and RNA metabolism. The degradation and loss of synapses account for the presence of TDP-43 in CSF (Nguyen et al., 2023).

Cyclic nucleotide phosphodiesterases (PDEs) are crucial regulators of cyclic nucleotide signaling because they hydrolyze the second messengers, cAMP and cGMP, into their inactive counterparts, 5′cAMP and 5′cGMP. This allows them to regulate the intracellular quantities of these molecules. They are associated with a number of neurodegenerative diseases, such as HD. PDE10A is mostly expressed in the brain, with the striatum expressing the highest amount of PDE10A. In the basal ganglia regions of HD mouse models and postmortem brains of patients, PDE10A becomes profoundly dysregulated, which may account for the deficits in cAMP-dependent signaling and related CREB transcriptional activity (Aqel et al., 2025).

9.6 SPINAL MUSCULAR ATROPHY

An autosomal recessively inherited neurodegenerative disease called spinal muscular atrophy (SMA) involves motor neuron cell loss with muscle weakening and atrophy (Meneri et al., 2023). A variety of potential molecular and physiological markers exist that assess the function of the neurological system or biological medium like blood or CSF. These biomarkers include creatine kinase (CK) and creatinine (Crn), neurofilament proteins (NFs), plasma protein analytes, SMN2 copy number, SMN mRNA and protein levels, and other electrophysiological and imaging measurements (Meneri et al., 2023).

9.7 CONCLUSIONS AND RECOMMENDATIONS FOR THE FUTURE

The significance of biomarkers in the diagnosis of neurodegenerative illnesses is an intriguing and actively researched subject. The first step toward effectively treating these diseases is identifying a trustworthy biomarker that might help to identify clinical-diagnostic units or differentiate the different disease presentations with additional proteinopathies. Presently, a variety of biomarkers that have been developed and regularly subjected to biochemical examinations are employed in diagnostic practice. Another approach emphasizes identifying blood biomarkers that exhibit the key traits of neurodegeneration and are substantially correlated with CSF biomarker levels. Because neurodegenerative illnesses are so heterogeneous, improved clinical management can only be achieved through the use of more precise and lucid biomarkers. This is the reason that numerous investigations have concentrated on their discovery. Long before clinical symptoms manifest, protein misfolding and aggregation processes start to occur. Given the necessity of early intervention for successful treatment, the capacity to estimate every component of disease in high resolution and blood test–based diagnoses of disease prior to the manifestation of first symptoms is a very significant feature of the study. Future research should also look into metabolite analysis, exosome analyses as active carriers, and chemicals involved in communications within neurons and neuronal–glial communications.

ACKNOWLEDGMENT

The author gratefully acknowledges the Koneru Lakshmaiah Education Foundation for providing the necessary resources and institutional support that facilitated the completion of this book chapter. Their commitment to fostering scientific research is deeply appreciated. We gratefully acknowledge the use of Inkscape, an open-source vector graphics editor, for creating the diagrams presented in this manuscript.

REFERENCES

Abukuri, D. N. (2024). Novel biomarkers for Alzheimer's disease: Plasma neurofilament light and cerebrospinal fluid. *International Journal of Alzheimer's Disease, 2024*, 1–6668159. https://doi.org/10.1155/2024/6668159

Agnello, L., Gambino, C. M., Sasso, B. Lo, Bivona, G., Milano, S., Ciaccio, A. M., Piccoli, T., La Bella, V., & Ciaccio, M. (2021). Neurogranin as a novel biomarker in Alzheimer's disease. *Lab Medicine, 52*(1), 188–196. https://doi.org/10.1093/labmed/lmaa062

Aqel, S., Ahmad, J., Saleh, I., Fathima, A., Al Thani, A. A., Mohamed, W. M. Y., & Shaito, A. A. (2025). Advances in Huntington's disease biomarkers: A 10-year bibliometric analysis and a comprehensive review. *Biology, 14*(2), 129. https://doi.org/10.3390/biology14020129

Aziz, N. A., Pijl, H., Frölich, M., Van Maurits Der Graaf, A. W., Roelfsema, F., & Roos, R. A. C. (2009). Increased hypothalamic-pituitary-adrenal axis activity in Huntington's disease. *Journal of Clinical Endocrinology and Metabolism, 94*(4), 1223–1228. https://doi.org/10.1210/jc.2008-2543

Battista, N., Bari, M., Tarditi, A., Mariotti, C., Bachoud-Lévi, A. C., Zuccato, C., Finazzi-Agrò, A., Genitrini, S., Peschanski, M., Di Donato, S., Cattaneo, E., & Maccarrone, M. (2007). Severe deficiency of the fatty acid amide hydrolase (FAAH) activity segregates with the Huntington's disease mutation in peripheral lymphocytes. *Neurobiology of Disease, 27*(1), 108–116. https://doi.org/10.1016/j.nbd.2007.04.012

Berge, T., Leikfoss, I. S., & Harbo, H. F. (2013). From identification to characterization of the multiple sclerosis susceptibility gene CLEC16A. *International Journal of Molecular Sciences, 14*(3), 4476–4497. https://doi.org/10.3390/ijms14034476

Bivona, G., Sammataro, S., & Ghersi, G. (2024). Nucleic acids-based biomarkers for Alzheimer's disease diagnosis and novel molecules to treat the disease. *International Journal of Molecular Sciences, 25*(14), 7893. https://doi.org/10.3390/ijms25147893

Boltjes, R., Knippenberg, S., Gerlach, O., Hupperts, R., & Damoiseaux, J. (2021). Vitamin D supplementation in multiple sclerosis: An expert opinion based on the review of current evidence. *Expert Review of Neurotherapeutics, 21*(6), 715–725. https://doi.org/10.1080/14737175.2021.1935878

Brooks, D. J., & Pavese, N. (2011). Imaging biomarkers in Parkinson's disease. *Progress in Neurobiology, 95*(4), 614–628. https://doi.org/10.1016/j.pneurobio.2011.08.009

Ciubotaru, A., Smihor, M. I., Grosu, C., Alexa, D., Covali, R., Anicăi, R. C., Păvăleanu, I., Cucu, A. I., Bobu, A. M., Ghiciuc, C. M., & Ignat, E. B. (2025). Neurodegenerative biomarkers in multiple sclerosis: At the interface between research and clinical practice. *Diagnostics, 15*(9), 1178. https://doi.org/10.3390/diagnostics15091178

Doxakis, E. (2022). Insights into the multifaceted role of circular RNAs: Implications for Parkinson's disease pathogenesis and diagnosis. *npj Parkinson's Disease, 8*(1). https://doi.org/10.1038/s41531-021-00265-9

Duan, W. X., Wang, F., Liu, J. Y., & Liu, C. F. (2024). Relationship between short-chain fatty acids and Parkinson's disease: A review from pathology to clinic. *Neuroscience Bulletin, 40*(4), 500–516. https://doi.org/10.1007/s12264-023-01123-9

Erkkinen, M. G., Kim, M. O., & Geschwind, M. D. (2018). Clinical neurology and epidemiology of the major neurodegenerative diseases. *Cold Spring Harbor Perspectives in Biology, 10*(4). https://doi.org/10.1101/cshperspect.a033118

Foster, E. M., Dangla-Valls, A., Lovestone, S., Ribe, E. M., & Buckley, N. J. (2019). Clusterin in Alzheimer's disease: Mechanisms, genetics, and lessons from other pathologies. *Frontiers in Neuroscience, 13*(FEB), 427607. https://doi.org/10.3389/fnins.2019.00164

Foulds, P. G., Diggle, P., Mitchell, J. D., Parker, A., Hasegawa, M., Masuda-Suzukake, M., Mann, D. M. A., & Allsop, D. (2013). A longitudinal study on α-synuclein in blood plasma as a biomarker for Parkinson's disease. *Scientific Reports, 3*. https://doi.org/10.1038/srep02540

Gao, R., & Ma, S. L. (2022). Is mitochondria DNA variation a biomarker for AD? *Genes, 13*(10), 1789. https://doi.org/10.3390/genes13101789

Gurram, P. C., Manandhar, S., Satarker, S., Mudgal, J., Arora, D., & Nampoothiri, M. (2022). Dopaminergic signaling as a plausible modulator of astrocytic toll-like receptor 4: A crosstalk between neuroinflammation and cognition. *CNS & Neurological Disorders – Drug Targets, 21*. https://doi.org/10.2174/1871527321666220413090541

Hansson, O. (2021). Biomarkers for neurodegenerative diseases. *Nature Medicine*, *27*(6), 954–963. https://doi.org/10.1038/s41591-021-01382-x

Hein (née Maier), K., Köhler, A., Diem, R., Sättler, M. B., Demmer, I., Lange, P., Bähr, M., & Otto, M. (2008). Biological markers for axonal degeneration in CSF and blood of patients with the first event indicative for multiple sclerosis. *Neuroscience Letters*, *436*(1), 72–76. https://doi.org/10.1016/j.neulet.2008.02.064

Hu, Y. (2022). Electroneurography abnormality in Parkinson's disease: A potential biomarker to help diagnosis. *Ageing and Neurodegenerative Diseases*, *2*(2), 9. https://doi.org/10.20517/and.2022.04

Iwaki, H., Blauwendraat, C., Leonard, H. L., Kim, J. J., Liu, G., Maple-Grødem, J., Corvol, J. C., Pihlstrøm, L., van Nimwegen, M., Hutten, S. J., Nguyen, K. D. H., Rick, J., Eberly, S., Faghri, F., Auinger, P., Scott, K. M., Wijeyekoon, R., Van Deerlin, V. M., Hernandez, D. G., … Nalls, M. A. (2019). Genomewide association study of Parkinson's disease clinical biomarkers in 12 longitudinal patients' cohorts. *Movement Disorders*, *34*(12), 1839–1850. https://doi.org/10.1002/mds.27845

Jansen, W. J., Janssen, O., Tijms, B. M., Vos, S. J. B., Ossenkoppele, R., Visser, P. J., Aarsland, D., Alcolea, D., Altomare, D., Von Arnim, C., Baiardi, S., Baldeiras, I., Barthel, H., Bateman, R. J., Van Berckel, B., Binette, A. P., Blennow, K., Boada, M., Boecker, H., … Zetterberg, H. (2022). Prevalence estimates of amyloid abnormality across the Alzheimer disease clinical spectrum. *JAMA Neurology*, *79*(3), 228–243. https://doi.org/10.1001/jamaneurol.2021.5216

Janssen Daalen, J. M., van den Bergh, R., Prins, E. M., Moghadam, M. S. C., van den Heuvel, R., Veen, J., Mathur, S., Meijerink, H., Mirelman, A., Darweesh, S. K. L., Evers, L. J. W., & Bloem, B. R. (2024). Digital biomarkers for non-motor symptoms in Parkinson's disease: The state of the art. *npj Digital Medicine*, *7*(1), 1–22. https://doi.org/10.1038/s41746-024-01144-2

Karikari, T. K., Pascoal, T. A., Ashton, N. J., Janelidze, S., Benedet, A. L., Rodriguez, J. L., Chamoun, M., Savard, M., Kang, M. S., Therriault, J., Schöll, M., Massarweh, G., Soucy, J. P., Höglund, K., Brinkmalm, G., Mattsson, N., Palmqvist, S., Gauthier, S., Stomrud, E., … Blennow, K. (2020). Blood phosphorylated tau 181 as a biomarker for Alzheimer's disease: A diagnostic performance and prediction modelling study using data from four prospective cohorts. *The Lancet Neurology*, *19*(5), 422–433. https://doi.org/10.1016/S1474-4422(20)30071-5

Karlsson, L., Vogel, J., Arvidsson, I., Åström, K., Janelidze, S., Blennow, K., Palmqvist, S., Stomrud, E., Mattsson-Carlgren, N., & Hansson, O. (2024). Cerebrospinal fluid reference proteins increase accuracy and interpretability of biomarkers for brain diseases. *Nature Communications*, *15*(1), 1–15. https://doi.org/10.1038/s41467-024-47971-5

Khalil, M., Teunissen, C. E., Otto, M., Piehl, F., Sormani, M. P., Gattringer, T., Barro, C., Kappos, L., Comabella, M., Fazekas, F., Petzold, A., Blennow, K., Zetterberg, H., & Kuhle, J. (2018). Neurofilaments as biomarkers in neurological disorders. *Nature Reviews Neurology*, *14*(10), 577–589. https://doi.org/10.1038/s41582-018-0058-z

Khoo, S. K., Petillo, D., Kang, U. J., Resau, J. H., Berryhill, B., Linder, J., Forsgren, L., & Neuman, L. A. (2012). Plasma-based circulating microRNA biomarkers for Parkinson's disease. *Journal of Parkinson's Disease*, *2*(4), 321–331. https://doi.org/10.3233/JPD-012144

Killoran, A., & Biglan, K. M. (2012). 8-OHdG: Its (limited) potential as a biomarker for Huntington's disease. *Biomarkers in Medicine*, *6*(6), 777–780. https://doi.org/10.2217/bmm.12.84

Klassen, B. T., Hentz, J. G., Shill, H. A., Driver-Dunckley, E., Evidente, V. G. H., Sabbagh, M. N., Adler, C. H., & Caviness, J. N. (2011). Quantitative EEG as a predictive biomarker for Parkinson disease dementia. *Neurology*, *77*(2), 118–124. https://doi.org/10.1212/WNL.0b013e318224af8d

Kremer, T., Taylor, K. I., Siebourg-Polster, J., Gerken, T., Staempfli, A., Czech, C., Dukart, J., Galasko, D., Foroud, T., Chahine, L. M., Coffey, C. S., Simuni, T., Weintraub, D., Seibyl, J., Poston, K. L., Toga, A. W., Tanner, C. M., Marek, K., Hutten, S. J., … Mollenhauer, B. (2021). Longitudinal analysis of multiple neurotransmitter metabolites in cerebrospinal fluid in early Parkinson's disease. *Movement Disorders*, *36*(8), 1972–1978. https://doi.org/10.1002/mds.28608

Lei, X., Cai, S., Chen, Y., Cui, J., Wang, Y., Li, Z., & Li, Y. (2017). Down-regulation of interleukin 7 receptor (IL-7R) contributes to central nervous system demyelination. *Oncotarget*, *8*(17), 28395–28407. https://doi.org/10.18632/oncotarget.16081

Leoni, V. (2011). The effect of apolipoprotein E (ApoE) genotype on biomarkers of amyloidogenesis, tau pathology and neurodegeneration in Alzheimer's disease. *Clinical Chemistry and Laboratory Medicine*, *49*(3), 375–383. https://doi.org/10.1515/CCLM.2011.088

Lind-Holm Mogensen, F., Scafidi, A., Poli, A., & Michelucci, A. (2023). PARK7/DJ-1 in microglia: Implications in Parkinson's disease and relevance as a therapeutic target. *Journal of Neuroinflammation*, *20*(1), 1–19. https://doi.org/10.1186/s12974-023-02776-z

Liu, C. Z., Guo, D. S., Ma, J. J., Dong, L. R., Chang, Q. Q., Yang, H. Q., Liang, K. K., Li, X. H., Yang, D. W., Fan, Y. Y., Gu, Q., Chen, S. Y., & Li, D. S. (2022). Correlation of matrix metalloproteinase 3 and matrix metalloproteinase 9 levels with non-motor symptoms in patients with Parkinson's disease. *Frontiers in Aging Neuroscience*, *14*, 889257. https://doi.org/10.3389/fnagi.2022.889257

Meneri, M., Abati, E., Gagliardi, D., Faravelli, I., Parente, V., Ratti, A., Verde, F., Ticozzi, N., Comi, G. P., Ottoboni, L., & Corti, S. (2023). Identification of novel biomarkers of spinal muscular atrophy and therapeutic response by proteomic and metabolomic profiling of human biological fluid samples. *Biomedicines*, *11*(5), 1254. https://doi.org/10.3390/biomedicines11051254

Michetti, F., D'Ambrosi, N., Toesca, A., Puglisi, M. A., Serrano, A., Marchese, E., Corvino, V., & Geloso, M. C. (2019). The S100B story: From biomarker to active factor in neural injury. *Journal of Neurochemistry*, *148*(2), 168–187. https://doi.org/10.1111/jnc.14574

Nampoothiri, M., Gurram, P. C., Bojja, S. L., & Satarker, S. (2023). Phytochemical nanoformulations for neurological disorders. *Advances in Phytonanotechnology for Treatment of Various Diseases*, 179–194. https://doi.org/10.1201/9781003231721-7

Nassar, A., Kodi, T., Satarker, S., Chowdari Gurram, P., Upadhya, D., SM, F., Mudgal, J., & Nampoothiri, M. (2022). Astrocytic MicroRNAs and transcription factors in Alzheimer's disease and therapeutic interventions. *Cells*, *11*(24). https://doi.org/10.3390/cells11244111

Ng, T. K. S., Ho, C. S. H., Tam, W. W. S., Kua, E. H., & Ho, R. C. M. (2019). Decreased serum brain-derived neurotrophic factor (BDNF) levels in patients with Alzheimer's disease (AD): A systematic review and meta-analysis. *International Journal of Molecular Sciences*, *20*(2), 257. https://doi.org/10.3390/ijms20020257

Nguyen, T. B., Miramontes, R., Chillon-Marinas, C., Maimon, R., Vazquez-Sanchez, S., Lau, A. L., McClure, N. R., England, W. E., Singha, M., Stocksdale, J. T., Jang, K.-H., Jung, S., McKnight, J. I., Ho, L. N., Faull, R. L. M., Steffan, J. S., Reidling, J. C., Jang, C., Lee, G., … Thompson, L. M. (2023). Aberrant splicing in Huntington's disease via disrupted TDP-43 activity accompanied by altered m6A RNA modification. *BioRxiv : The Preprint Server for Biology*, *28*(2), 280–292. https://doi.org/10.1101/2023.10.31.565004

Niu, M., Li, Y., Li, G., Zhou, L., Luo, N., Yao, M., Kang, W., & Liu, J. (2020). A longitudinal study on α-synuclein in plasma neuronal exosomes as a biomarker for Parkinson's disease development and progression. *European Journal of Neurology*, *27*(6), 967–974. https://doi.org/10.1111/ene.14208

Olsson, B., Lautner, R., Andreasson, U., Öhrfelt, A., Portelius, E., Bjerke, M., Hölttä, M., Rosén, C., Olsson, C., Strobel, G., Wu, E., Dakin, K., Petzold, M., Blennow, K., & Zetterberg, H. (2016). CSF and blood biomarkers for the diagnosis of Alzheimer's disease: A systematic review and meta-analysis. *The Lancet Neurology*, *15*(7), 673–684. https://doi.org/10.1016/S1474-4422(16)00070-3

Pérez, M. M. G., & Eisele, S. J. G. (2022). MicroRNAs as a possible biomarker in the treatment of multiple sclerosis. *IBRO Neuroscience Reports*, *13*, 492–499. https://doi.org/10.1016/j.ibneur.2022.11.001

Pittock, S. J., Lennon, V. A., Bakshi, N., Shen, L., McKeon, A., Quach, H., Briggs, F. B. S., Bernstein, A. L., Schaefer, C. A., & Barcellos, L. F. (2014). Seroprevalence of aquaporin-4-IgG in a Northern California population representative cohort of multiple sclerosis. *JAMA Neurology*, *71*(11), 1433–1436. https://doi.org/10.1001/jamaneurol.2014.1581

Plinta, K., Plewka, A., Wójcik-Pędziwiatr, M., Zmarzły, N., Rudziński, M., & Rudzińska-Bar, M. (2021). Is TGF-β1 a biomarker of Huntington's disease progression? *Journal of Clinical Medicine*, *10*(13), 3001. https://doi.org/10.3390/jcm10133001

Podbielska, M., Szulc, Z. M., Ariga, T., Pokryszko-Dragan, A., Fortuna, W., Bilinska, M., Podemski, R., Jaskiewicz, E., Kurowska, E., Yu, R. K., & Hogan, E. L. (2020). Distinctive sphingolipid patterns in chronic multiple sclerosis lesions. *Journal of Lipid Research*, *61*(11), 1464–1479. https://doi.org/10.1194/jlr.RA120001022

Qiu, Y., Jacobs, D. M., Messer, K., Salmon, D. P., Wellington, C. L., Stukas, S., Revta, C., Brewer, J. B., Léger, G. C., Askew, B., Donahue, L., Kaplita, S., Coric, V., Qureshi, I. A., & Feldman, H. H. (2025). Prognostic value of plasma biomarkers for informing clinical trial design in mild-to-moderate Alzheimer's disease. *Alzheimer's Research & Therapy*, *17*(1), 97. https://doi.org/10.1186/s13195-025-01745-3

Rideout, H. J., Chartier-Harlin, M. C., Fell, M. J., Hirst, W. D., Huntwork-Rodriguez, S., Leyns, C. E. G., Mabrouk, O. S., & Taymans, J. M. (2020). The current state-of-the art of LRRK2-based biomarker assay development in Parkinson's disease. *Frontiers in Neuroscience*, *14*, 566602. https://doi.org/10.3389/fnins.2020.00865

Russell, A. C., Šimurina, M., Garcia, M. T., Novokmet, M., Wang, Y., Rudan, I., Campbell, H., Lauc, G., Thomas, M. G., & Wang, W. (2017). The N-glycosylation of immunoglobulin G as a novel biomarker of Parkinson's disease. *Glycobiology*, *27*(5), 501–510. https://doi.org/10.1093/glycob/cwx022

Satue, M., Obis, J., Rodrigo, M. J., Otin, S., Fuertes, M. I., Vilades, E., Gracia, H., Ara, J. R., Alarcia, R., Polo, V., Larrosa, J. M., Pablo, L. E., & Garcia-Martin, E. (2016). Optical coherence tomography as a biomarker for diagnosis, progression, and prognosis of neurodegenerative diseases. *Journal of Ophthalmology*, *2016*(1), 8503859. https://doi.org/10.1155/2016/8503859

Shu, Y., Li, R., Qiu, W., Chang, Y., Sun, X., Fang, L., Chen, C., Yang, Y., Lu, Z., Hu, X., & Kermode, A. G. (2017). Association of serum gamma-glutamyltransferase and C-reactive proteins with neuromyelitis optica and multiple sclerosis. *Multiple Sclerosis and Related Disorders*, *18*, 65–70. https://doi.org/10.1016/j.msard.2017.09.021

Shusharina, N., Yukhnenko, D., Botman, S., Sapunov, V., Savinov, V., Kamyshov, G., Sayapin, D., & Voznyuk, I. (2023). Modern methods of diagnostics and treatment of neurodegenerative diseases and depression. *Diagnostics*, *13*(3), 573. https://doi.org/10.3390/diagnostics13030573

Suárez-Calvet, M., Kleinberger, G., Araque Caballero, M. Á., Brendel, M., Rominger, A., Alcolea, D., Fortea, J., Lleó, A., Blesa, R., Gispert, J. D., Sánchez-Valle, R., Antonell, A., Rami, L., Molinuevo, J. L., Brosseron, F., Traschütz, A., Heneka, M. T., Struyfs, H., Engelborghs, S., … Haass, C. (2016). sTREM 2 cerebrospinal fluid levels are a potential biomarker for microglia activity in early-stage Alzheimer's disease and associate with neuronal injury markers. *EMBO Molecular Medicine*, *8*(5), 466–476. https://doi.org/10.15252/emmm.201506123

Talaat, F., Abdelatty, S., Ragaie, C., & Dahshan, A. (2023). Chitinase-3-like 1-protein in CSF: A novel biomarker for progression in patients with multiple sclerosis. *Neurological Sciences*, *44*(9), 3243–3252. https://doi.org/10.1007/s10072-023-06764-2

Tsao, H. H., Huang, C. G., & Wu, Y. R. (2022). Detection and assessment of alpha-synuclein in Parkinson disease. *Neurochemistry International*, *158*, 105358. https://doi.org/10.1016/j.neuint.2022.105358

Vafaeyan, H., Ebrahimzadeh, S. A., Rahimian, N., Alavijeh, S. K., Madadi, A., Faeghi, F., Harirchian, M. H., & Rad, H. S. (2015). Quantification of diagnostic biomarkers to detect multiple sclerosis lesions employing 1H-MRSI at 3T. *Australasian Physical and Engineering Sciences in Medicine*, *38*(4), 611–618. https://doi.org/10.1007/s13246-015-0390-1

van den Berg, M. M. J., Krauskopf, J., Ramaekers, J. G., Kleinjans, J. C. S., Prickaerts, J., & Briedé, J. J. (2020). Circulating microRNAs as potential biomarkers for psychiatric and neurodegenerative disorders. *Progress in Neurobiology*, *185*. https://doi.org/10.1016/j.pneurobio.2019.101732

Varma, V. R., Varma, S., An, Y., Hohman, T. J., Seddighi, S., Casanova, R., Beri, A., Dammer, E. B., Seyfried, N. T., Pletnikova, O., Moghekar, A., Wilson, M. R., Lah, J. J., O'Brien, R. J., Levey, A. I., Troncoso, J. C., Albert, M. S., & Thambisetty, M. (2017). Alpha-2 macroglobulin in Alzheimer's disease: A marker of neuronal injury through the RCAN1 pathway. *Molecular Psychiatry*, *22*(1), 13–23. https://doi.org/10.1038/mp.2016.206

Wankhede, N. L., Koppula, S., Umare, P. D., Umare, M. D., Lokhande, S. K., Taksande, B. G., Upaganlawar, A. B., Umekar, M. J., Kopalli, S. R., & Kale, M. B. (2025). Neuroimaging biomarkers: A critical review of their role in neurological disease diagnosis and management. *Brain Disorders*, *18*, 100222. https://doi.org/10.1016/j.dscb.2025.100222

Xiang, C., Cong, S., Tan, X., Ma, S., Liu, Y., Wang, H., & Cong, S. (2022). A meta-analysis of the diagnostic utility of biomarkers in cerebrospinal fluid in Parkinson's disease. *npj Parkinson's Disease*, *8*(1), 1–8. https://doi.org/10.1038/s41531-022-00431-7

Zheng, D., Chen, C., Song, W. C., Yi, Z. Q., Zhao, P. W., Zhong, J. G., Dai, Z. Y., Shi, H. C., & Pan, P. L. (2019). Regional gray matter reductions associated with mild cognitive impairment in Parkinson's disease: A meta-analysis of voxel-based morphometry studies. *Behavioural Brain Research*, *371*, 111973. https://doi.org/10.1016/j.bbr.2019.111973

10 Cutting-Edge Research and Technological Advancements in Neurodegenerative Diseases

Puneet K. Samaiya
Department of Pharmacy, Shri G.S. Institute of Technology & Science, Indore, Madhya Pradesh, India

10.1 INTRODUCTION

Neurodegenerative diseases (NDDs) are a heterogeneous group of age-related diseases of the central nervous system (CNS) and the peripheral nervous system (PNS). A battery of events, such as genetic, environmental, and sedentary lifestyle, are the major contributors to NDDs. The execution of NDDs results in motor, cognitive, behavioral, and social anomalies. These diseases are very common in elderly individuals, and their frequency increases with age nowadays (Figure 10.1). Coincidentally, the increasing life expectancy and declining fertility rate are resulting in a hike in the median age pervasiveness of these disorders and aggravated health and economic burdens globally (Ricci, 2024).

10.1.1 Alzheimer's Disease (AD)

Alzheimer's disease (AD) is a kind of NDD with a global incidence of 70% of identified cases out of 55 million cases of dementia. Pathological alterations in the cleavage of amyloid precursor protein (*APP*) and the generation of the *APP* fragment amyloid-beta (Aβ) and accumulation of hyperphosphorylated tau collectively impart their role in diminishing synaptic function, synapse loss, and neurodegeneration. Additionally, metabolic, vascular, and inflammatory changes, along with other co-occurring pathologies, are crucial factors in the progression of the disease (Figure 10.1). A few cases of early-onset AD (EOAD) are governed by the mutation in three genes, namely *APP*, *PSEN1* (presenilin 1), and *PSEN2* (presenilin 2), with the clinical symptoms arising before the age of 65. A wide number of patients show late-onset AD (LOAD), which appears later in life, is sporadic and non-hereditary. However, current evidence shows the existence of several genetic risk factors, among which is the presence of the E4 allele in the *ApoE* (apolipoprotein E) gene occurring in about 16% of the population (Rostagno, 2022). The details of *ApoE* have been discussed in another section.

10.1.1.1 Clinical Diagnostics for AD Detection

Earlier in the 21st century, the diagnosis of AD was likely to be based on clinical features in postmortem subjects. After this, techniques like cerebrospinal fluid (CSF) biomarkers and positron emission tomography (PET) were introduced that significantly improved the clinical diagnosis of molecular and cellular signatures (amyloid plaques and tau) of AD pathogenesis. Later in 2002, the groundbreaking emergence of Pittsburgh compound B (PiB) facilitated the visualization of Aβ plaques in living individuals (Hampel et al., 2023; Krishnadas et al., 2021). At present, florbetapir, florbetaben, and flutemetamol are used as PET ligands for diagnosis but are limited in use due to their higher expense.

 DOI: 10.1201/9781003530510-10

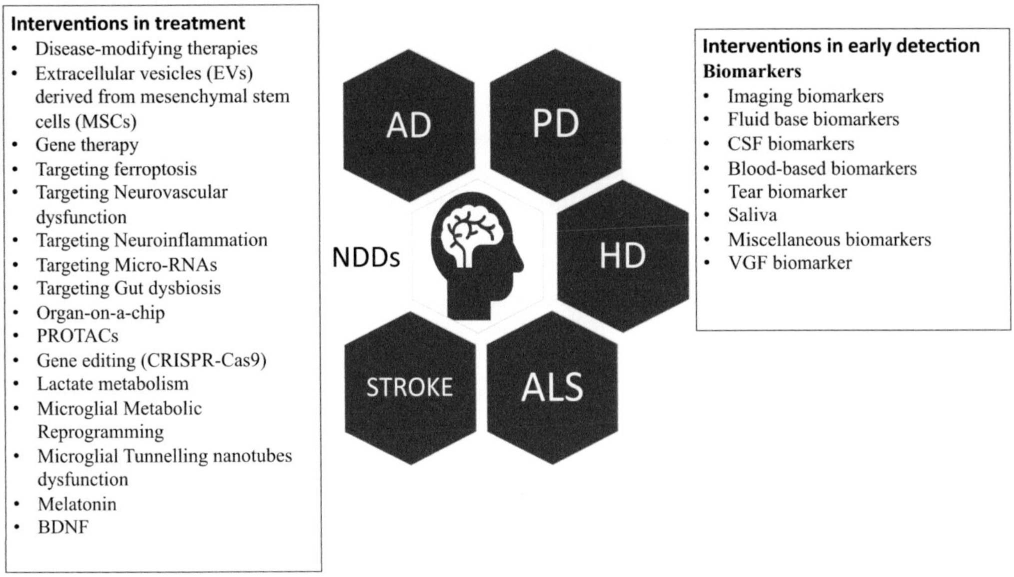

FIGURE 10.1 Interventions for early detection as well as treatment for NDDs.

10.1.1.2 Biomarkers

10.1.1.2.1 Imaging Biomarkers

The three best validated neuroimaging biomarkers for AD are magnetic resonance imaging (MRI), 18-fluorodeoxyglucose (18FDG)-PET, and amyloid-PET, of which medial temporal lobe atrophy on MRI and posterior cingulate and temporoparietal hypometabolism on 18FDG-PET are measures of neurodegeneration, and cortical Aβ deposition on amyloid-PET imaging. These three biomarkers have nearly accomplished the clinical validity of phases 1–3. However, evidence for their clinical utility for phases 4 and 5 is yet to be achieved (Frisoni et al., 2017).

10.1.1.3 Fluid Base Biomarkers

10.1.1.3.1 CSF Biomarkers

Among the various phosphorylated tau (p-tau) forms, p-tau18, p-tau217, and p-tau231—along with phosphorylation occupancies, which represent the ratio of phosphorylated to non-phosphorylated mid-region tau (np-tau) fragments—have been extensively studied. These forms display significant correlations with Aβ pathology and moderate correlations with tau, as measured by both PET and neuropathological assessments. These biomarkers are valuable in enhancing the diagnostic process for AD and in predicting disease progression (Salvadó et al., 2024).

10.1.1.3.2 Blood-Based Biomarkers (BBB)

FDA has already approved PET and CSF biomarkers for clinical diagnosis of AD research. However, these biomarkers are not practiced routinely. According to a 2023 report, 8.8 million Indians older than 60 years live with dementia. The number of individuals will reach 16.9 million in 2036 due to the growth in the older Indian population. Plasma Aβ 42/40, phosphorylated-tau (pTau) isoforms, NFL, and glial fibrillary acidic protein (GFAP) are a few blood-based biomarkers (BBB) biomarkers that are closest to clinical implementation for early diagnosis of AD-associated dementia (Altuna-Azkargorta & Mendioroz-Iriarte, 2021; Hampel et al., 2023).

A new CSF proteomic analysis on 3416 individuals from six prospective AD case–control cohorts identified synapse proteins as key predictors, leading to the development of the YWHAG:

NPTX2 ratio as an improved biomarker. This ratio strongly correlates with AD progression and is shown as a blood test (Oh et al., 2024).

10.1.1.3.3 Tear Biomarker

Due to their extended shelf life and non-invasiveness, tears can represent a fruitful source of information relevant to various illnesses like AD. The levels of antibacterial and immunomodulatory proteins (AMP) present in tears (like lipocalin-1, lactotransferrin, and lysozyme-C) play a defensive role against various pathogens and tear composition. These AMPs with their levels may also represent the biomarkers for NDDs like AD. Apart from these, certain other components like Aβ38, Aβ40, Aβ42, T-tau, P-tau, and microRNAs can also be detected in tear fluid (Kenny et al., 2019).

10.1.1.4 Saliva

The potential AD biomarkers found in saliva include Aβ peptides, T-tau and P-tau, acetylcholinesterase, lactoferrin, and trehalose, each associated with distinct AD-related pathophysiological mechanisms (Ashton et al., 2021).

10.1.1.4.1 Miscellaneous Biomarkers

Neural injury and synaptic dysfunction biomarkers like neurofilament light (NFL), neurogranin, BACE1, synaptotagmin, SNAP-25, GAP-43, and synaptophysin have also been developed. Apart from this some biomarkers of neuroinflammation such as sTREM2 and YKL-40 have also been extensively studied to understand the underlying mechanism of AD (Bălașa et al., 2020; Chang et al., 2021).

10.1.1.4.2 Microbiome and EEG as Early Biomarkers for AD

The microbiome and EEG show potential non-invasive biomarkers for early detection of AD. Early diagnostic methods can be developed based on the unique microbiome signature and EEG patterns seen in people with MCI. Combining microbiome and EEG data, early AD diagnosis could be more accurate and reliable, which could improve patient outcomes by allowing for immediate intervention (Krothapalli et al., 2024).

10.1.1.5 Disease-Modifying Therapies of AD (DMAD)

Currently, only the FDA-approved first-choice drugs for the management of AD Cholinesterase Inhibitors like donepezil, rivastigmine, galantamine, and N-methyl D-aspartate (NMDA) antagonist memantine are available. However, these drugs only provide symptomatic treatment. Therefore, to prevent the progression of disease, new and specific DMAD therapies are in the pipeline. Recently, the FDA has approved some monoclonal antibodies under DMAD intervention, which are aducanemab (approved by the FDA in June 2020)(Budd Haeberlein et al., 2022), lecanemab (recently approved), and donanemab (expected to receive FDA approval soon). These monoclonal antibodies act by removing amyloid from the brain as immunotherapy. Remternetug is another one of these promising amyloid-targeting immunotherapy drugs in the pipeline. Aducanemab received accelerated FDA approval in June 2020. It was shown to reduce Aβ plaque in the brain. However, it missed the primary phase III trial endpoint of clinical improvement. Lecanemab received accelerated FDA approval in January 2023. It reduced the Aβ burden in the brain. The phase III trial showed a 27% slowing of disease progression. Kisunla (Donanemab-azbt) is a humanized IgG1 monoclonal antibody targeted against an epitope at the N-terminal of a specific type of Aβ-pyroglutamate, which is found only in brain amyloid plaques associated with AD. It has just been approved by the FDA for early symptomatic AD. It slowed the cognitive and functional decline by up to 35% compared to the placebo at 18 months of the phase 3 clinical trial (*https://investor.lilly.com/news-releases/news-release-details/lillys-kisunlatm-donanemab-azbt-approved-fda-treatment-early*; Mintun et al., 2021).

10.1.1.6 Extracellular Vesicles (EVs) Derived from Mesenchymal Stem Cells (MSCs) and CNS

Recently, inflammation-related activated astrocytes have gained increasing attention concerning NDDs. It has been shown that the presence of type 1 astrocytes (neurotoxic phenotype) that secrete proinflammatory cytokines has been associated with NDDs. Similarly, CNS-derived EVs are the next-level neural circuit biomarkers (Goetzl et al., 2019).

10.1.1.7 Apolipoprotein E (APoE)

According to the post-multiple large-scale genome-wide association studies (GWAS) and GWAS meta-analyses, the ε4 allele of the APoE gene continues to be the most prevalent genetic risk factor associated with sporadic or late-onset AD (LOAD). While the APoE ε2 allele has an estimated two times decreased lifetime risk of AD. ApoE, a 299 amino acid glycoprotein with a molecular weight of 34 kDa, acts as a lipid transporter aiming to facilitate cholesterol and phospholipids in the periphery and CNS. Its synthesis in the periphery is governed by hepatocytes and macrophages, and in the CNS by astrocytes, activated microglia, vascular mural cells, choroid plexus cells, and to a lesser extent in stressed neurons. Recently, it has been known that APoE interacts and co-deposits with Aβ as well as tau, which can be the strongest genetic risk factor for sporadic or late-onset AD (LOAD) and the principal mechanism of APoE in AD risk (Raulin et al., 2022). Gene-therapy-mediated transition of APoE ε4 isoform to protective APoE ε2 isoform can be a fruitful approach. Another committed APoE-targeted therapeutic approach is the use of immunotherapies against ApoE–Aβ complexes using anti-ApoE antibodies. Recently, an anti-APoE monoclonal antibody has been reported to modulate the APoE Christchurch mutation by downregulating ApoE–heparin sulfate proteoglycan (HSPG) interaction and reducing tau pathology pre-clinically, providing a novel therapeutic intervention compounding genetics as well as antibodies for an ApoE-targeted therapy (Narasimhan et al., 2024).

10.1.2 Parkinson's Disease

The primary Parkinson's disease (PD) neuropathology seen at autopsy is that of alpha-synuclein (α-syn) aggregates, which is a 140 amino acid long intrinsically disordered protein (IDP) encoded by the *SNCA* gene located on chromosome 4. The α-syn promotes the loss of pre-synaptic dopaminergic neurons in the striatal pathway. The basic test for predicting PD is electroencephalography (EEG), which is a cost-effective, non-invasive, and convenient technique for detecting PD pathogenesis. The PET scan is not a suitable technique for detecting α-syn in PD non-invasively, as in the case with Aβ and tau. Even though the nigro-striatal dopaminergic synaptic endings can be estimated with SPECT and PET imaging (Brücke et al., 2000).

10.1.2.1 Genetic Test for PD

Detection of mutations in genes like LRRK2 (leucine-rich repeat kinase-2), α-syn, and GBA1 (glucocerebrosidase) has revealed a better understanding of the molecular and pathophysiological mechanisms of the hereditary forms and PD in general (Munhoz et al., 2024).

10.1.2.2 Clinical Diagnosis for PD Detection

10.1.2.2.1 VGF Biomarker

It has been observed that the levels of VGF are elevated in the substantia nigra (SN) and peripheral circulation (Plasma) of PD patients. Therefore, VGF can play a useful role as a biomarker in detecting early stages of PD. Expression differences of VGF-derived peptides have also been associated with frontotemporal lobar dementias, amyotrophic lateral sclerosis (ALS), Lewy body diseases, Huntington's disease (HD), pain, schizophrenia, bipolar disorder, depression, and antidepressant response (Quinn et al., 2021).

10.1.2.2.2 Imaging Biomarkers

For the clinical detection of pre-synaptic dopaminergic neuron dysfunction, dopamine transporter DAT-SPECT is a reliable test. However, DAT SPECT is *not* a reliable test to differentiate PD from atypical Parkinsonism. SPECT-TRODAT on the other hand has a precise sensitivity for estimating the decrement of DAT in PD subjects as compared to other imaging techniques. More recently, nigrosome and neuromelanine evaluation through MRI may act as an in vivo marker for the progression of nigral degeneration from early to advanced stages of PD (Pavese & Tai, 2018).

10.1.2.2.3 Cerebrospinal Fluid (CSF) Biomarkers

A few potential CSF biomarkers have been investigated in patients with PD, including total αsynuclein, oligomeric α-synuclein, lysosomal enzyme activities, and neurofilament light chain. CSF α-synuclein seed amplification assays (SAA) have been recently used for early identification of people at risk of PD and detection of disease heterogeneity (Saw et al., 2024).

10.1.3 Huntington's Disease (HD)

HD is an autosomal dominant neurodegenerative disorder caused by an expanded polyglutamine-coding (CAG) trinucleotide repeat in the huntingtin (*HTT*) gene. A progressive course of motor, cognitive, and psychiatric symptoms that have been observed in HD has been attributed to the extensive cellular degeneration of the medium spiny neurons of the caudate nucleus and putamen (Dhingra & Gaidhane, 2023).

10.1.3.1 Fluid-Based HD Biomarkers

Recently, quantification of mutant huntingtin protein levels and neurofilament light chain (NFL) in the CSF is emerging as an early biochemical biomarker of HD. Previous studies reported increased plasma levels of IL-6, IL-10, and YKL-40 in HD patients (Rodrigues et al., 2016; Soltani Khaboushan et al., 2023).

10.1.3.2 Neuroimaging Biomarkers

Various studies conducted on imaging biomarkers assess brain atrophy by employing volumetric MRI alongside various other MRI techniques, such as functional MRI, magnetic resonance spectroscopy (MRS), transcranial magnetic stimulation, and brain iron analysis. PET scans have also been used to examine brain glucose metabolism, PDE10A activity, as well as dopamine D1 and D2 receptor activity in the striatum (van Eimeren et al., 2023).

10.1.4 Amyotrophic Lateral Sclerosis (ALS)

ALS, the leading cause of motor neuron disease (MND), is characterized by muscle weakness and atrophy. Approximately 50% of ALS patients experience some sort of cognitive or behavioral impairment, and around 15% meet the criteria for a diagnosis of frontotemporal dementia (FTD). These conditions also overlap genetically, with the hexanucleotide repeat expansion (HRE) in the C9ORF72 gene being the most common genetic link to ALS. Another factor contributing to the varied presentation of ALS symptoms is the extent of neuronal degeneration in the cortex (upper motor neurons), brainstem, and spinal cord (lower motor neurons). Neurological examination typically reveals both upper and lower motor neuron dysfunction, with some patients showing more upper or lower motor neuron signs. Validated screening tests specific to ALS, such as the Edinburgh Cognitive and Behavioral ALS Screen or ALS Cognitive Behavioral Screen, are recommended in all patients. The majority of ALS cases are sporadic (sALS), and only about 10% are familial (fALS) and caused by genetic mutations. The discovery of TDP-43 in 2006 as the primary component of ubiquitinated aggregates in motor neurons of sporadic and most familial ALS, as well as in cortical neurons in a subset of frontotemporal dementia cases, marked a new era in ALS pathology (Suk & Rousseaux, 2020).

10.1.4.1 Biomarkers for ALS

10.1.4.1.1 Imaging

MRI plays a crucial role in the exploration of neurological biomarkers, with its capabilities further enhanced through the combination of multiple imaging sequences, the development of new protocols, the integration of machine learning, and its pairing with other modalities like PET and proton spectroscopy. Both task-based and resting-state functional MRI (fMRI) enable the identification of distinct blood oxygen level-dependent (BOLD) activity patterns, helping to investigate neural network connectivity. Additionally, MRS offers the ability to quantify key neuronal and glial metabolites, such as N-acetyl aspartate (a marker of neuronal integrity), creatine (an indicator of energy metabolism), choline (a marker of cell membrane turnover), and myoinositol (a glial marker). Complementing this, PET imaging can reveal localized changes in brain metabolism by utilizing various receptor ligands to map specific regions of interest (Witzel et al., 2022).

10.1.4.2 Electrophysiology

Advances in EEG have improved spatial resolution and signal quality. Motor unit number estimation (MUNE) tracks ALS progression by estimating motor units in a muscle, with the newer MScanFit MUNE method providing objective estimates from stimulus-response data. Motor unit number index (MUNIX) measures motor unit number and size using maximum CMAP and surface electromyography at different force levels. Transcranial magnetic stimulation (TMS) non-invasively assesses corticospinal motor neurons, while electrical impedance myography (EIM) is a research tool for evaluating muscle health (Verber & Shaw, 2020).

10.1.4.3 Fluid Based Biomarkers

Neurofilaments, the structural proteins of the cytoskeleton found in neurons, are the biomarkers that have been extensively studied in the CSF and blood for ALS and various neurological disorders. The build-up of these proteins has been associated with motor neuron impairment. A previous report recognized the elevated expression of phosphorylated neurofilament heavy chain (pNFH) in the CSF of patients. Additionally, research on pheno-converters (individuals who carry ALS-related genes and later develop ALS or frontotemporal dementia symptoms) revealed elevated NFL levels at least a year before the onset of clinical symptoms (Menke et al., 2015).

10.1.4.4 Advances in ALS Therapy

10.1.4.4.1 Disease-Modifying Therapies

The US Food and Drug Administration (FDA) has approved three disease-modifying drugs for the treatment of ALS: (1) Riluzole has anti-glutaminergic action, increases survival, and delays the decline in muscle testing score. However, the drug has major adverse effects like asthenia, gastrointestinal symptoms, and an increase in liver enzymes. (2) Edaravone works through a free radical scavenging mechanism, decreases oxidative stress, and retards ALS progression. (3) The combination of sodium phenylbutyrate and taurursodiol improves mitochondrial function, reduces endoplasmic reticulum stress, and prevents cell death (Ilieva et al., 2023; Paganoni et al., 2020).

10.1.5 Stroke

Stroke is an acute, focal neurological deficit with no other explanation than a cerebrovascular cause. Common symptoms include hemiparesis, dysarthria, sensory deficits, aphasia, and visual deficits. It is the second leading cause of death globally. The primary pathological mechanism of ischemic stroke is intravascular thrombosis, which can lead to the death of brain tissue and localized neurological deficits. There are three main contributors to ischemic strokes: 50% arise from arteriosclerotic plaques in cerebral vessels and plaque rupture, 20% are due to cardiogenic cerebral infarctions, and 25% result from Lacunar infarcts caused by small vessel disease. The remaining 5% occur due to rarer conditions, including vasculitis and extracranial arterial dissection (Tan et al., 2024).

10.1.5.1 Biomarkers

IL-6 and high-sensitive CRP (hs-CRP) have been identified in a higher recurrence risk after ischemic stroke by both an individual participant data meta-analysis and a Mendelian randomization study (Chaturvedi & De Marchis, 2024).

10.2 CURRENT APPROACHES IN THE PATHOGENESIS AND TREATMENT OF NDDs

10.2.1 Targeting Ferroptosis

Ferroptosis takes place when there is a redox imbalance between the iron-mediated production of lipid hydroperoxides and several antioxidant defense systems, principally glutathione-dependent glutathione peroxidase 4 (GPX4), which is known to detoxify free radicals and lipid oxidation products. There are elevated levels of iron as well as lipid peroxidation products in the AD brain, which confirms the role of ferroptosis in the pathogenesis of AD. Some recent studies have reported that iron is raised in the cerebral cortex (medial frontal and temporal gyrus) in the AD post-mortem brain. Most recently, the prognosis of PD is attributed to ferroptosis (an iron-mediated cell death). The cascade of ferroptosis is initiated by α-syn aggregates, promoting the production of ROS and lipid peroxides within cell membranes. Furthermore, a bidirectional relationship has been observed in dopamine depletion and ferroptosis. The early pathogenesis of certain NDDs like PD, AD, HD, and multiple sclerosis (MS) involves altered iron metabolism (Sun et al., 2023).

10.2.2 Neurovascular Coupling (NVC) Dysfunction in AD, PD, and Ischemic Stroke

The neurovascular coupling (NVC) unit (NVCU) consists of neurons, endothelial and other vascular cells, astrocytes, and microglial cells, which play a pivotal role in neuronal protection. It has been observed that the NVC establishes a bidirectional transport of metabolites, fluid, and other substances across the periphery and the brain (Rowsthorn et al., 2023). The reduction in NVC can play a pivotal role in the early detection of several neurodegenerative disorders like AD, PD, HD, and ischemic stroke.

10.2.3 Neuroinflammation

Increasing evidence suggests that neuroinflammation plays a key role in the early pathogenesis of AD, PD, HD, ALS, and stroke. Astrocytes and microglia are proposed to be the main players in neuroinflammation (Morales et al., 2014). Microglia have been found in two activated subsets, i.e., classic (M1) or selectively activated state (M2 state). Activated microglial cells release potentially harmful proinflammatory cytokines in the M1 state. However, in the M2 state, these cells release neurotrophic and anti-inflammatory cytokines such as IL-10, brain-derived neurotrophic factor (BDNF), and glial cell-derived neurotrophic factor (GDNF), which help control inflammation and promote the health of neurons. One of the hallmarks of AD is the accumulation of Aβ protein in the brain, which can trigger an inflammatory response. Aβ activates microglia, the immune cells of the brain, to produce pro-inflammatory cytokines and chemokines. These cytokines and chemokines can lead to the recruitment of peripheral immune cells to the brain, further amplifying the inflammatory response. Neuroinflammation can also contribute to the development of tau pathology in AD.

Similarly, recent evidence of neuroinflammation in the pathogenesis of PD has gained noteworthy attention. Some of these are elevated levels of proinflammatory cytokines like TNF-α and IL-1β in the CSF and blood, activation of microglia and macrophages in the substantia nigra, T lymphocyte infiltration in affected brain regions, and altered T and B lymphocyte activity in PD patients. Toll-like receptors (TLRs), predominantly expressed on microglia, can be activated by molecules such as misfolded α-synuclein, which plays a key role in the pathogen- or damage-associated molecular pattern (PAMP/DAMP)-mediated dysregulation of TLR2 and TLR4 signaling. This dysregulation triggers the MyD88 and NFκB pathways, leading to the production of proinflammatory cytokines.

Unlike other NDDs, HD does not exhibit significant infiltration of peripheral immune cells (T cells) into the brain. Instead, the primary inflammatory responses are driven by activated microglia and astrocytes. Neuroinflammation is evident by the presence of translocator protein (TSPO), which is normally expressed at low levels in a healthy CNS but becomes upregulated by microglia and astrocytes during neuroinflammatory processes. PET imaging reveals increased TSPO tracer levels in the HD striatum, correlating with disease severity and linked to microglial activation. In postmortem, striatal tissue from HD patients, elevated levels of IL-6, IL-8, and tumor necrosis factor (TNF)-α have been observed (O'Regan et al., 2021).

The potential role in neuroinflammation in ALS has drawn attention as it has been observed that inflammatory mediators are correlated to neuronal loss even in the early pre-synaptic phase of ALS. Moreover, there is evidence to point out that there is a cross-talk between the CNS and peripheral immune cells during ALS progression. A previous report has depicted that during ALS progression, the disruption of the BBB takes place. Subsequently, peripheral immune cells like dendritic cells (DC), natural killer (NK) cells, macrophages, monocytes, and neutrophils could infiltrate into the CNS, as demonstrated in mSOD1 ALS both practically and clinically. Therefore, the immune cells and the BBB can be prime targets to inhibit the progression of ALS.

In the condition of stroke, neuroinflammation plays a crucial role in disease progression. In stroke reperfusion injury is the prime cause of ROS generation as well as activation of immune cells (including microglia, astrocytes, and leukocytes), triggering inflammatory chemokines like CCL2, CCL3, and CXCL8, and adhesion molecules, and causing neuroinflammation. Various neuroprotective therapeutic interventions have been developed, but their clinical trials show disappointing results due to insignificant efficacy or undesirable side effects. Targeting adhesion molecules and integrin proteins, ILs, and TNF alpha can be a game-changer and a promising therapeutic intervention to improve outcomes after ischemic stroke (Alsbrook et al., 2023).

10.2.4 Micro-RNAs (miRNAs)

miRNAs are single-stranded non-coding RNAs that regulate gene expression post-transcriptionally via RNA interference. Some recent studies have demonstrated that a population of these miRNAs plays a pivotal role in NDDs. It has been observed that microRNA-485-3p (miR-485-3p) is overexpressed in CSF of patients with AD. In a transgenic mouse model, the antisense oligonucleotide (ASO) of miR-485-3p reduced Aβ and tauopathy and improved cognitive function. In another study, dysregulation of miRNA-485 was involved in the pathogenesis of AD and PD. In a different study, miRNA-124-3p-loaded small extracellular vesicles have shown promising therapy in treating 6-hydroxydopamine-induced PD in a pre-clinical study. In the context of ischemic stroke, miRNAs play a mechanistic role in the pathogenesis of stroke. In a clinical study using 74 patients, 55 thrombi were obtained, and analysis of 2578 microRNAs showed that 50 were altered, in quantification of miR-378f and miR-450b-5p showing a significant elevation among the cardioembolic thrombi. In a pre-clinical study, it has been reported that mi-R-21 decreased post-stroke brain damage. Another study carried out on animal models of HD depicted many microRNAs, like miR-9/9*, miR-29b, miR-124a, miR-132, miR-128, miR-139, miR-122, miR-138, miR-23b, miR-135b, miR-181 (all downregulated), and miR-448 (upregulated), had altered expression levels in the striatum and cerebral cortex. In a study carried out in bulbar-onset ALS patients, it was observed that the expression of miR-16-5p, miR-23a-3p, miR-22-3p, and miR-93-5p was significantly up-regulated in plasma exosomes compared with healthy control individuals (Liu et al., 2023). Therefore, the emergence of miRNAs provides a platform for the diagnosis and treatment of NDDs.

10.2.5 Gut Dysbiosis

It has been widely accepted that there is a bidirectional correlation between the gut microbiota and the CNS. The gut and brain crosstalk with each other through various aspects, like the enteric nervous system (ENS), neuroendocrine system, immune system, and vagus nerve, via the production of neuroactive substances, metabolites, and hormones. Gut dysbiosis has been observed in many NDDs, such as AD, PD, HD, ALS, and ischemic stroke, suggesting a possible role in disease pathogenesis. It has been observed that the pathogenesis of AD, i.e., accumulation of Aβ, begins in the pre-clinical phase, i.e., 10–20 years before the onset of the clinical phase. Gut dysbiosis is seen in 80% of cases, which is attributed to the early nonmotor symptom of PD. Constipation is one of the symptoms that occurs decades before the motor symptoms are exacerbated in PD. Earlier studies have shown that α-syn was present in the enteric tissues of PD patients. Another study has proposed that α-syn inclusion may initiate in the gut. An earlier report has depicted that vagotomy may reduce the risk of developing PD. Therefore, identifying early pathogenesis of AD in the gut can prevent the advancement of this disease. Studies have shown the implication of pre-clinical and clinical evidence of probiotics, prebiotics, and fecal microbiota transplantation (FMT) in NDDs.

10.2.6 Organ-on-a-chip (OOC)

NDDs (AD, PD, ALS, and HD) despite differing in their manifestations share a battery of common phenotypes like mitochondrial dysfunction, neuroinflammation, neuronal loss, oxidative damage, and vascular dysfunction. Due to the lack of précised pre-clinical animal models of the prognosis of NDDs, there are more chances of failing the new chemical entities in clinical trials. Therefore, there is an urge for such pre-clinical models that mimic NDD-associated phenotypes with more accuracy. In context of this demand, the US Food and Drug Administration (FDA) recently approved the use of alternative models that bypass the pre-clinical phase and involve clinical studies as part of the FDA Modernization Act 2.0. One such approach is organ-on-a-chip (OOC) technology, which offers a promising alternative, closely mimicking human tissue niches in vitro (Table 10.1) (Stewart et al., 2023).

10.2.7 Proteolysis-targeting Chimeras (PROTACs)

Proteolysis-targeting chimeras (PROTACs) are heterobifunctional chemical entities having the potential to overcome the constraints of conventional small-molecule therapies. A typical PROTAC consists of two ligands: one that binds to the protein of interest (POI) and the other that binds to an E3 ubiquitin ligase, connected via a linker (Figure 10.2). PROTAC-mediated degradation occurs through the ubiquitin-proteasome system (UPS), where the ubiquitinated target protein is recognized and degraded by the 26S proteasome. PROTAC has a unique characteristic of ambushing "undruggable" targets (Yao et al., 2022). PROTAC can be a pioneering entity in the early diagnosis and treatment of NDDs, as the aggregation of misfolded proteins, such as tau protein and α-synuclein protein, which are the main cause of such diseases, can be precisely targeted.

In 2019, Kargbo's group reported the first small-molecule tau degrader, revealing that tau-targeting PROTACs successfully degraded tau in human tau-p301L and tau-a152T neurons (Kargbo, 2019). Another report regarding PROTAC 60 showed that it preferentially degrades tau in FTD neurons of frontotemporal dementia compared to normal cells (Silva et al., 2019). Another study carried

TABLE 10.1
Various OOC Models for NDDs

NDDs	OOC model	Reference
AD	Decreased interstitial fluid flow on aβ_{42}-mediated neurotoxicity	Park et al. (2015)
	Perusable microvascular network of the BBB adjacent to pre-differentiated neuronal microtissues replicating key pathological phenotypes in vitro	Ko et al. (2023)
	Tri-culture system, the authors replicated critical phenotypes of the NDD in vitro	Park et al. (2018)
	3D human AD tri-culture model by using neurons, astrocytes, and microglia in a 3D microfluidic platform (3D hNeuroGliAD). Model provided key representative AD features: Aβ aggregation, p-tau accumulation, and neuroinflammatory activity	Alavi-Moghadam et al. (2024)
	Matrigel® into a conventional Campenot-like chamber to study the correlation between mitochondria-associated endoplasmic reticulum membranes, which have been shown to be upregulated in AD, and β-secretase activity	Bhattacharyya et al. (2021)
	A physiologically relevant 3D human neural cell culture microfluidic model has been developed, having a brain endothelial cell monolayer with a BBB-like phenotype. This model is shown to recapitulate several key aspects of BBB dysfunction observed in AD patients	Shin et al. (2019)
PD	Culturing PD patient-specific dopaminergic neurons within a Matrigel-loaded commercial microfluidic platform	Bolognin et al. (2019)
	Replicated vascular dysfunction in the context of PD within a membrane-based platform	Pediaditakis et al. (2021)
	Mimetas OrganoPlate to investigate PD-associated astrocyte dysfunction	de Rus Jacquet et al. (2023)
ALS	Stem cell (iPSC)-derived and light-sensitive channelrhodopsin-2-induced MN spheroids from a patient with sporadic ALS	Osaki et al. (2018)
	Microfluidic neuromuscular junction model to examine the correlation between the familial ALS mutation superoxide dismutase (SOD)	Machado et al. (2019)
	Neuromuscular junction-on-a-chip	Alavi-Moghadam et al. (2024)
HD	iPSC-based BBB-chip (HD-iBMECs)	Vatine et al. (2019)

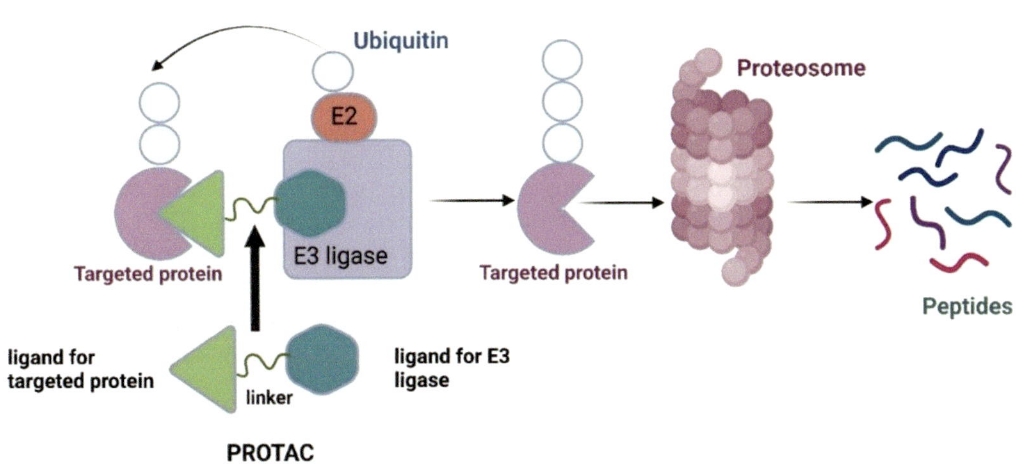

FIGURE 10.2 Mechanism of PROTAC-mediated protein degradation.

out on PROTAC 61 showed its effectiveness toward tau degradation under both physiological and pathological conditions (Wang et al., 2021).

In another report, PROTAC 62 was targeted against glycogen synthase kinase 3 (GSK-3), a class of serine/threonine protein kinases that boost the phosphorylation of tau protein and the production of the Aβ peptide to induce AD (Jiang et al., 2021).

In PD pathogenesis, neuronal cytoplasmic aggregation of Lewy bodies is composed of aggregates of α-synuclein protein, resulting in neuronal degeneration. In 2020, a PROTAC that induces α-synuclein protein degradation was reported by Kargbo's group (Kargbo, 2020).

In a different research conducted in the year 2017 by Ishikawa's group on PROTACs 63 and 64 to target HD. Interestingly, PROTAC 63 and PROTAC 64 effectively downregulated mHtt protein levels in HD patients and mHtt-transfected HeLa cells (Tomoshige et al., 2017). In continuation of this, they synthesized a novel PROTAC 65 that showed an improved affinity compared to the previous two PROTACs, i.e., PROTAC 63 and PROTAC 64 (Tomoshige et al., 2018).

However, the PROTACs targeting strategies in the field of stroke are in the initial stage. In a previous report in 2018, Duris et al. identified a protein "BET protein" that poses histone acetyltransferase activity, regulates gene transcription after ischemic stroke, and promotes rehabilitation. Therefore, PROTAC dBET1 can be a hopeful target for treating stroke (Ma et al., 2021).

10.2.8 Clustered Regularly Interspaced Short Palindromic Repeats and CRISPR-associated Protein 9 (CRISPR-Cas9)

CRISPR/Cas9 is a pioneer gene-editing method having great potential for correcting specific targeted gene sequences and offering new insights for the treatment of NDDs. Regardless of its genetic inheritance, an altered Aβ metabolism is generally detected in both familial and sporadic AD cases. As already reported, mutations in the PSEN-1, PSEN-2, and APP genes contribute to familial AD, CRISPR/Cas9 technique can effectively target excessive Aβ production or correct mutations in these genes. In summary, CRISPR/Cas9 stands out as one of the most powerful gene-editing tools available for advancing research into NDDs like AD. More precise and newer models of AD have been developed using CRISPR-Cas9 technology to study the pathogenesis, phenotype, and identify harmful genes and simultaneously find better treatment for the disease. Serenels and his team used CRISPR-Cas9 to develop new animal models of AD by introducing three specific point mutations—F681Y, G676R, and R684H—into the endogenous APP gene of experimental animals (rat and mice), converting it to the human form. This approach resulted in humanized animal models, which will be valuable for studying APP processing and uncovering new pathways in disease development. In another study using the CRISPR-Cas9 technique, Vermilyea and Babinski successfully induced monogenic mutations in common marmosets and supported the use of this species for generating a novel genetically based model of PD that expresses physiological levels of LRRK2 G2019S (Vermilyea et al., 2020). Furthermore, by the application of the CRISPR-Cas9 technique, Yan and Tu developed a knock-in (KI) pig model of HD that endogenously expresses full-length mutant huntingtin (HTT) (Yan et al., 2018). In a different study on ALS using the CRISPR/Cas9 method and human induced pluripotent stem cells (iPSCs), Kim and Ryu generated highly pure, iPSC-derived MNs with a SOD1-G93A missense mutation and identified pathological phenotypes relevant to ALS. Moreover, they elucidated spontaneous disease phenotypes relevant to ALS in mutant SOD1 MNs from genome-edited and patient-derived iPSCs. Their observations revealed that SOD1 mutations in human MNs cause cell-autonomous proteinopathy, axonopathy, synaptic pathology, and aberrant neurotransmission. TTR buildup in the CNS, potentially leading to stroke and cognitive decline, continues to be a significant unresolved concern. A CRISPR-Cas9-based therapy aimed at editing TTR genes is currently under investigation in a phase 1 clinical trial (Table 10.2).

TABLE 10.2
This Table Showing Therapeutic Application of CRISPR/Cas9 in Managing NDDs

NDDs	Point Mutation in Genetic Target	Organism/Cell Line	Outcome	Ref.
AD	G676R, F681Y, and R684H of APP	Mouse	Humanized animal models	Serneels et al. (2020)
	3′-UTR of APP	APP-KI mice	Reduced Aβ pathology	Nagata et al. (2018)
	C-terminus of APP	Tg2576 mice	Downregulated Aβ production	Sun et al. (2019)
	Deletion in the transcriptional start codon of MAPT gene	Tau knock-out mice	More Background resistant to excitotoxicity of C57Bl/6J	Tozzo et al. (2020)
	APoE-E3/E4	Human-induced pluripotent stem cells	Downregulation in APoE-E3/E4 and Tau protein hyper-phosphorylation	Wadhwani et al. (2019)
	P522R of Plcγ2-P522R variant	Mouse	Plcγ2-P522R knock-in mouse mode	György et al. (2018)
	Downregulation of the expression of BACE-1	Mouse	Reduced Aβ pathology	Park et al. (2019)
	Upregulation of the expression of ADAM10	Mouse	Reduced Aβ pathology	Park et al. (2021)
PD	Leucine-rich repeat kinase 2 (LRRK2) G2019S	Marmoset stem cells	Genetic-based model of PD	Vermilyea et al. (2020)
	DNAJC6, encoding HSP40 auxilin	Human embryonic stem cells	Genetic-based model of PD	Wulansari et al. (2021)
	Vacuolar protein sorting 35 gene (VPS35) D620N	Mouse	Age-related alterations in the nigrostriatal dopamine system	Ishizu et al. (2016)
	PINK1 and DJ-1 genes deletion	Monkey	Transgenic monkey model for future PD studies	Li et al. (2021)
	SNCA (JARID1A)	PD-iPSCs	Downregulation of α-synuclein	Guhathakurta et al. (2021)
	A53T-SNCA	Rat	Downregulation of α-synuclein	Yoon et al. (2022)
HD	Downregulated huntingtin (HTT) gene	Ex-vivo	Mutant allele-specific inactivation of the HD mutant allele	Shin et al. (2016)
	HTT gene	Pig	Pig model of HD	Yan et al. (2018)
ALS	SOD1	iPSC-derived motor neurons	Features associated with ALS	Kim et al. (2020)
Stroke	Hereditary transthyretin (TTR)	Human	Phase 1 clinical trial	Adams et al. (2023)

10.2.9 Lactate Metabolism

Apart from being a metabolic substrate to provide energy as a primary role physiologically, lactate under pathophysiological conditions can also act as a signaling molecule to modulate cellular functions. It is clear from the Astrocyte-Neuron Lactate Shuttle that lactate plays a crucial role in the CNS. Moreover, protein lactylation highlights the novel role of lactate in regulating transcription, cellular functions, and disease development (Yang et al., 2024).

In AD patients, along with Aβ and tau deposition, an alteration in aerobic glycolysis has also been reported. A technique, 2-deoxy-2-[18F] fluoro-D-glucose PET, which is intended for the analysis of cerebral glucose metabolism, has shown that glucose uptake is decreased in the brain of AD patients, whereas lactate level is increased in the CSF of AD and mild cognitive impairment patients, suggesting that a glucose metabolism disorder is prevalent in the advancement of AD. Microglia transform into a glucose-hypermetabolic state in AD patients. Therefore, targeting compromised lactate metabolism may be a new AD intervention approach. Similarly, in the CSF of late-onset patients with PD, it has been observed that lactate levels were unusually raised. Upregulated hexokinase and lactate dehydrogenase with increased lactate levels lead to the apoptosis of dopaminergic neurons. These findings imply that targeting glucose metabolism may be an important PD therapeutic strategy. Likewise, in MS patients, CSF lactate was higher. Therefore, measuring CSF or serum lactate could be a low-cost and reliable laboratory test to gauge an MS patient's response to treatment and the severity of their condition. The research demonstrated that ascorbic acid's inhibition of glucose transport results in the promotion of lactate absorption, but overexpressing glucose transporter in HD cells restored lactate. These findings imply that lactate can be employed as a diagnostic marker and therapeutic target for NDDs. Molecular mechanism studies have revealed that aerobic exercise can promote cerebral blood circulation and improve cerebral glucose metabolism.

10.2.10 Microglial Metabolic Reprogramming

Microglia, the brain's resident immune cells, are determinant in maintaining CNS homeostasis. Nevertheless, in NDDs, these cells undergo metabolic changes in response to pathological factors like Aβ plaques, Tau tangles, and α-synuclein aggregates. This transition involves moving from oxidative phosphorylation (OXPHOS) to glycolysis, increased glucose uptake, higher production of lactate, lipids, and succinate, and elevated levels of glycolytic enzymes. These metabolic alterations lead to changes in microglial function, such as misleading inflammation and reduced phagocytic ability, which contribute to the worsening of neurodegeneration (Miao et al., 2023). Therefore, therapies targeting microglial metabolic reprogramming can open a new avenue in the field of NDDs.

10.2.11 Microglial Tunneling Nanotubes Dysfunction

In a recent report, it has been reported that microglia and neurons get connected via tunneling nanotubes (TNTs) for quick exchange of organelles, vesicles, and proteins in both physiological and pathological conditions. In NDDs like AD and PD, toxic aggregates of Aβ, tau, and α-syn accumulate within neurons and exhibit "prion-like" characteristics. These prion-like features disperse among cells and foster misfolding of native proteins in neighboring neurons, thereby propagating neuronal damage and neurodegeneration across interconnected brain regions. Apart from other mechanisms (endocytosis, exocytosis, direct penetration of aggregates, exosome-carried proteins), TNTs are one of the prime mechanisms for this pathogenesis. However, in physiological conditions, naive microglia donate healthy mitochondria to rescue cells burdened with inflammation, thereby enhancing overall cell survival within the microglial population (Scheiblich et al., 2024).

10.2.12 Melatonin

Disruption in the circadian rhythm (CR) of aging individuals is a prime reason for developing NDDs like AD and PD. Melatonin is a hormone for managing proper CR. The disruption of CR leads to an imbalance between the production and clearance of Aβ as well as alpha-synuclein, leading to the pathogenesis of AD and PD. Therefore, clinical trials with melatonin treatment have shown promising effects on maintaining the Aβ production/clearance ratio by maintaining normal sleep-wakefulness and improving cognitive function in AD patients. In another study, it has been shown that melatonin can increase the clearance of alpha-synuclein by inducing ubiquitination, thus inducing proteasomal degradation of alpha-synuclein. In stroke patients, melatonin exerts its neuroprotective effects through inducing autophagy and antioxidant pathways. Similarly, in HD patients, the biosynthesis of melatonin is hindered. Therefore, melatonin treatment can impart a neuroprotective effect in HD. In another report, it has been shown that melatonin can slow the progression of ALS through autophagy and antioxidant potential (Bald et al., 2021).

10.2.13 Brain-Derived Neurotrophic Factors (BDNF)

BDNF is a well-known growth factor in mammalian brain, which plays a pivotal role in synaptic plasticity and long-term potentiation for learning and memory function. Therefore, therapeutic targets that can elevate the brain levels of BDNF could provide neuroprotection against various NDDs (Gao et al., 2022; Keifer, 2022; Palasz et al., 2020).

10.3 CONCLUSION

The rapid evolution of research and technological advancements in NDDs represents a ray of hope for millions of individuals affected by these debilitating conditions. As we have explored in this chapter, breakthroughs in areas such as biomarker discovery, genetic research, and neuroimaging have paved the way for earlier detection, more precise diagnoses, and the development of targeted therapies.

REFERENCES

Adams, D., Sekijima, Y., Conceição, I., Waddington-Cruz, M., Polydefkis, M., Echaniz-Laguna, A., & Reilly, M. M. (2023). Hereditary transthyretin amyloid neuropathies: Advances in pathophysiology, biomarkers, and treatment. *Lancet Neurol*, *22*(11), 1061–1074. https://doi.org/10.1016/s1474-4422(23)00334-4

Alavi-Moghadam, S., Kokabi-Hamidpour, S., Rezaei-Tavirani, M., Larijani, B., Arjmand, R., Rahim, F., … Arjmand, B. (2024). Neuromuscular junction-on-a-chip for amyotrophic lateral sclerosis modeling. *Methods in Molecular Biology*, *2736*, 139–150. https://doi.org/10.1007/7651_2022_474

Alsbrook, D. L., Di Napoli, M., Bhatia, K., Biller, J., Andalib, S., Hinduja, A., … Divani, A. A. (2023). Neuroinflammation in acute ischemic and hemorrhagic stroke. *Curr Neurol Neurosci Rep*, *23*(8), 407–431. https://doi.org/10.1007/s11910-023-01282-2

Altuna-Azkargorta, M., & Mendioroz-Iriarte, M. (2021). Blood biomarkers in Alzheimer's disease. *Neurologia (Engl Ed)*, *36*(9), 704–710. https://doi.org/10.1016/j.nrleng.2018.03.006

Ashton, N. J., Blennow, K., & Zetterberg, H. (2021). Spitting image: Can saliva biomarkers reflect Alzheimer's disease? *EBioMedicine*, *68*, 103437. https://doi.org/10.1016/j.ebiom.2021.103437

Bălașa, A. F., Chircov, C., & Grumezescu, A. M. (2020). Body fluid biomarkers for Alzheimer's disease-an up-to-Date overview. *Biomedicines*, *8*(10). https://doi.org/10.3390/biomedicines8100421

Bald, E. M., Nance, C. S., & Schultz, J. L. (2021). Melatonin may slow disease progression in amyotrophic lateral sclerosis: Findings from the Pooled Resource Open-Access ALS Clinic Trials database. *Muscle Nerve*, *63*(4), 572–576. https://doi.org/10.1002/mus.27168

Bhattacharyya, R., Black, S. E., Lotlikar, M. S., Fenn, R. H., Jorfi, M., Kovacs, D. M., & Tanzi, R. E. (2021). Axonal generation of amyloid-β from palmitoylated APP in mitochondria-associated endoplasmic reticulum membranes. *Cell Rep*, *35*(7), 109134. https://doi.org/10.1016/j.celrep.2021.109134

Bolognin, S., Fossépré, M., Qing, X., Jarazo, J., Ščančar, J., Moreno, E. L., … Schwamborn, J. C. (2019). 3D cultures of Parkinson's disease-specific dopaminergic neurons for high content phenotyping and drug testing. *Adv Sci (Weinh)*, *6*(1), 1800927. https://doi.org/10.1002/advs.201800927

Brücke, T., Djamshidian, S., Bencsits, G., Pirker, W., Asenbaum, S., & Podreka, I. (2000). SPECT and PET imaging of the dopaminergic system in Parkinson's disease. *J Neurol*, *247*(Suppl 4), Iv/2-7. https://doi.org/10.1007/pl00007769

Budd Haeberlein, S., Aisen, P. S., Barkhof, F., Chalkias, S., Chen, T., Cohen, S., … Sandrock, A. (2022). Two randomized phase 3 studies of aducanumab in early Alzheimer's disease. *J Prev Alzheimers Dis*, *9*(2), 197–210. https://doi.org/10.14283/jpad.2022.30

Chang, C. H., Lin, C. H., & Lane, H. Y. (2021). Machine learning and novel biomarkers for the diagnosis of Alzheimer's disease. *Int J Mol Sci*, *22*(5). https://doi.org/10.3390/ijms22052761

Chaturvedi, S., & De Marchis, G. M. (2024). Inflammatory biomarkers and stroke subtype: An important new frontier. *Neurology*, *102*(2), e208098. https://doi.org/10.1212/wnl.0000000000208098

de Rus Jacquet, A., Alpaugh, M., Denis, H. L., Tancredi, J. L., Boutin, M., Decaestecker, J., … Cicchetti, F. (2023). The contribution of inflammatory astrocytes to BBB impairments in a brain-chip model of Parkinson's disease. *Nat Commun*, *14*(1), 3651. https://doi.org/10.1038/s41467-023-39038-8

Dhingra, H., & Gaidhane, S. A. (2023). Huntington's disease: Understanding its novel drugs and treatments. *Cureus*, *15*(10), e47526. https://doi.org/10.7759/cureus.47526

Frisoni, G. B., Boccardi, M., Barkhof, F., Blennow, K., Cappa, S., Chiotis, K., … Winblad, B. (2017). Strategic roadmap for an early diagnosis of Alzheimer's disease based on biomarkers. *Lancet Neurol*, *16*(8), 661–676. https://doi.org/10.1016/s1474-4422(17)30159-x

Gao, L., Zhang, Y., Sterling, K., & Song, W. (2022). Brain-derived neurotrophic factor in Alzheimer's disease and its pharmaceutical potential. *Transl Neurodegener*, *11*(1), 4. https://doi.org/10.1186/s40035-022-00279-0

Goetzl, E. J., Nogueras-Ortiz, C., Mustapic, M., Mullins, R. J., Abner, E. L., Schwartz, J. B., & Kapogiannis, D. (2019). Deficient neurotrophic factors of CSPG4-type neural cell exosomes in Alzheimer disease. *FASEB J*, *33*(1), 231–238. https://doi.org/10.1096/fj.201801001

Guhathakurta, S., Kim, J., Adams, L., Basu, S., Song, M. K., Adler, E., … Kim, Y. S. (2021). Targeted attenuation of elevated histone marks at SNCA alleviates α-synuclein in Parkinson's disease. *EMBO Mol Med*, *13*(2), e12188. https://doi.org/10.15252/emmm.202012188

György, B., Lööv, C., Zaborowski, M. P., Takeda, S., Kleinstiver, B. P., Commins, C., … Ingelsson, M. (2018). CRISPR/Cas9 mediated disruption of the Swedish APP allele as a therapeutic approach for early-onset Alzheimer's disease. *Mol Ther Nucleic Acids*, *11*, 429–440. https://doi.org/10.1016/j.omtn.2018.03.007

Hampel, H., Hu, Y., Cummings, J., Mattke, S., Iwatsubo, T., Nakamura, A., … Schindler, S. E. (2023). Blood-based biomarkers for Alzheimer's disease: Current state and future use in a transformed global healthcare landscape. *Neuron*, *111*(18), 2781–2799. https://doi.org/10.1016/j.neuron.2023.05.017

Ilieva, H., Vullaganti, M., & Kwan, J. (2023). Advances in molecular pathology, diagnosis, and treatment of amyotrophic lateral sclerosis. *BMJ*, *383*, e075037. https://doi.org/10.1136/bmj-2023-075037

Ishizu, N., Yui, D., Hebisawa, A., Aizawa, H., Cui, W., Fujita, Y., … Watase, K. (2016). Impaired striatal dopamine release in homozygous Vps35 D620N knock-in mice. *Hum Molecul Genet*, *25*(20), 4507–4517. https://doi.org/10.1093/hmg/ddw279

Jiang, X., Zhou, J., Wang, Y., Liu, X., Xu, K., Xu, J., … Sun, H. (2021). PROTACs suppression of GSK-3β, a crucial kinase in neurodegenerative diseases. *Eur J Med Chem*, *210*, 112949. https://doi.org/10.1016/j.ejmech.2020.112949

Kargbo, R. B. (2019). Treatment of Alzheimer's by PROTAC-Tau protein degradation. *ACS Med Chem Lett*, *10*(5), 699–700. https://doi.org/10.1021/acsmedchemlett.9b00083

Kargbo, R. B. (2020). PROTAC compounds targeting α-synuclein protein for treating neurogenerative disorders: Alzheimer's and Parkinson's diseases. *ACS Med Chem Lett*, *11*(6), 1086–1087. https://doi.org/10.1021/acsmedchemlett.0c00192

Keifer, J. (2022). Regulation of AMPAR trafficking in synaptic plasticity by BDNF and the impact of neurodegenerative disease. *J Neurosci Res*, *100*(4), 979–991. https://doi.org/10.1002/jnr.25022

Kenny, A., Jiménez-Mateos, E. M., Zea-Sevilla, M. A., Rábano, A., Gili-Manzanaro, P., Prehn, J. H. M., … Hernández, F. (2019). Proteins and microRNAs are differentially expressed in tear fluid from patients with Alzheimer's disease. *Sci Rep*, *9*(1), 15437. https://doi.org/10.1038/s41598-019-51837-y

Kim, B. W., Ryu, J., Jeong, Y. E., Kim, J., & Martin, L. J. (2020). Human motor neurons with SOD1-G93A mutation generated from CRISPR/Cas9 Gene-edited iPSCs develop pathological features of amyotrophic lateral sclerosis. *Front Cell Neurosci*, *14*, 604171. https://doi.org/10.3389/fncel.2020.604171

Ko, E. C., Spitz, S., Pramotton, F. M., Barr, O. M., Xu, C., Pavlou, G., … Kamm, R. D. (2023). Accelerating the in vitro emulation of Alzheimer's disease-associated phenotypes using a novel 3D blood-brain barrier neurosphere co-culture model. *Front Bioeng Biotechnol*, *11*, 1251195. https://doi.org/10.3389/fbioe.2023.1251195

Krishnadas, N., Villemagne, V. L., Doré, V., & Rowe, C. C. (2021). Advances in brain amyloid imaging. *Seminars in Nuclear Medicine*, *51*(3), 241–252. https://doi.org/https://doi.org/10.1053/j.semnuclmed.2020.12.005

Krothapalli, M., Buddendorff, L., Yadav, H., Schilaty, N. D., & Jain, S. (2024). From gut microbiota to brain waves: The potential of the microbiome and EEG as biomarkers for cognitive impairment. *Int J Mol Sci*, *25*(12). https://doi.org/10.3390/ijms25126678

Li, H., Wu, S., Ma, X., Li, X., Cheng, T., Chen, Z., … Hu, X. (2021). Co-editing PINK1 and DJ-1 genes via adeno-associated virus-delivered CRISPR/Cas9 system in adult monkey brain elicits classical Parkinsonian phenotype. *Neurosci Bull*, *37*(9), 1271–1288. https://doi.org/10.1007/s12264-021-00732-6

Liu, Y., Ding, M., Pan, S., Zhou, R., Yao, J., Fu, R., … Lu, Z. (2023). MicroRNA-23a-3p is upregulated in plasma exosomes of bulbar-onset ALS patients and targets ERBB4. *Neuroscience*, *524*, 65–78. https://doi.org/10.1016/j.neuroscience.2023.05.030

Ma, K., Han, X. X., Yang, X. M., & Zhou, S. L. (2021). Proteolysis targeting chimera technology: A novel strategy for treating diseases of the central nervous system. *Neural Regen Res*, *16*(10), 1944–1949. https://doi.org/10.4103/1673-5374.308075

Machado, C. B., Pluchon, P., Harley, P., Rigby, M., Gonzalez Sabater, V., Stevenson, D. C., … Lieberam, I. (2019). In vitro modelling of nerve-muscle connectivity in a compartmentalised tissue culture device. *Adv Biosyst*, *3*(7). https://doi.org/10.1002/adbi.201800307

Menke, R. A., Gray, E., Lu, C. H., Kuhle, J., Talbot, K., Malaspina, A., & Turner, M. R. (2015). CSF neurofilament light chain reflects corticospinal tract degeneration in ALS. *Ann Clin Transl Neurol*, *2*(7), 748–755. https://doi.org/10.1002/acn3.212

Miao, J., Chen, L., Pan, X., Li, L., Zhao, B., & Lan, J. (2023). Microglial metabolic reprogramming: Emerging insights and therapeutic strategies in neurodegenerative diseases. *Cellular and Molecular Neurobiology*, *43*(7), 3191–3210. https://doi.org/10.1007/s10571-023-01376-y

Mintun, M. A., Lo, A. C., Duggan Evans, C., Wessels, A. M., Ardayfio, P. A., Andersen, S. W., … Skovronsky, D. M. (2021). Donanemab in early Alzheimer's disease. *New England Journal of Medicine*, *384*(18), 1691–1704. https://doi.org/10.1056/NEJMoa2100708

Morales, I., Guzmán-Martínez, L., Cerda-Troncoso, C., Farías, G. A., & Maccioni, R. B. (2014). Neuroinflammation in the pathogenesis of Alzheimer's disease. A rational framework for the search of novel therapeutic approaches. *Front Cell Neurosci*, *8*, 112. https://doi.org/10.3389/fncel.2014.00112

Munhoz, R. P., Tumas, V., Pedroso, J. L., & Silveira-Moriyama, L. (2024). The clinical diagnosis of Parkinson's disease. *Arquivos de Neuro-Psiquiatria*, *82*(6), 1–10. https://doi.org/10.1055/s-0043-1777775 (O diagnóstico clínico da doença de Parkinson.)

Nagata, K., Takahashi, M., Matsuba, Y., Okuyama-Uchimura, F., Sato, K., Hashimoto, S., … Saido, T. C. (2018). Generation of app knock-in mice reveals deletion mutations protective against Alzheimer's disease-like pathology. *Nat Commun*, *9*(1), 1800. https://doi.org/10.1038/s41467-018-04238-0

Narasimhan, S., Holtzman, D. M., Apostolova, L. G., Cruchaga, C., Masters, C. L., Hardy, J., … Hampel, H. (2024). Apolipoprotein E in Alzheimer's disease trajectories and the next-generation clinical care pathway. *Nat Neurosci*. https://doi.org/10.1038/s41593-024-01669-5

O'Regan, G. C., Farag, S. H., Casey, C. S., Wood-Kaczmar, A., Pocock, J. M., Tabrizi, S. J., & Andre, R. (2021). Human Huntington's disease pluripotent stem cell-derived microglia develop normally but are abnormally hyper-reactive and release elevated levels of reactive oxygen species. *J Neuroinflamm*, *18*(1), 94. https://doi.org/10.1186/s12974-021-02147-6

Oh, H. S.-H., Urey, D. Y., Karlsson, L., Zhu, Z., Shen, Y., Farinas, A., … Wyss-Coray, T. (2024). Synapse protein signatures in cerebrospinal fluid and plasma predict cognitive maintenance versus decline in Alzheimer's disease. *bioRxiv*, 2024.2007.2022.604680. https://doi.org/10.1101/2024.07.22.604680

Osaki, T., Uzel, S. G. M., & Kamm, R. D. (2018). Microphysiological 3D model of amyotrophic lateral sclerosis (ALS) from human iPS-derived muscle cells and optogenetic motor neurons. *Sci Adv*, *4*(10), eaat5847. https://doi.org/10.1126/sciadv.aat5847

Paganoni, S., Macklin, E. A., Hendrix, S., Berry, J. D., Elliott, M. A., Maiser, S., … Cudkowicz, M. E. (2020). Trial of sodium phenylbutyrate-taurursodiol for amyotrophic lateral sclerosis. *N Engl J Med*, *383*(10), 919–930. https://doi.org/10.1056/NEJMoa1916945

Palasz, E., Wysocka, A., Gasiorowska, A., Chalimoniuk, M., Niewiadomski, W., & Niewiadomska, G. (2020). BDNF as a promising therapeutic agent in Parkinson's disease. *Int J Mol Sci*, *21*(3). https://doi.org/10.3390/ijms21031170

Park, H., Hwang, Y., & Kim, J. (2021). Transcriptional activation with Cas9 activator nanocomplexes rescues Alzheimer's disease pathology. *Biomaterials*, *279*, 121229.

Park, H., Oh, J., Shim, G., Cho, B., Chang, Y., Kim, S., … Choi, H. (2019). In vivo neuronal gene editing via CRISPR–Cas9 amphiphilic nanocomplexes alleviates deficits in mouse models of Alzheimer's disease. *Nat Neurosci*, *22*(4), 524–528.

Park, J., Lee, B. K., Jeong, G. S., Hyun, J. K., Lee, C. J., & Lee, S. H. (2015). Three-dimensional brain-on-a-chip with an interstitial level of flow and its application as an in vitro model of Alzheimer's disease. *Lab Chip*, *15*(1), 141–150. https://doi.org/10.1039/c4lc00962b

Park, J., Wetzel, I., Marriott, I., Dréau, D., D'Avanzo, C., Kim, D. Y., … Cho, H. (2018). A 3D human triculture system modeling neurodegeneration and neuroinflammation in Alzheimer's disease. *Nat Neurosci*, *21*(7), 941–951. https://doi.org/10.1038/s41593-018-0175-4

Pavese, N., & Tai, Y. F. (2018). Nigrosome imaging and neuromelanin sensitive MRI in diagnostic evaluation of Parkinsonism. *Mov Disord Clin Pract*, *5*(2), 131–140. https://doi.org/10.1002/mdc3.12590

Pediaditakis, I., Kodella, K. R., Manatakis, D. V., Le, C. Y., Hinojosa, C. D., Tien-Street, W., … Karalis, K. (2021). Modeling alpha-synuclein pathology in a human brain-chip to assess blood-brain barrier disruption. *Nat Commun*, *12*(1), 5907. https://doi.org/10.1038/s41467-021-26066-5

Quinn, J. P., Kandigian, S. E., Trombetta, B. A., Arnold, S. E., & Carlyle, B. C. (2021). VGF as a biomarker and therapeutic target in neurodegenerative and psychiatric diseases. *Brain Commun*, *3*(4), fcab261. https://doi.org/10.1093/braincomms/fcab261

Raulin, A. C., Doss, S. V., Trottier, Z. A., Ikezu, T. C., Bu, G., & Liu, C. C. (2022). ApoE in Alzheimer's disease: Pathophysiology and therapeutic strategies. *Mol Neurodegener*, *17*(1), 72. https://doi.org/10.1186/s13024-022-00574-4

Ricci, C. (2024). Neurodegenerative disease: From molecular basis to therapy. *Int J Mol Sci*, *25*(2). https://doi.org/10.3390/ijms25020967

Rodrigues, F. B., Byrne, L. M., McColgan, P., Robertson, N., Tabrizi, S. J., Zetterberg, H., & Wild, E. J. (2016). Cerebrospinal fluid inflammatory biomarkers reflect clinical severity in Huntington's disease. *PLoS ONE*, *11*(9), e0163479. https://doi.org/10.1371/journal.pone.0163479

Rostagno, A. A. (2022). Pathogenesis of Alzheimer's disease. *Int J Mol Sci*, *24*(1). https://doi.org/10.3390/ijms24010107

Rowsthorn, E., Pham, W., Nazem-Zadeh, M. R., Law, M., Pase, M. P., & Harding, I. H. (2023). Imaging the neurovascular unit in health and neurodegeneration: A scoping review of interdependencies between MRI measures. *Fluids Barriers CNS*, *20*(1), 97. https://doi.org/10.1186/s12987-023-00499-0

Salvadó, G., Horie, K., Barthélemy, N. R., Vogel, J. W., Pichet Binette, A., Chen, C. D., … Hansson, O. (2024). Disease staging of Alzheimer's disease using a CSF-based biomarker model. *Nat Aging*, *4*(5), 694–708. https://doi.org/10.1038/s43587-024-00599-y

Saw, A., Bari, S., & Robertson, N. P. (2024). Novel biomarkers and neuroimaging techniques in Parkinson's disease. *Journal of Neurology*, *271*(2), 1056–1058. https://doi.org/10.1007/s00415-024-12187-6

Scheiblich, H., Eikens, F., Wischhof, L., Opitz, S., Jüngling, K., Cserép, C., … Heneka, M. T. (2024). Microglia rescue neurons from aggregate-induced neuronal dysfunction and death through tunneling nanotubes. *Neuron*. https://doi.org/10.1016/j.neuron.2024.06.029

Serneels, L., T'Syen, D., Perez-Benito, L., Theys, T., Holt, M. G., & De Strooper, B. (2020). Modeling the β-secretase cleavage site and humanizing amyloid-beta precursor protein in rat and mouse to study Alzheimer's disease. *Mol Neurodegener*, *15*(1), 60. https://doi.org/10.1186/s13024-020-00399-z

Shin, J. W., Kim, K. H., Chao, M. J., Atwal, R. S., Gillis, T., MacDonald, M. E., … Lee, J. M. (2016). Permanent inactivation of Huntington's disease mutation by personalized allele-specific CRISPR/Cas9. *Hum Mol Genet*, *25*(20), 4566–4576. https://doi.org/10.1093/hmg/ddw286

Shin, Y., Choi, S. H., Kim, E., Bylykbashi, E., Kim, J. A., Chung, S., … Tanzi, R. E. (2019). Blood-brain barrier dysfunction in a 3D in vitro model of Alzheimer's disease. *Adv Sci (Weinh)*, *6*(20), 1900962. https://doi.org/10.1002/advs.201900962

Silva, M. C., Ferguson, F. M., Cai, Q., Donovan, K. A., Nandi, G., Patnaik, D., … Haggarty, S. J. (2019). Targeted degradation of aberrant tau in frontotemporal dementia patient-derived neuronal cell models. *Elife*, *8*. https://doi.org/10.7554/eLife.45457

Soltani Khaboushan, A., Moeinafshar, A., Ersi, M. H., Teixeira, A. L., Majidi Zolbin, M., & Kajbafzadeh, A. M. (2023). Circulating levels of inflammatory biomarkers in Huntington's disease: A systematic review and meta-analysis. *J Neuroimmunol*, *385*, 578243. https://doi.org/10.1016/j.jneuroim.2023.578243

Stewart, A., Denoyer, D., Gao, X., & Toh, Y. C. (2023). The FDA modernisation act 2.0: Bringing non-animal technologies to the regulatory table. *Drug Discov Today*, *28*(4), 103496. https://doi.org/10.1016/j.drudis.2023.103496

Suk, T. R., & Rousseaux, M. W. C. (2020). The role of TDP-43 mislocalization in amyotrophic lateral sclerosis. *Mol Neurodegener*, *15*(1), 45. https://doi.org/10.1186/s13024-020-00397-1

Sun, J., Carlson-Stevermer, J., Das, U., Shen, M., Delenclos, M., Snead, A. M., … Roy, S. (2019). CRISPR/Cas9 editing of APP C-terminus attenuates β-cleavage and promotes α-cleavage. *Nat Commun*, *10*(1), 53. https://doi.org/10.1038/s41467-018-07971-8

Sun, S., Shen, J., Jiang, J., Wang, F., & Min, J. (2023). Targeting ferroptosis opens new avenues for the development of novel therapeutics. *Signal Transduct Target Ther*, *8*(1), 372. https://doi.org/10.1038/s41392-023-01606-1

Tan, K. S., Pandian, J. D., Liu, L., Toyoda, K., Leung, T. W. H., Uchiyama, S., … Venketasubramanian, N. (2024). Stroke in Asia. *Cerebrovasc Dis Extra*, *14*(1), 58–75. https://doi.org/10.1159/000538928

Tomoshige, S., Nomura, S., Ohgane, K., Hashimoto, Y., & Ishikawa, M. (2017). Discovery of small molecules that induce the degradation of huntingtin. *Angewandte Chemie Int Ed In English*, *56*(38), 11530–11533. https://doi.org/10.1002/anie.201706529

Tomoshige, S., Nomura, S., Ohgane, K., Hashimoto, Y., & Ishikawa, M. (2018). Degradation of huntingtin mediated by a hybrid molecule composed of IAP antagonist linked to phenyldiazenyl benzothiazole derivative. *Bioorg Med Chem Lett*, *28*(4), 707–710. https://doi.org/10.1016/j.bmcl.2018.01.012

Tozzo, P., Zullo, S., & Caenazzo, L. (2020). Science runs and the debate brakes: Somatic Gene-editing as a new tool for gender-specific medicine in Alzheimer's disease. *Brain Sci*, *10*(7). https://doi.org/10.3390/brainsci10070421

van Eimeren, T., Giehl, K., Reetz, K., Sampaio, C., & Mestre, T. A. (2023). Neuroimaging biomarkers in Huntington's disease: Preparing for a new era of therapeutic development. *Parkinsonism Relat Disord*, *114*, 105488. https://doi.org/10.1016/j.parkreldis.2023.105488

Vatine, G. D., Barrile, R., Workman, M. J., Sances, S., Barriga, B. K., Rahnama, M., ... Svendsen, C. N. (2019). Human iPSC-derived blood-brain barrier chips enable disease modeling and personalized medicine applications. *Cell Stem Cell*, *24*(6), 995–1005.e1006. https://doi.org/10.1016/j.stem.2019.05.011

Verber, N., & Shaw, P. J. (2020). Biomarkers in amyotrophic lateral sclerosis: A review of new developments. *Curr Opin Neurol*, *33*(5), 662–668. https://doi.org/10.1097/wco.0000000000000854

Vermilyea, S. C., Babinski, A., Tran, N., To, S., Guthrie, S., Kluss, J. H., ... Golos, T. G. (2020). In vitro CRISPR/Cas9-directed Gene editing to model LRRK2 G2019S Parkinson's disease in common marmosets. *Sci Rep*, *10*(1), 3447. https://doi.org/10.1038/s41598-020-60273-2

Wadhwani, A. R., Affaneh, A., Van Gulden, S., & Kessler, J. A. (2019). Neuronal apolipoprotein E4 increases cell death and phosphorylated tau release in Alzheimer disease. *Ann Neurol*, *85*(5), 726–739. https://doi.org/10.1002/ana.25455

Wang, W., Zhou, Q., Jiang, T., Li, S., Ye, J., Zheng, J., ... Wang, J. Z. (2021). A novel small-molecule PROTAC selectively promotes tau clearance to improve cognitive functions in Alzheimer-like models. *Theranostics*, *11*(11), 5279–5295. https://doi.org/10.7150/thno.55680

Witzel, S., Mayer, K., & Oeckl, P. (2022). Biomarkers for amyotrophic lateral sclerosis. *Curr Opin Neurol*, *35*(5), 699–704. https://doi.org/10.1097/wco.0000000000001094

Wulansari, N., Darsono, W. H. W., Woo, H. J., Chang, M. Y., Kim, J., Bae, E. J., ... Lee, S. H. (2021). Neurodevelopmental defects and neurodegenerative phenotypes in human brain organoids carrying Parkinson's disease-linked DNAJC6 mutations. *Sci Adv*, *7*(8). https://doi.org/10.1126/sciadv.abb1540

Yan, S., Tu, Z., Liu, Z., Fan, N., Yang, H., Yang, S., ... Li, X. J. (2018). A huntingtin knockin pig model recapitulates features of selective neurodegeneration in Huntington's disease. *Cell*, *173*(4), 989–1002.e1013. https://doi.org/10.1016/j.cell.2018.03.005

Yang, C., Pan, R. Y., Guan, F., & Yuan, Z. (2024). Lactate metabolism in neurodegenerative diseases. *Neural Regen Res*, *19*(1), 69–74. https://doi.org/10.4103/1673-5374.374142

Yao, T., Xiao, H., Wang, H., & Xu, X. (2022). Recent advances in PROTACs for drug targeted protein research. *Int J Mol Sci*, *23*(18). https://doi.org/10.3390/ijms231810328

Yoon, H. H., Ye, S., Lim, S., Jo, A., Lee, H., Hong, F., ... Jeon, S. R. (2022). CRISPR-Cas9 gene editing protects from the A53T-SNCA overexpression-induced pathology of Parkinson's disease in vivo. *Crispr J*, *5*(1), 95–108. https://doi.org/10.1089/crispr.2021.0025

11 Challenges Faced by Patients and Caregivers in Combating Neurodegenerative Diseases

*Sanjoli Srivastav and Hemlata Nimesh**
Amity Institute of Pharmacy, Amity University, Noida, Sector-125, Uttar Pradesh, India

11.1 INTRODUCTION

Neurodegenerative diseases (ND) are characterized by the loss of function or death of the cells of the central nervous system, such as Alzheimer's disease (AD), Parkinson's disease (PD), epilepsy, Huntington's, amyotrophic lateral sclerosis (ALS), etc. In a majority of cases, these tend to appear in geriatric patients, but in some rare cases, they can also be observed in younger patients [1]. They are chronic and progressive, but there are multiple approaches that delay the progression of the disease and assist in the management of the symptoms but do not cure the underlying cause [2, 3]. Chronic fatigue, paralysis, convulsions, dementia, altered perception, and pain are some of the symptoms of ND [1]. Though linked with distinct genes, the main cause of ND is the formation of protein aggregates that alter the chemical and physical characteristics of nerve cells. Knowledge of molecular genetics helps in discovering the specific genes linked to the disorder and also helps in the determination of the most suitable approach for treatment [2, 4, 5]. The World Health Organization (WHO) estimates that in the coming years, neurological disorders will surpass cancer and become the second most prevalent cause of mortality, after cardiovascular diseases [6]. The anticipated trend for dementia patients by 2050 has been illustrated in Table 11.1.

Despite having different presentations, all neurodegenerative diseases are ultimately fatal, but it has been demonstrated that supportive palliative care improves the standard of living, patient outcomes, and often extends a patient's life [7]. Patients suffering from neurodegenerative diseases often require an informal caregiver to assist them with their daily chores, with the progression of their disease, which leads to a gradual loss of autonomy and decreased quality of life (QoL). These caregivers face a multifaceted burden due to which they experience feelings of stress and neglect, eventually leading to burnout or breakdown [8, 9]. It is recognized that providing care for people with neurological disorders presents numerous obstacles that include problems with cognition, communication, behavior, altered sensory perception, and pain [10, 11]. The rates of suffering and exhaustion experienced by caregivers for neurology patients are comparable to those of cancer patients' caregivers [10, 12]. For effective treatment of the condition, professionals who have been trained in recognizing and developing solutions associated with human psychology, emotions, and life cycle phases should be employed [10]. Figure 11.1 illustrates the relation between a patient and their caregiver.

Patients, families, and caregivers bear a tremendous burden and cost as a result of neurodegenerative illnesses. They raise the expense of healthcare and create a lot of moral and ethical conundrums, particularly as people age [13]. Everyone encounters difficulties when caring for another person. Some caretakers manage the experience well, while others suffer from diminishing mental health. Caregivers' study is increasingly focused on understanding their variability [14].

*Corresponding Author. hemlatanimesh@gmail.com

 DOI: 10.1201/9781003530510-11

TABLE 11.1
Projected Number of Dementia Patients by the Year 2050 across Different Continents, Based on World Health Organization Estimates. It Highlights the Growing Global Burden of Neurodegenerative Diseases, with Asia Expected to Experience the Highest Increase in Patient Numbers

	Patients in 2050 (in million)
America	27.08
Europe	18.65
Africa	3.92
Asia	60.92
World	115.38

Source: World Health Organization and Regional Office for Europe (2013).

11.2 VARIOUS CHALLENGES FACED BY PATIENTS SUFFERING FROM NEURODEGENERATIVE DISEASES

Patients suffering from neurodegenerative diseases face a wide range of multidimensional challenges. These conditions not only impair cognitive and physical functioning but also significantly affect emotional stability and social interactions. These challenges are experienced by both patients and their caregivers, highlighting the complexity of care and emphasizing the need for timely and holistic support systems. The challenges faced by patients and caregivers are depicted in Figure 11.2.

11.2.1 Quality of Life (QoL)

QoL is a holistic notion that encompasses the extent of necessity and contentment in the domains of the body, cognition, social interaction, activities, materials, and structure [15, 16]. As the condition progresses, patients with neurodegenerative diseases frequently experience dementia, which makes

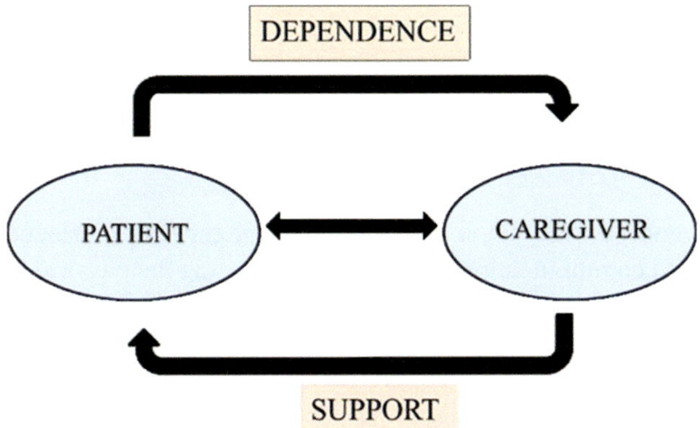

FIGURE 11.1 Dynamic and interdependent relationship between patients with neurodegenerative diseases and their caregivers. It highlights how the patient's declining physical, emotional, and cognitive functions increase their reliance on the caregiver, while the caregiver's well-being is equally influenced by the patient's condition, forming a cyclical support system that requires balance and empathy.

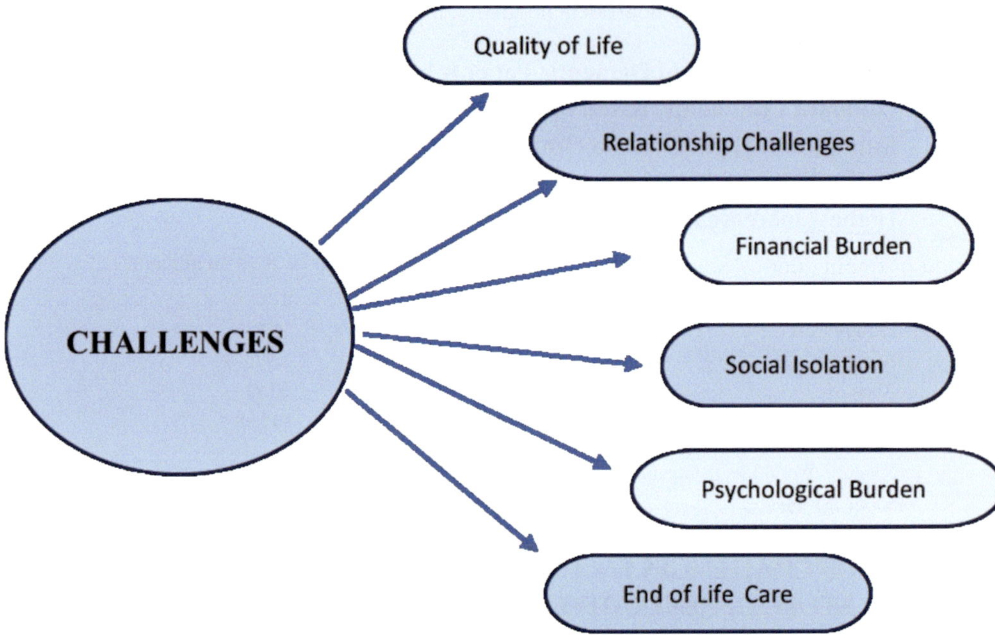

FIGURE 11.2 Various challenges encountered by both patients and caregivers in the context of neurodegenerative diseases. It visually categorizes the burdens, such as emotional stress, financial strain, social isolation, and physical exhaustion, faced by caregivers, alongside the patients' declining autonomy, communication issues, and cognitive impairments. The diagram emphasizes the need for comprehensive support systems.

it more difficult for them to carry out daily tasks independently and necessitates caregiver support. This wears out the caregiver and makes the patient feel ashamed and low in self-worth.

The QoL of caregivers is directly impacted by several factors such as health and sociodemographic traits, diagnosis, symptom manifestation, and the subsequent need for care [17]. The current findings demonstrated and validated the significant burden that late-stage PD places on informal caregivers. A total of 60%–70% of caregivers for Parkinson's patients attested to phrases like "situation wears me down," "always on my mind," and "feel guilty if not there" [18]. Additionally, a correlation between patient age and caregiver QoL was found, indicating that informal caregivers' QoL typically increases as patients age. This might be explained to some extent by a rise in home health and social care services, as well as a potential rise in situational acceptability.

11.2.2 Relationship Challenges

When a patient is diagnosed with ND, it might impair their capacity to execute their responsibilities due to cognitive and communication difficulties, influencing the physical and emotional health of their family members. About 17% of children between the ages of 12 and 24 whose parents had PD were diagnosed with mild to moderate depression [19]. When the wards of Multiple Sclerosis (MS) patients were compared with their counterparts of the same age and gender having healthy parents, the results indicated that the children of patients suffering from MS had greater psychological and behavioral issues. Certain patients have personal aids with them at all times, due to which their children are forced to spend time with people they are not familiar with [19]. Children of patients who are now above the age of 18 acknowledged a desire for practical advice on managing the illness.

Taking care of a spouse with a neurodegenerative disease has a quadrupled detrimental impact on mood as compared to similar-aged non-caregivers. This is mostly because of the disease's effects on social isolation, physical health, and sleep [17, 20, 21]. Losing a supporting spouse can lead to increased stress and loneliness, both of which are known to contribute to depression development over time [14]. From the early to late stages, the spousal care partner's sorrow grew linearly. In residential care facilities, spousal care partners expressed that they needed emotional support to keep their feelings of mutuality and connection with their partner intact, as well as support to enable shared moments and organize visits. In an unusual manner, siblings of dementia patients were interested in finding out more about the genetic components of the illness [22, 23]. The illness of their life partner impacts many aspects of family life, including socio-professional, financial, and psychological aspects. Research indicates that caring for patients with early-onset dementia carries an even more significant burden than caring for elderly patients with similar illnesses, and this may be because younger caregivers are unlikely to be adequately prepared for the role, possibly as a result of a decreased level of acceptance and knowledge of the illness [13].

11.2.3 Financial Burden

In 2020, the combined economic burden of PD, ALS, and spinal muscular atrophy in the United States was estimated at $655 billion. This figure includes direct medical and non-medical expenses, as well as indirect costs such as lost productivity and the value of unpaid caregiving hours [24, 25]. From age 65 until death, individuals with dementia incur an additional $45,940 in medical and long-term care costs, with nearly 98% of this amount attributed to nursing facility expenses [25, 26]. As the disease progresses, the financial responsibility often shifts to the patient's family. Family caregivers are particularly affected, as many reduce their working hours or leave their jobs entirely to provide care for their loved ones. A closer analysis shows that 63% of surveyed caregivers were below retirement age. Among this group, half reported having stopped working to provide care. Of those previously employed full-time, 61% reduced their working hours, while 71% of part-time workers limited their schedules to no more than 20 hours per week. These findings are consistent with other research that links informal caregiving to reduced workforce participation [16].

11.2.4 Social Isolation

Neuropsychiatric symptoms, including mood disturbances and behavioral abnormalities, are frequently used for identification and during neurodegenerative illnesses [27, 28]. The National Institute on Aging-Alzheimer's Association workgroups (NIA-AA) recognizes changes in personality, conduct, and demeanor as a key factor in diagnosing dementia [27, 29]. Impaired social cognition can be indicated by changes in attitude and behavior, including social disengagement, lack of social etiquette, or failure to respond to others emotionally. Neurodegenerative research is now increasingly concentrating on social cognition, which involves processing and reacting to social information from others, including emotions and conclusions [27, 30].

Apathy, disinhibition, and awkwardness in social settings have been identified as the first few symptoms in 2.0%, 5.0%, and 6.9% of AD patients, respectively [27]. According to studies, people with Idiopathic Parkinson's Disease (IPD) have trouble recognizing faces and identifying emotions from visual and aural clues [29, 31, 32]. Although IPD patients are recognized to be at a greater risk of impulse control disorders (ICDs), such as compulsive engagement in gambling, buying, sexual activities, and eating, these behavioral characteristics are assumed to be predominantly related to the use of PD medicines, notably dopamine agonists [27, 33].

Caregivers often experience discomfort due to the patient's lack of empathy and impulsive actions [34], disintegration in temperament and interpersonal behavior. Attending to a patient with a low sense of empathy is connected with increased loneliness, tension, and hardship for caregivers, along with depression [14].

11.2.5 Psychological Burden

People with neurodegenerative diseases frequently experience depression, which significantly affects their QoL. Depression in these individuals has been associated with earlier institutionalization, increased mortality, greater caregiver burden, and a more rapid decline in physical and functional abilities. Descriptive population studies suggest that approximately 80% of individuals with AD may experience depression at some point during the illness. While cognitive symptoms typically persist and worsen as the disease progresses, depressive symptoms can fluctuate and, in some cases, may even remit [35]. For a variety of reasons, it may be difficult to diagnose depression in Alzheimer's patients, as there is currently no validated tool to identify and measure the illness, and the depression may go unnoticed, as dementia signs like apathy might be mistaken for common depressive symptoms like melancholy or anhedonia. Lastly, these individuals' cognitive impairment suggests that they may have trouble expressing basic emotional sentiments like melancholy and hopelessness [19]. The most prevalent behavioral symptoms are agitation, irritability, sadness, anxiety, and apathy [35–37]. Hallucinations, excitement, disinhibition, and abnormal motor behavior are some of the less common symptoms. It is believed that there is a higher chance that apathy alone or in combination with depression may eventually lead to AD [35, 38].

Studies have shown that patients having Huntington's disease have suicide rates that are more than four times those of healthy individuals [39, 40]. Up to 40% of patients are diagnosed with depression. Feelings such as apathy, hostility, and disinhibition are also prevalent [41–43]. Neuropsychiatric symptoms, particularly depression, are associated with the majority of neurodegenerative diseases. Depressive symptoms are a major factor in rising healthcare expenses, morbidity, disability, and caregiver load. Appropriate diagnosis and evaluation of these symptoms are necessary to start the right early non-pharmacological and pharmacological treatment and restrict their extent [35]. Insensitive patients can leave caregivers feeling unsupported, thus impacting the quality of patient-caregiver interactions [14, 35].

11.2.6 End-of-Life Care

End-of-life decision-making can be challenging, especially for patients with neurological diseases who may experience cognitive impairment, necessitating decision-making by family members or surrogates, which can be challenging due to a lack of advance directives and care plans [44–47].

Hospital ethics committees are frequently consulted by patients with ethical dilemmas pertaining to the withholding or withdrawal of therapy in the latter stages of life [13, 42]. Although withdrawing therapy is an act and withholding treatment is an omission, both are morally equivalent in that one does not bear a greater burden of guilt for either a commission or an omission [13]. Cardiopulmonary resuscitation and other life-sustaining procedures may look fruitless for most patients with end-stage neurodegenerative diseases and should be weighed carefully in accordance with the patient's previous preferences [13, 43]. It then becomes a challenge for the family member to respect the wishes of the patient, despite whatever they feel.

When the primary goal of therapy shifts to comfort and support, a palliative care approach becomes the most appropriate option for patients in the terminal stages of neurodegenerative diseases. The presence of behavioral and psychiatric symptoms—such as aggression, anger, or disruptive behavior—can exacerbate the burden on caregivers and families. In such cases, neuroleptics, which function as pharmacological restraints, may be used; however, they are associated with serious side effects, including stroke, movement disorders, instability, and falls, and should be employed only as a last resort. Whether pharmacological or physical, restraints deprive patients

TABLE 11.2
Common Challenges Encountered by Caregivers of Patients with Neurodegenerative Diseases and Suggests Practical Solutions to Manage Them. It Includes Physical, Emotional, Financial, and Social Aspects of Caregiving, Aiming to Support Improved Caregiver Well-Being

S. No	Challenges	Solutions
1	Physical exhaustion	Regular breaks. Hiring home aids
2	Emotional stress	Counseling, meditation, support groups
3	Financial strain	Budgeting, insurance
4	Sleep deprivation	Establishing sleep routine
5	Social isolation	Seeking community support

of autonomy and dignity and should only be applied when there is a clear risk of the patient harming themselves [14]. Some of the caregiver challenges and their potential solutions are given in Table 11.2.

11.3 AVAILABLE SUPPORT FOR PATIENTS AND CAREGIVERS

11.3.1 Palliative Care for Patients

The WHO refers to palliative therapy as a strategy that enhances the QoL for individuals and their families dealing with issues related to life-threatening diseases by preventing and alleviating suffering through early detection, precise assessment, and treatment of pain and other issues associated with the physical, mental, and spiritual aspects of life [44].

The rehabilitative approach inherent in palliative care might easily be combined with contemporary neurorehabilitation methods for the long-term treatment of elderly patients suffering from stroke or neurodegenerative diseases [41].

Palliative care can benefit children with progressive neurodegenerative diseases (PNDI) from the time of diagnosis. Pediatric palliative care is defined as an active and holistic approach to treatment that addresses the physical, psychological, and spiritual needs of both the child and the family. The primary goal is to enhance the QoL for the patient while providing ongoing support to the family. This approach includes effective management of distressing symptoms, emotional and psychological support, and bereavement care after the child's death. Palliative care can be provided alongside curative or disease-modifying treatments, or it may become the primary focus when curative options are no longer effective or appropriate [48].

11.3.2 Support for Caregivers

A condition of physical, emotional, and mental tiredness and anguish that might include sadness, misery, worry, and other similar symptoms is known as caregiver burnout [49].

Female sex, low levels of education, living with the care recipient, a greater number of hours spent providing care, depressive symptoms, social exclusion, monetary strain, and a lack of choice in providing care are risk factors for caregiver burden are some of the risk factors of neurodegenerative diseases [50].

Denial of the severity of the condition, anger at the situation or patient, social isolation, disruptive sleep, and anxiety are some of the signs of caregiver burden. When feeling overwhelmed

TABLE 11.3
A Curated List of Organizations and Online Platforms Offering Assistance to Patients with Neurodegenerative Diseases and Their Caregivers. It Includes Services Such as Emotional Support, Medical Information, Community Engagement, and Educational Resources

Support Group	Beneficiary	Assistance Offered	Weblink
ALZConnected	Alzheimer's patients and caregivers	Accelerating global research, risk reduction, early detection, and maximizing quality care and support.	https://alzwa.org
Atrium Health	ND patients and caregivers	Development of an individualized treatment plan to help restore QoL for patients and caregivers and mental health support	https://atriumhealth.org/
ACRM Neurodegenerative Diseases Networking Group	ND patients	Promoting and facilitating neurodegenerative rehabilitation research and scholarship.	https://acrm.org/neurodegenerative-diseases-networking-group
UCSF Huntington's Disease Support Group	Huntington's patients	Offers support and education in a caring, safe, and relaxed environment.	https://hdsa.org/find-help/community-social-support/hdsa-support-groups/
Centre for Brain Research, Bengaluru	ND patients	Minimize the burden of neurodegenerative diseases through prevention, early diagnosis, and scientifically proven intervention	office.cbr@iisc.ac.in
The Indian Association of Palliative Care (IAPC)	Caregivers	A national forum to connect, support and motivate individuals, and institutions involved in palliative care.	https://www.palliativecare.in/
Caregiver Saathi	Caregivers	Support to improve the QoL, well-being of caregivers of patients with terminal illness or chronic conditions	https://caregiversaathi.co.in/home info@caregiversaathi.co.in
White Swan Foundation	Caregivers	Mental health and psychological support strategies	https://www.mhinnovation.net/
National Parkinson's Foundation	Parkinson patients	Identifying community resources	http://www.parkinson.org/
National Multiple Sclerosis Society	MS patients	Source of information on multiple sclerosis	https://www.nationalmssociety.org/
Family Caregiver Alliance, National Centre on Caregiving	ND Caregivers	Education, advocacy and improving QoL of family caregivers	https://www.caregiver.org/
National Rehabilitation Information Centre	ND patients	Medical and vocational rehabilitation, employment, education, assistive technology information	http://www.naric.com/

by having to take care of everything on their own, caregivers must acknowledge when they need support, which includes unpaid assistance, pertinent training, peer support groups, stress alleviation methods, organizational tools, and well-informed care plans. Improved clinical treatment and focused programming will have the twin effect of increasing caregiver well-being while decreasing the chance of care recipient institutionalization. Tailored healthcare services to assist caregivers in managing challenges associated with neurodegenerative illness are crucial for a higher QoL [17]. A selection of the support groups available for ND patients and their caregivers is included in Table 11.3.

11.4 CONCLUSION

In this chapter, we discussed the various challenges navigated by patients suffering from neurodegenerative diseases and their caregivers, which include relationship challenges, social isolation, decreased QoL, and financial burden. From the initial shock of the diagnosis to the myriad symptoms as the disease progresses, each stage brings new challenges. Daily tasks become challenging due to a deterioration in physical, cognitive, and emotional abilities.

The research literature suggests that the stress experienced by the caregiver is often overlooked, which eventually leads to burnout, isolation, and poor mental health. Feelings such as denial, guilt, anger, resentment, and anticipatory fear are some of the prevalent emotions but against the background of struggle is a story of resilience, innovation, and hope. At present, caregivers receive limited support; however, research continues.

The long-term struggle against neurodegenerative illnesses ultimately demonstrates how resilient individuals who become patients and caregivers are. Their tales, hardships, and accomplishments serve as a strong reminder of the necessity of empathy, support, and continuous innovation in the face of such challenging circumstances.

REFERENCES

1. Chaki, J., & Woźniak, M. (2023). Deep learning for neurodegenerative disorder (2016 to 2022): A systematic review. *Biomedical Signal Processing and Control*, *80*, 104223.
2. Ciurea, A. V., et al. (2023). Unraveling molecular and genetic insights into neurodegenerative diseases: Advances in understanding Alzheimer's, Parkinson's, and Huntington's diseases and amyotrophic lateral sclerosis. *International Journal of Molecular Sciences*, *24*(13), 10809.
3. Brettschneider, J., et al. (2015). Spreading of pathology in neurodegenerative diseases: A focus on human studies. *Nature Reviews Neuroscience*, *16*(2), 109–120.
4. Castellani, R. J., Rolston, R. K., & Smith, M. A. (2010). Alzheimer disease. *Disease-a-Month*, *56*(9), 484.
5. Kovacs, G. G. (2018). Concepts and classification of neurodegenerative diseases. In *Handbook of Clinical Neurology* (pp. 301–307). Elsevier.
6. Zaib, S., et al. (2023). Neurodegenerative diseases: Their onset, epidemiology, causes and treatment. *ChemistrySelect*, *8*(20), e202300225.
7. Armitage, A. E., & Fonkem, E. (2023). Supportive care of neurodegenerative patients. *Frontiers in Oncology*, *13*.
8. Klietz, M. (2022). Caregiver burden in movement disorders and neurodegenerative diseases: Editorial. *Brain Sciences*, *12*(9), 1184.
9. Martinez-Martin, P., et al. (2008). Burden, perceived health status, and mood among caregivers of Parkinson's disease patients. *Movement Disorders*, *23*(12), 1673–1680.
10. Lithin, Z., et al. (2020). Palliative care needs and caregiver burden in neurodegenerative diseases: A cross sectional study. *Annals of Indian Academy of Neurology*, *23*(3), 313–317.
11. Bhatnagar, S., & Prabhakar, H. (2012). Palliative care beyond oncology! *Indian Journal of Palliative Care*, *18*(2), 85.
12. Boersma, I., et al. (2014). Palliative care and neurology. *Neurology*, *83*(6), 561–567.
13. Low, J. A., & Ho, E. (2017). Managing ethical dilemmas in end-stage neurodegenerative diseases. *Geriatrics*, *2*(1), 8.

14. Brown, C. L., et al. (2018). Empathic accuracy deficits in patients with neurodegenerative disease: Association with caregiver depression. *The American Journal of Geriatric Psychiatry*, *26*(4), 484–493.
15. Marston, H. R., & van Hoof, J. (2019). Who doesn't think about technology when designing urban environments for older people? A case study approach to a proposed extension of the WHO's age-friendly cities model. *International Journal of Environmental Research and Public Health*, *16*(19), 3525.
16. Maresova, P., et al. (2020). Activities of daily living and associated costs in the most widespread neurodegenerative diseases: A systematic review. *Clinical Interventions in Aging*, 15, 1841–1862.
17. Roland, K. P., & Chappell, N. L. (2019). Caregiver experiences across three neurodegenerative diseases: Alzheimer's, Parkinson's, and Parkinson's with dementia. *Journal of Aging and Health*, *31*(2), 256–279.
18. Rosqvist, K., et al. (2022). Caregiver burden and quality of life in late stage Parkinson's disease. *Brain Sciences*, *12*(1), 111.
19. Ferm, U. N., Nolbris, S. M. J., Linnsand, P., & Jonsson, A. (2018). Impact of a parent's neurodegenerative disease and care on the daily life of children. In M. Mollaoglu (Ed.), *Caregiving and Home Care* (p. 238). BoD – Books on Demand.
20. Richardson, T. J., et al. (2013). Caregiver health: Health of caregivers of Alzheimer's and other dementia patients. *Current Psychiatry Reports*, *15*(7), 367.
21. Fried, E. I., et al. (2015). From loss to loneliness: The relationship between bereavement and depressive symptoms. *Journal of Abnormal Psychology*, *124*(2), 256.
22. Kinchin, I., et al. (2022). Care partner needs of people with neurodegenerative disorders: What are the needs, and how well do the current assessment tools capture these needs? A systematic meta-review. *International Journal of Geriatric Psychiatry*, *37*(7).
23. Millenaar, J. K., et al. (2016). The care needs and experiences with the use of services of people with young-onset dementia and their caregivers: A systematic review. *International Journal of Geriatric Psychiatry*, *31*(12), 1261–1276.
24. Thorpe, K. E., Levey, A. I., & Thomas, J. (2021). US burden of neurodegenerative disease. *Partnership to Fight Chronic Disease (PFCD) (2021): 1–13.*
25. Alzheimer's Association. (2021). 2021 Alzheimer's disease facts and figures. *Alzheimer's & Dementia*, *17*(3), 327–406.
26. Hudomiet, P., Hurd, M. D., & Rohwedder, S. (2019). The relationship between lifetime out-of-pocket medical expenditures, dementia, and socioeconomic status in the US. *The Journal of the Economics of Ageing*, *14*, 100181.
27. Desmarais, P., et al. (2018). Social inappropriateness in neurodegenerative disorders. *International Psychogeriatrics*, *30*(2), 197–207.
28. Teipel, S. J., et al. (2015). Neuropsychiatric symptoms in people screened positive for dementia in primary care. *International Psychogeriatrics*, *27*(1), 39–48.
29. McKhann, G. M., et al. (2011). The diagnosis of dementia due to Alzheimer's disease: Recommendations from the National Institute on Aging-Alzheimer's Association workgroups on diagnostic guidelines for Alzheimer's disease. *Alzheimer's & Dementia*, *7*(3), 263–269.
30. Henry, J. D., et al. (2016). Clinical assessment of social cognitive function in neurological disorders. *Nature Reviews Neurology*, *12*(1), 28–39.
31. Ariatti, A., Benuzzi, F., & Nichelli, P. (2008). Recognition of emotions from visual and prosodic cues in Parkinson's disease. *Neurological Sciences*, *29*(4), 219–227.
32. Bodden, M. E., Dodel, R., & Kalbe, E. (2010). Theory of mind in Parkinson's disease and related basal ganglia disorders: A systematic review. *Movement Disorders*, *25*(1), 13–27.
33. Weintraub, D., et al. (2015). Clinical spectrum of impulse control disorders in Parkinson's disease. *Movement Disorders*, *30*(2), 121–127.
34. Hsieh, S., et al. (2013). When one loses empathy: Its effect on carers of patients with dementia. *Journal of Geriatric Psychiatry and Neurology*, *26*(3), 174–184.
35. Baquero, M., & Martín, N. (2015). Depressive symptoms in neurodegenerative diseases. *World Journal of Clinical Cases*, *3*(8), 682–693.
36. Ready, R. E., et al. (2003). Apathy and executive dysfunction in mild cognitive impairment and Alzheimer disease. *The American Journal of Geriatric Psychiatry*, *11*(2), 222–228.
37. Robert, P. H., et al. (2006). Neuropsychological performance in mild cognitive impairment with and without apathy. *Dementia and Geriatric Cognitive Disorders*, *21*(3), 192–197.
38. Modrego, P. J., & Ferrández, J. (2004). Depression in patients with mild cognitive impairment increases the risk of developing dementia of Alzheimer type: A prospective cohort study. *Archives of Neurology*, *61*(8), 1290–1293.

39. van Duijn, E., et al. (2014). Neuropsychiatric symptoms in a European Huntington's disease cohort (REGISTRY). *Journal of Neurology, Neurosurgery & Psychiatry*, *85*(12), 1411.
40. Di Maio, L., et al. (1993). Suicide risk in Huntington's disease. *Journal of Medical Genetics*, *30*(4), 293.
41. Paulsen, J. S., et al. (2001). Neuropsychiatric aspects of Huntington's disease. *Journal of Neurology, Neurosurgery & Psychiatry*, *71*(3), 310.
42. Hurst, S. A., et al. (2007). Ethical difficulties in clinical practice: Experiences of European doctors. *Journal of Medical Ethics*, *33*(1), 51.
43. Volicer, L. (2005). End-of-life care for people with dementia in residential care settings. *Alzheimer's Association.*
44. Borasio, G. D. (2013). The role of palliative care in patients with neurological diseases. *Nature Reviews Neurology*, *9*(5), 292–295.
45. Jox, R. J., et al. (2012). Surrogate decision making for patients with end-stage dementia. *International Journal of Geriatric Psychiatry*, *27*(10), 1045–1052.
46. Vandervoort, A., et al. (2012). Advance directives and physicians' orders in nursing home residents with dementia in Flanders, Belgium: Prevalence and associated outcomes. *International Psychogeriatrics*, *24*(7), 1133–1143.
47. Seeber, A. A., et al. (2012). Discussions about treatment restrictions in chronic neurologic diseases. *Neurology*, *78*(8), 590–597.
48. Rallison, L. B., & Raffin-Bouchal, S. (2013). Living in the in-between: Families caring for a child with a progressive neurodegenerative illness. *Qualitative Health Research*, *23*(2), 194–206.
49. Yu, H., et al. (2015). Measuring the caregiver burden of caring for community-residing people with Alzheimer's disease. *PLoS One*, *10*(7), e0132168.
50. Adelman, R. D., et al. (2014). Caregiver burden: A clinical review. *JAMA*, *311*(10), 1052–1060.

Index

Note: **Bold** and *italic* page numbers refer to **tables** and figures, respectively.

Q

R

S

For Product Safety Concerns and Information please contact our
EU representative GPSR@taylorandfrancis.com Taylor & Francis
Verlag GmbH, Kaufingerstraße 24, 80331 München, Germany